AMINOGLYCOSIDE ANTIBIOTICS

BICENTENNIAL

1807

WILEY

2007

BICENTENNIAL

THE WILEY BICENTENNIAL–KNOWLEDGE FOR GENERATIONS

Each generation has its unique needs and aspirations. When Charles Wiley first opened his small printing shop in lower Manhattan in 1807, it was a generation of boundless potential searching for an identity. And we were there, helping to define a new American literary tradition. Over half a century later, in the midst of the Second Industrial Revolution, it was a generation focused on building the future. Once again, we were there, supplying the critical scientific, technical, and engineering knowledge that helped frame the world. Throughout the 20th Century, and into the new millennium, nations began to reach out beyond their own borders and a new international community was born. Wiley was there, expanding its operations around the world to enable a global exchange of ideas, opinions, and know-how.

For 200 years, Wiley has been an integral part of each generation's journey, enabling the flow of information and understanding necessary to meet their needs and fulfill their aspirations. Today, bold new technologies are changing the way we live and learn. Wiley will be there, providing you the must-have knowledge you need to imagine new worlds, new possibilities, and new opportunities.

Generations come and go, but you can always count on Wiley to provide you the knowledge you need, when and where you need it!

William J. Pesce
President and Chief Executive Officer

Peter Booth Wiley
Chairman of the Board

AMINOGLYCOSIDE ANTIBIOTICS

From Chemical Biology to Drug Discovery

DEV P. ARYA
Clemson University

BICENTENNIAL 1807 WILEY 2007 BICENTENNIAL

WILEY-INTERSCIENCE
A John Wiley & Sons, Inc., Publication

Published by John Wiley & Sons, Inc., Hoboken, New Jersey.
Published simultaneously in Canada.

Wiley Bicentennial Logo: Richard J. Pacifico

Library of Congress Cataloging-in-Publication Data:

ISBN 978-0-471-74302-6

Printed in the United States of America.

10 9 8 7 6 5 4 3 2 1

CONTENTS

PREFACE

Since Selman Waksman's discovery of streptomycin in 1944, aminoglycosides have been at the forefront of antibacterial drug treatment. Advances in carbohydrate chemistry and biochemistry, coupled with other technological advances in nucleic acid synthesis, biochemistry and structure analysis have led to a substantial increase in aminoglycoside research among both chemists and biochemists. While by no means exhaustive, this endeavor attempts to chronicle the advances made in aminoglycoside-related work over the last two decades to assist new researchers in the field of aminoglycoside research while also offering a reference guide for the expert. Any omission of work is unintentional and due partly to the time constraints always present in such a venture.

In the introduction, Dr. Davies' narrative of aminoglycoside research provides us with an elegant historical perspective on aminoglycoside discovery and the mechanism of action over the past 50 years. Chapter 2 illustrates the progress made in the elucidation of the biosynthetic pathways for aminoglycoside synthesis. Piepersburg and coworkers have outlined the major pathways and their work should serve as an important reference guide for years to come. A major problem in aminoglycoside therapy has been drug resistance and in Chapter 3 are the findings of Shakya and Wright regarding their articulation of the major mechanisms of aminoglycoside resistance and the possible methods for overcoming these pathways. To get past the resistance mechanisms, chemistry and biology must come together and in Chapter 4, Wang and Chang summarize the synthetic advances made in aminoglycoside chemistry over the past decade and their uses in the development of improved antibiotics.

A large interest among chemists and biochemists in aminoglycosides stems also from the fact that their molecular RNA target of therapeutic action has been identified. Chapters 5–7 provide a broad overview of the interaction of aminoglycoside with RNA focusing particularly on the A-site, their ribosomal target. In Chapter 5, Marshall and Puglisi first summarize the advances made in the NMR-based structural investigation of aminoglycosides with RNA, including the A-site. In Chapter 6, Kondo and Westhof detail the progress made in the crystallographic investigations of RNA–aminoglycoside interactions and how that has been used to decipher the differences between eukaryotic and prokaryotic ribosomal drug targets. Kling, Chow and Mobashery summarize the different approaches regarding the further comprehension of this important nucleic-acid interaction. Together, these accounts should provide an excellent reference guide for students and established researchers interested in the field of aminoglycoside biochemistry, biophysics, chemical biology or medicinal chemistry.

In Chapter 8, Cowan and coworkers review the state of the art on the metal-mediated aminoglycoside research resulting in possible metallozymes that can be developed as novel drugs and tools for DNA and RNA cleavage. A major issue preventing the development of aminoglycoside therapies has to do with their toxicity. As the mechanisms causing these aminoglycoside toxicities (ototoxicity and nephrotoxicity) become clearer, molecular mechanisms to circumvent the propagation of these toxicities are being proposed. Talaska and Schacht discuss these proposed mechanisms and the possible role of iron-mediated free radical damage leading to oxidative stress. For other molecular mechanisms in the genetic basis of aminoglycoside ototoxicity, the reader is advised to consult the references at the end of Chapter 1. In Chapter 10, Elson-Schwab and Tor describe the use of aminoglycosides in other medicinally relevant RNA targets, specifically those involved in HIV therapy. The work elegantly illustrates how the congruence of chemistry and biology can lead to novel paradigms in drug development. Finally, Arya and coworkers summarize the recent work in some non-RNA targets discovered for aminoglycoside-based recognition and how chemical manipulations can be used to tailor aminoglycoside specificities.

Aminoglycoside research is now a legitimate aspect of scientific endeavor; with much knowledge, yet unknown, that we can learn. Aminoglycosides have brought together a number of different disciplines (and many more disciplines are expected to be positively affected by aminoglycoside research). It is my hope that this book will assist in both the continuing progress and expansion of aminoglycoside research into drug development, chemical biology, biophysics, microbiology, toxicology, molecular recognition and carbohydrate chemistry. I wish to thank all the authors for their wholehearted collaboration in this endeavor. Without them this book would not have been possible.

DEV PRIYA ARYA
CLEMSON

CONTRIBUTORS

Khaled M. Aboshanab, BU Wuppertal, Chemical Microbiology, D-42097 Wuppertal, Germany

Dev P. Arya, Laboratory of Medicinal Chemistry, Department of Chemistry, Clemson University, Clemson, SC 29634

Cheng-Wei Tom Chang, Department of Chemistry and Biochemistry, Utah State University, Logan, UT 84322

Christine Chow, Department of Chemistry, Wayne State University, Detroit, MI 48202

J. A. Cowan, Department of Chemistry, Ohio State University, Columbus, OH 43210

Julian Davies, Department of Microbiology and Immunology, University of British Columbia[1] Life Sciences Institute,[2] Vancouver BC V6T 1Z3, Canada

Lev Elson-Schwab, Department of Chemistry and Biochemistry, University of California, San Diego, La Jolla, CA 92093

Nikhil Gokhale, Department of Chemistry, Ohio State University, Columbus, OH 43210

Dale Kling, Department of Chemistry and Biochemistry, University of Notre Dame, Notre Dame, IN 46556

Jiro Kondo, Institut de Biologie Moléculaire et Cellulaire, UPR9002 CNRS, Université Louis Pasteur, 67084 Strasbourg, France

R. Andrew Marshall, Department of Chemistry, Stanford University, Stanford, CA 94305

Shahriar Mobashery, Department of Chemistry and Biochemistry, University of Notre Dame, Notre Dame, IN 46556

Anjali Patwardhan, Department of Chemistry, Ohio State University, Columbus, OH 43210

Wolfgang Piepersberg, BU Wuppertal, Chemical Microbiology, D-42097 Wuppertal, Germany

Joseph D. Puglisi, Department of Structural Biology, Stanford University, Stanford, CA 94305

Jochen Schacht, Kresge Hearing Research Institute, Department of Otolaryngology, University of Michigan, Ann Arbor, MI 48109

Heike Schmidt-Beißner, BU Wuppertal, Chemical Microbiology, D-42097 Wuppertal, Germany

Tushar Shakya, Antimicrobial Research Centre, Department of Biochemistry and Biomedical Sciences, McMaster University, Hamilton, Ontario, Canada, L8N 3Z5

Nicholas Shaw, Laboratory of Medicinal Chemistry, Department of Chemistry, Clemson University, Clemson, SC 29634.

Andra E. Talaska, Kresge Hearing Research Institute, Department of Otolaryngology, University of Michigan, Ann Arbor, MI 48109

Yitzhak Tor, Department of Chemistry and Biochemistry, University of California, San Diego, La Jolla, CA 92093

Jinhua Wang, Department of Chemistry and Biochemistry, Utah State University, Logan, UT 84322

Udo F. Wehmeier, BU Wuppertal, Chemical Microbiology, D-42097 Wuppertal, Germany

Eric Westhof, Institut de Biologie Moléculaire et Cellulaire, UPR9002 CNRS, Université Louis Pasteur, 67084 Strasbourg, France

Gerard D. Wright, Antimicrobial Research Centre, Department of Biochemistry and Biomedical Sciences, McMaster University, Hamilton, Ontario, Canada, L8N 3Z5

Hongjuan Xi, Laboratory of Medicinal Chemistry, Department of Chemistry, Clemson University, Clemson, SC 29634.

1

IN THE BEGINNING THERE WAS STREPTOMYCIN

JULIAN DAVIES

Department of Microbiology and Immunology[1], *Life Sciences Institute*[2], *University of British Columbia, Vancouver BC V6T 1Z3, Canada*

1.1. INTRODUCTION

Although streptomycin was not the first antibiotic (penicillin, a fungal product, had been isolated some years earlier), its discovery was a landmark in antibiotic history. It was the first effective therapeutic for tuberculosis, a disease that had terrorized humans for centuries and a cause of human morbidity and mortality unmatched by wars or any other pestilence. Streptomycin was the first aminoglycoside to be identified and characterized and is noteworthy in being the first useful antibiotic isolated from a bacterial source.[1] At the present time, the use

I dedicate this article to the memory of Kenneth Rinehart (1929–2005), who contributed much to structural studies and developed the mutasynthetic approach to novel aminoglycosides.

Aminoglycoside Antibiotics: From Chemical Biology to Drug Discovery, Second Edition, by Dev P. Arya

Figure 1.1. The founders of the antibiotic era: Selman Waksman (discoverer of streptomycin) and Alexander Fleming (discoverer of penicillin).

of streptomycin in infectious disease therapy has largely been replaced by less toxic and equally effective compounds, but it still has significant applications as a second-line treatment for TB and occasionally for the treatment of nosocomial multidrug-resistant gram-positive infections. Streptomycin's preeminent place in the history of antibiotics is assured!

Selman Waksman's commitment to the isolation and screening of soil bacteria in the search for bioactive small molecules, especially potential antibiotics, was validated by the discovery of streptomycin.[2] This led to the creation of the modern biopharmaceutical industry and the subsequent isolation of tens of thousands of bioactive small molecules from soil bacteria and other environments. A proportion of these compounds have become highly successful therapeutics, not only for all types of infectious diseases, but also in the treatment of many other human and animal ailments and as anticancer, immuno-modulatory, and cardiovascular agents.[3] Waksman and Fleming could be considered the fathers of chemical biology (Figure 1.1).

Following on the discovery of streptomycin and its streptamine-based relatives (Figure 1.2), a new generation of the aminoglycosides derived from 2–deoxystreptamine (DOS) was not long in coming (Figure 1.3). For a variety of reasons, many of these compounds have not been employed as human therapeutics; for

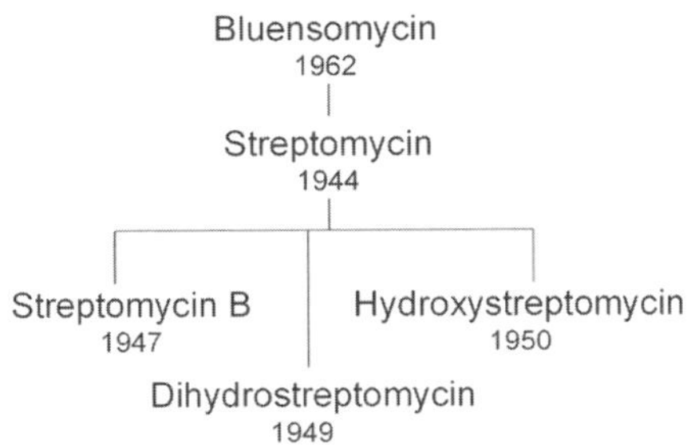

Figure 1.2. The structural relationships of the streptamine family of antibiotics.

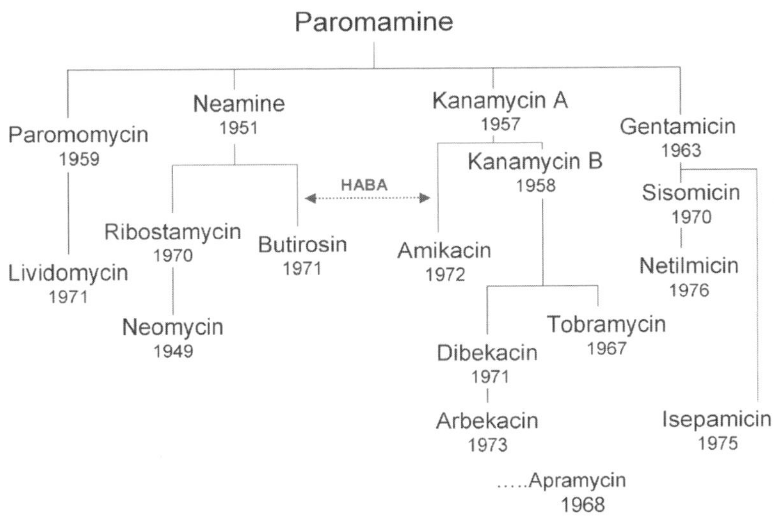

Figure 1.3. The structural relationships of the 2-deoxystreptamine family of antibiotics.

example, neomycin has rarely been used in the clinic because of its extreme toxicity. Surprisingly, paromomycin, a naturally occurring 6′-desaminoderivative of neomycin, is receiving increasing interest in the treatment of a variety of tropical diseases, including leishmaniasis and certain types of fungal infection.[4] This serves to illustrate that the aminoglycosides (and the related aminocyclitols, such as spectinomycin) have a broad range of biological activities and have found use in a wide variety of applications as indicated in Table 1.1. In addition to these compounds, there is a large group of atypical aminoglycosides, compounds that are of diverse microbial origin, structure, and biological activity (Table 1.2).

Many applications of the aminoglycosides have been of historical significance in genetics and microbiology. For example, mutations to streptomycin resistance were employed as counterselective genetic markers in the historic experiments of William Hayes that demonstrated the existence of bacterial conjugation and the requirement of donor (Hfr or F+) and receptor (F−) species.[5]

TABLE 1.1. Some of the Myriad Properties and Applications of the Aminoglycosides

Protein synthesis inhibition—prokaryotes
Protein synthesis inhibition—eukaryotes
Mistranslation on ribosomes
Nonsense mutation suppression
DNA translation
Phenotypic suppression
Membrane leakiness
Nucleic acid binding/precipitation
Probing ribosome structure
Allosteric activation of enzyme activity
Ribozyme/intron binding, inhibition, activation
Antibacterial
Antiprotozoal
Antiviral
Genetic markers for ribosome function
Broad-spectrum selective agents for gene transfer
Promoter-reporters (resistance genes)

TABLE 1.2. Other Classes of Aminoglycoside-Aminocyclitol Antibiotics

Ashimycin
Astromicin/Istamycin
Boholmycin
Kasugamycin
Myomycin
Spectinomycin
Trehalosamine
Validamycin

These experiments showed that conjugal gene transfer occurs with directional polarity and led to the subsequent characterization of sex factors that were ultimately shown to be extrachromosomal DNA elements, or plasmids. The finding by Ruth Sager that streptomycin interferes with chlorophyll production in *Chlamydomonas* and that high-level streptomycin resistance mutants exhibit cytoplasmic rather than Mendelian inheritance (due to alteration of the chloroplast genome)[6] provided evidence in support of the bacterial (endosymbiotic) origin of chloroplasts in *Chlamydomonas* and plants.[7] In the early 1970s, kanamycin resistance (encoded by a resistance plasmid) was used as a dominant selective genetic marker for heterologous gene transfer in the seminal recombinant DNA studies of Herbert Boyer, Stanley Cohen, and their colleagues.[8] The later observation that kanamycin and certain other aminoglycosides have inhibitory activity against some types of eukaryotic cells led to their application in the genetic manipulation in higher organisms, including plants.[9] In particular, two antibiotics have been widely used for eukaryotic gene cloning: G-418 (Geneticin™)[10] and hygromycin B.[11] G-418, related to gentamicin, has become the preferred selective agent for mammalian cell studies, and large amounts of the compound are currently employed for this purpose. The neomycin phosphotransferase gene was the first bacterial gene approved in tests of human gene therapy in 1982[12] and remains the genetic marker of choice for all types of eukaryotic cloning.

1.2. MODE OF ACTION

The biochemical mode of action of the aminoglycosides as antibacterials has long been a topic of great interest. Early experiments carried out soon after the introduction of streptomycin suggested a variety of modes of action, but these conclusions were based largely on symptomatic analyses of antibiotic-treated bacterial cultures. One important experiment done in 1948 showed that streptomycin blocks enzyme induction in susceptible bacteria[13]; this was the closest that anyone came to identifying the mechanism of action at the time.

A series of genetic and biochemical studies in the late 1950s and early 1960s led to the definitive identification of protein synthesis as the primary target for the antibacterial action of streptomycin. Initially, Erdos and Ullmann employed the incorporation of radioactive amino acids to show that production of labeled protein by cell-free extracts of *Mycobacterium tuberculosis* is effectively blocked by streptomycin.[14] The results of these experiments were subsequently confirmed by others, using defined cell-free translation systems from other bacteria and with synthetic polynucleotides as messenger RNAs. A seminal paper by Spotts and Stanier proposed the ribosome as the probable target for streptomycin action and came up with a plausible biochemical mechanism for the phenomenon of streptomycin dependence.[15]

Within the next few years (1962–1965) a flurry of research activity in a number of laboratories confirmed this model, and *in vitro* translation studies employing hybrids of sensitive and resistant ribosome subunits showed that streptomycin acts by binding to the 30S ribosome subunit.[16,17] This led to the investigation of the effects of streptomycin and other aminoglycosides on coding fidelity during translation, providing evidence for the active role of the 30S subunit in protein synthesis and the important finding that streptomycin and other aminoglycosides induce errors in translation.[18] These studies provided the first evidence that the ribosome is not simply an inert support in the process of peptide bond formation but plays an active role in the selection of aminoacylated tRNAs by the ribosome-bound messenger RNA.

Current work on the three-dimensional structure of ribosome complexes has amply confirmed the dynamic role of the ribosome in translation and the mechanism by which this process is perturbed by the binding of aminoglycosides to specific sites on the 30S subunit.[19] There is now strong genetic and phenotypic evidence for translation misreading by aminoglycosides in living cells.[20] While it has been shown that a number of other translation inhibitors also provoke mistranslation, this may be a symptom of protein synthesis inhibition and not a direct effect on codon reading, as with the aminoglycoside antibiotics.[21] Surprisingly, in the presence of some aminoglycosides, DNA can be accurately read as a messenger on the ribosome and to generate polypeptides *in vitro*; it is not known if this occurs *in vivo*[22]. Parenthetically, the ability of aminoglycosides to cause mistranslation has itself been applied recently to an "indirect" form of gene therapy; the administration of gentamicin or related compounds to patients with hereditary diseases such as severe hemophilia or cystic fibrosis can result in partial suppression of the disease.[23,24] Aminoglycoside-induced read-through of nonsense mutations leads to the production of small amounts of the missing protein and prevents nonsense-mediated decay of messenger RNA.

As is the case with most antibiotics, at subinhibitory concentrations the aminoglycosides induce significant changes in the transcription of some 5% of the genes in susceptible bacteria.[25] The mechanism responsible is not known but may be due to some form of coupling between translation and transcription not previously identified. We can assume that transcription modulation is associated with antibiotic activity in therapeutic use and may contribute to some of the side effects. On the other hand, at low concentrations in the environment, the aminoglycosides and other antibiotics may be acting as cell-signaling molecules.[26]

The use of streptomycin or spectinomycin resistance as a genetic marker was critical to the cloning and identification of the gene clusters encoding structural elements of the ribosome in bacteria. Once it had been demonstrated that resistance to streptomycin and spectinomycin is associated with amino acid changes in ribosomal proteins, bacteriophage P1 transduction studies showed that the associated genes are linked in clusters on the bacterial chromosome.[27] Masayasu Nomura and others then used disruption and reconstitution of ribosome particles from 16S rRNA and isolated R proteins to demonstrate the roles of the proteins RpsL (str) and RpsE (spc) in the determination of antibiotic resistance[28]; this confirmed the earlier genetic and phenotypic studies and ratified the role of R proteins in ribosome function. There followed a decade of argument as to the relative importance of ribosomal RNA *versus* ribosomal proteins in the structure and function of the particle, and a paradigm change occurred when it was shown by numerous sequence and functional studies that the two major rRNA molecules are the structural basis of ribosome function in translation. The fact that these RNA molecules are the targets for the binding and interaction of different antibiotics on the ribosome, resulting in interruption of the translation process, provides strong confirmation of their roles in translation. The spectacularly successful rRNA footprinting studies and X–ray structure analyses carried out by the groups of Noller,[29] Ramakrishnan,[30] and others have amply confirmed this dominant role of rRNA and the consequences of antibiotic binding to the ribosome, initially with the aminoglycosides but subsequently with most ribosomal inhibitors. However, although the primordial template for peptide bond formation is likely to have been RNA alone, the involvement of both RNA and protein is essential in the dynamic role of the "modern" ribosome in translation; this is a topic of continuing interest.[31] To date, it is only in the case of streptomycin that three-dimensional structure analysis of the antibiotic/ribosome complex identifies an interaction of the drug with both R proteins and rRNA.[19] There is increasing evidence for the existence of nonribosomal functions of the protein components of the ribosome. Studies using antibiotics such as the aminoglycosides will undoubtedly continue to play important roles in developing this story.

During these years of exciting revelations concerning aminoglycoside activity and the ribosome, one question relative to the therapeutic use of aminoglycosides has remained unsolved. Unlike most antibiotic inhibitors of protein synthesis in bacteria that lead to bacteriostasis, the aminoglycosides are rapidly bactericidal. The ability of the aminoglycosides to kill bacterial pathogens is an important attribute in their therapeutic use. This action is somewhat surprising when we consider that most inhibitors of ribosome function act in a similar fashion to the aminoglycosides, by binding to target sequences within the 16S or 23S rRNAs (as described above). For example, the aminocyclitol spectinomycin is bactericidal in action. The difference between cidal and static action has been the topic of much discussion and many publications; this work has been largely physiological in nature, and a satisfactory biochemical explanation for the lethal action of the aminoglycosides still eludes us.[32] The possibility that aminoglycosides (as distinct from other translation inhibitors) induce a process of programmed cell death (apoptosis) in bacteria[33] could provide an explanation.

1.3. RESISTANCE AND AMINOGLYCOSIDE EVOLUTION

Antibiotic resistance (both endogenous and acquired) is an important determining factor in the historical development of the aminoglycosides as therapeutic agents. After streptomycin was introduced for the treatment of tuberculosis, it was found that bacterial resistance to the drug often developed; this was shown to be due to spontaneous mutants arising during the course of therapy with the antibiotic, although the biochemical mechanism was not known at the time. Kanamycin, the first useful DOS aminoglycoside, was isolated in Japan in 1957 and rapidly became an antibiotic of choice in that country. However, the appearance of strains resistant to both streptomycin and kanamycin increasingly interfered with their therapeutic use; in addition, hospital infections of *Pseudomonas aeruginosa,* a bacterium that is naturally less susceptible to antibiotics, were on the rise. A major breakthrough came with the discovery of a novel class of 2–DOS compounds, the gentamicins.[34] These are extremely effective antibiotics with good activity against the pseudomonads and other problem pathogens, such as *Proteus* and *Serratia* species, that were being increasingly encountered as nosocomial infections.

Gentamicin and related compounds lack the 3′ OH group, and the absence eliminates the modification by phosphorylation at this site and confers activity against pathogens possessing aminoglycoside 3′ OH phosphotransferases; gentamicin, being a mixture, contains one component with a modified 6′ amino group and has reasonable potency against strains harboring plasmid-encoded 6′ acetytransferases that inactivate kanamycin. By this time it was known that resistance to antibiotics by enzymic modification could be acquired by plasmid transfer. Gentamicin was also effective for the treatment of staphylococcal and enterococcal infections, frequently being used in combination with a β-lactam antibiotic in these circumstances.

In spite of its nephrotoxicity, gentamicin was the treatment of choice for gram-negative nosocomial infections for many years, and its success led to the introduction of tobramycin, a related compound. However, novel antibiotic resistance mechanisms began to appear on the scene; of particular concern was the adenylylation of the 2′ OH of gentamicin and related compounds that appeared on the scene in 1971 and conferred high-level resistance to the newest generation of aminoglycosides.[35,36] The increasing, worldwide use of different aminoglycosides led to the appearance of many different types of resistant strains; the local use of specific classes of aminoglycoside often led to the selection of distinct local classes of resistance.[37]

Fortunately, the discovery of a novel DOS derivative in 1971 provided the next breakthrough. This compound, butirosin, related to ribostamycin and produced by a *Bacillus* species (not an actinomycete), inhibits a variety of aminoglycoside-resistant hospital pathogens, including those inactivating the drugs by 3′ phosphorylation and 2″ adenylylation.[38] This property is due to the presence of a 4-hydroxy-2-aminobutyric acid (HABA) substituent on the 1-position of the DOS of butirosin. The latter antibiotic lacked good pharmaceutical properties, but synthetic insertion of a HABA or related group on the 1-position of the DOS of kanamycin (and subsequently of gentamicin-derived compounds) provided a novel series of potent semisynthetic aminoglycoside antibiotics with improved activity against a number of types of resistant strains. In particular, amikacin, (Figure 1.3), a kanamycin derivative with a broad spectrum of activity against resistant strains, has had considerable clinical and commercial success.[39] Since its discovery in 1976, no chemical modifications of substance have been reported, in spite of the fact that the spread of resistance has continued unabated and new resistance enzymes and efflux systems have appeared,[40] in particular a great variety of 6′ acetyltransferases.[41] Effort has been channeled primarily to tinkering with the DOS core.[42]

A variety of bacterial genera have been shown to produce aminoglycoside–aminocyclitol antibiotics. These include *Streptomyces, Micromonospora, Bacillus*, and so on. Only those compounds emanating from Streptomyces are named "-mycins" (e.g., tobramycin) while others are "-micins" (gentamicin), "-osins," "-asins," or "-acins." The biosynthetic pathways for the aminoglycosides and the control of their expression are not well-studied. Streptomycin is the exception, and Piepersberg's group has contributed significantly to this effort.[43] The intricacy of the biosynthesis is evident from the fact that upwards of 30 enzymatic steps are required for the formation of streptomycin from D-glucose.

The therapeutic use of aminoglycosides has diminished somewhat, but they are still important potent and widely used antibiotics in hospitals; most of the class are now generics. There is no question that a novel aminoglycoside derivative with demonstrated activity against the current generation of resistant pathogens, and preferably with reduced toxicity, would be a welcome addition for the treatment of infectious diseases. Attempts have been made to produce inhibitors of one or more of the aminoglycoside-modifying enzymes, and a number of different small molecule inhibitors have been described.[44–47] In principal, such inhibitors could be used in combination with an aminoglycoside for the treatment of resistant infections, much like the successful combination of a β-lactam antibiotic with a lactamase inhibitor. However, none of the inhibitors of aminoglycoside resistance enzymes have been employed in serious clinical trials. Given the increasing problems of antibiotic resistance in hospitals worldwide, it is surprising that this approach has not been pursued with more purpose.

1.4. TOXICITY

As previously mentioned, another drawback limiting an expanded therapeutic use of the aminoglycosides is their toxicity, which varies in form and intensity with the different types of molecules; the main toxic responses are ototoxicity and renal toxicity.[48] Streptomycin and other aminoglycosides target sensory hair cells of the inner ear and can lead to hair-cell degeneration and permanent loss; this occurs by an as yet undetermined mechanism and leads to irreparable hearing loss in up to 5% of patients on extended treatment with aminoglycosides.[49] A variety of dosing regimens have been employed and shown to reduce the incidence of toxicity.[50] On the positive side, significant advances in understanding of the general mechanisms of drug-induced ototoxicity in recent years have provided important information on the genetic and structural elements of hearing loss in humans; it would appear that mutations affecting mitochondrial rRNA predispose to aminoglycoside ototoxicity.[51]

From a therapeutic point of view, however, relatively little effort has been put into attempts to redesign aminoglycoside structure to reduce toxic responses, probably because good *in vitro* testing models have not been available. The largely random analyses of structure–activity relationships between the inhibitory and toxicity responses of the aminoglycosides have provided few significant insights into the problem. One has the impression that, because the two responses are so closely related in structure–activity terms, a less toxic, equipotent aminoglycoside is unattainable! An interesting series of experiments on the relationship between activity against eukaryotic cells and the role of the various functional groups of the DOS aminoglycosides has provided some valuable clues concerning antibiotic/ribosome/rRNA interactions,[52,53] but this work has not yet been exploited with reference to toxic responses during aminoglycoside therapy. Obviously, such information would be of great value in the design of new aminoglycosides for use as antimicrobials or in other therapeutic applications. To date, the development of semisynthetic aminoglycosides has been largely driven by the goal of finding compounds active against evolving resistant or recalcitrant bacterial pathogens.

1.5. CONCLUSIONS AND COMMENTS

The aminoglycosides have wide-ranging properties and are known to enhance or interfere with many cellular processes; (Table 1.1) significant and diverse research on potential medical and industrial applications is ongoing (see relevant chapters in this volume). The fact that this class of compounds interacts specifically with different types of nucleic acids has long been known and explored extensively. For example, the aminoglycosides interact with ribozymes and other forms of catalytic RNA, and the possibility of using small molecule effectors to modulate RNA reactions has been the subject of many investigations.[54,55] Much effort has gone into work on aminoglycoside-based inhibitors of the replication and function of viral RNA genomes.[56,57]

It has been suggested that bioactive small molecules, including the aminoglycosides, may have acted as naturally occurring allosteric effectors of catalytic RNAs during the "RNA world" stage of chemical evolution;[58] they may have been among the original riboswitches.[59] Aminoglycosides (and related compounds) could equally well have been involved as effector molecules in primordial DNA-based reactions; as has been mentioned, in the presence of aminoglycosides, DNA directs polypeptide synthesis on ribosomes.[22]

Finally, it should be clear to the reader that the aminoglycosides are a biologically and chemically diverse class of molecules with great therapeutic potential. Their exploitation as molecular tools in chemical biology applications is of continuing interest. The structures of these relatively simple natural products can be manipulated synthetically and, in principle, by methods of combinatorial biology. The latter molecular genetic-based approach has been successfully applied to achieve novel structural modification of the polyketides[60] and nonribosomal peptides.[61] Since the biosynthetic gene clusters of a number of aminoglycosides have recently been cloned and sequenced,[62] the stage is set for the use of molecular genetic approaches to develop aminoglycoside biology in greater depth. This approach holds great promise for the discovery and development of novel molecules that are not be readily available by chemical synthesis. The aminoglycosides are still very much alive!

ACKNOWLEDGMENTS

I am indebted to Bernard D. Davis, who introduced me to streptomycin in 1962. Thank you to all my laboratory colleagues from the University of Wisconsin

(1967–1980), especially Masayasu Nomura, for friendship, scientific contributions, and good times. Financial support was provided by the National Institutes of Health and the National Science Foundation. I am very grateful to Bristol-Myers, Eli Lilly, the Upjohn Company, and Schering Plough for unrestricted supplies of materials, much-needed information, and excellent technical support. Finally I wish to thank Dorothy Davies for her patient editorial efforts.

REFERENCES

1. Schatz, A.; Bugie, E.; Waksman, S. A. *Proc. Soc. Exp. Biol. Med.* **1944,** *55*, 66–69.
2. Waksman, S. A. *Antimicrob. Agents Chemother.* **1965,** *5*, 9–19.
3. Demain, A. *Appl. Microbiol. Biotechnol.* **1999,** *52*, 455–463.
4. Maarouf, M.; Lawrence, F.; Croft, S. L.; Robert-Gero, M. *Parasitol. Res.* **1995,** *81*, 421–425.
5. Hayes, W. *Br. Med. Bull.* **1962,** *18*, 36–40.
6. Sager, R. *Science* **1960,** *132*, 1459–1465.
7. Woese, C. R. *J. Mol. Evol.* **1977,** *10*, 93–96.
8. Cohen, S. N.; Chang, A. C.; Boyer, H. W.; Helling, R. B. *Proc. Natl. Acad. Sci. USA* **1973,** *70*, 3240–3244.
9. Valvekens, D.; Van Montagu, M.; Van Lijsebettens, M. *Proc. Natl. Acad. Sci. USA* **1988,** *85*, 5536–5540.
10. Davies, J.; Jimenez, A. *Am. J. Trop. Med. Hyg.* **1980,** *29*, 1089–1092.
11. Gritz, L.; Davies, J. *Gene* **1983,** *25*, 179–188.
12. Kasid, A.; Morecki, S.; Aebersold, P.; Cornetta, K.; Culver, K.; Freeman, S.; Director, E.; Lotze, M. T.; Blaese, R. M.; Anderson, W. F.; Rosenberg, S. A. *Proc. Natl. Acad. Sci. USA* **1990,** *87*, 473–477.
13. Fitzgerald, R. J.; Bernheim, F.; Fitzgerald, D. B. *J. Biol. Chem.* **1948,** *175*, 195–200.
14. Erdos, T.; Ullmann, A. *Nature* **1959,** *183*, 618–619.
15. Spotts, C. R.; Stanier, R. Y. *Nature* **1961,** *192*, 633–637.
16. Davies, J. E. *Proc. Natl. Acad. Sci. USA* **1964,** *51*, 659–664.
17. Cox, E. C.; White, J. R.; Flaks, J. G. *Proc. Natl. Acad. Sci. USA* **1964,** *51*, 703–709.
18. Davies, J.; Gilbert, W.; Gorini, L. *Proc. Natl. Acad. Sci. USA* **1964,** *51*, 883–890.
19. Carter, A.; Clemons, W.; Broderson, D.; Morgan-Warren, R.; Wimberley, B.; Ramakrishnan, V. *Nature* **2000,** *407*, 340–348.
20. Edelmann, P.; Gallant, J. *Cell* **1977,** *10*, 131–137.
21. Thompson, J.; O'Connor, M.; Mills, J. A.; Dahlberg, A. E. *J. Mol. Biol.* **2002,** *322*, 273–279.
22. Bretscher, M. S. *Nature* **1968,** *220*, 1088–1091.
23. Barton-Davis, E. R.; Cordier, L.; Shoturma, D. I.; Leland, S. E.; Sweeney, H. L. *J. Clin. Invest.* **1999,** *104*, 375–381.
24. James, P.; Raut, S.; Rivard, G.; Poon, M.-C.; McKenna, S.; Leggo, J.; Lillicrap, D. *Blood* **2005,** *106*, 3043–3048.
25. Goh, E.; Yim, G.; Tsui, W.; McClure, J.; Surette, M.; Davies, J. *Proc. Natl. Acad. Sci. USA* **2002,** *99*, 17025.

26. Yim, G.; Wang, H.; Davies, J. *Int. J. Mol. Microbiol.* **2006,** *296.*
27. Nomura, M.; Morgan, E. A. *Annu. Rev. Genet.* **1977,** *11*, 297–347.
28. Nomura, M. *Trends Biochem. Sci.* **1997,** *22*, 275–279.
29. Noller, H. *Science* **2005,** *309*, 1508–1514.
30. Ogle, J.; Ramakrishnan, V. *Annu. Rev. Biochem.* **2005,** *74*, 129–177.
31. Wilson, D. N.; Nierhaus, K. H. *Crit. Rev. Biochem. Mol. Biol.* **2005,** *40*, 243–267.
32. Magnet, S.; Blanchard, J. *Chem. Rev.* **2005,** *105*, 477–497.
33. Engelberg-Kulka, H.; Sat, B.; Hazan, R. *ASM News* **2001,** *67*, 617–624.
34. Weinstein, M. J.; Luedemann, G. M.; Oden, E. M.; Wagman, G. H. *Antimicrob. Agents Chemother.* **1963,** *161*, 1–7.
35. Martin, C. M.; Ikari, N. S.; Zimmerman, J.; Waitz, J. A. *J. Infect. Dis.* **1971,** *124*, Suppl 124:24–29.
36. Benveniste, R.; Davies, J. *FEBS Lett.* **1971,** *14*, 293–296.
37. Miller, G. H.; Sabatelli, F. J.; Naples, L.; Hare, R. S.; Shaw, K. J. *J. Chemother.* **1995,** *7 Suppl 2*, 31–44.
38. Woo, P.; Dion, H.; Bartz, Q. *Tetrahedron Lett.* **1971**, 2625–2628.
39. Kawaguchi, H. *J. Infect. Dis.* **1976,** *134 Suppl*, S242–S248.
40. Poole, K. *J. Antimicrob. Chemother.* **2005,** *56*, 20–51.
41. Miller, G. H.; Sabatelli, F. J.; Hare, R. S.; Glupczynski, Y.; Mackey, P.; Shlaes, D.; Shimizu, K.; Shaw, K. J. *Clin. Infect. Dis.* **1997,** *24 Suppl 1*, S46–S62.
42. Busscher, G. F.; Rutjes, F. P.; van Delft, F. L. *Chem. Rev.* **2005,** *105*, 775–791.
43. Piepersberg, W. *Biotechnology* **1995,** *28*, 531–570.
44. Daigle, D. M.; McKay, G. A.; Wright, G. D. *J. Biol. Chem.* **1997,** *272*, 24755–24758.
45. Liu, M.; Haddad, J.; Azucena, E.; Kotra, L. P.; Kirzhner, M.; Mobashery, S. *J. Org. Chem.* **2000,** *65*, 7422–7431.
46. Kotra, L. P.; Haddad, J.; Mobashery, S. *Antimicrob. Agents Chemother.* **2000,** *44*, 3249–3256.
47. Boehr, D. D.; Draker, K. A.; Koteva, K.; Bains, M.; Hancock, R. E.; Wright, G. D. *Chem. Biol.* **2003,** *10*, 189–196.
48. Rougier F. C. D.; Maurin M,; Sedoglavic A.; Ducher M.; Corvaisier S.; Jeliffe R.; Maire P. *Antimicrob. Agents Chemother.* **2003,** *47*, 1010–1016.
49. Waguespack, J. R.; Ricci, A. J. *J. Physiol.* **2005,** *567*, 359–360.
50. Peloquin, C. A.; Berning, S. E.; Nitta, A. T.; Simone, P. M.; Goble, M.; Huitt, G. A.; Iseman, M. D.; Cook, J. L.; Curran-Everett, D. *Clin. Infect. Dis.* **2004,** *38*, 1538–1544.
51. Guan, M.-X.; Fischel-Ghodsian, N.; Attardi, G. *Hum. Mol. Genet.* **2000,** *9*, 1787–1793.
52. Hobbie, S.; Pfister, P.; C, B.; Westhof, E.; Bottger, E. *Antimicrob. Agents Chemother.* **2005,** *49*, 5112–5118.
53. Pfister, P.; Hobbie, S.; Brull, C.; Corti, N.; Vasella, A.; Westhof, E.; Bottger, E. C. *J. Mol. Biol.* **2005,** *346*, 467–475.
54. Pearson, N. D.; Prescott, C. D. *Chem. Biol.* **1997,** *4*, 409–414.
55. Schroeder, R.; Waldsich, C.; Wank, H. *Embo J.* **2000,** *19*, 1–9.
56. Zapp, M. L.; Stern, S.; Green, M. R. *Cell* **1993,** *74*, 969–978.

57. Riguet, E.; Desire, J.; Boden, O.; Ludwig, V.; Gobel, M.; Bailly, C.; Decout, J. L. *Bioorg. Med. Chem. Lett.* **2005,** *15*, 4651–4655.
58. Davies, J.; von Ahsen, U.; Schroeder, R. In *The RNA World;* Gesteland, R. F.; Atkins, J. F. Eds.; Cold Spring Harbor, NY: Cold Spring Harbor Laboratory Press, 1993; pp. 185–204.
59. Winkler, W. C. *Curr. Opin. Chem. Biol.* **2005,** *9*, 594–602.
60. Cane, D. E.; Walsh, C. T.; Khosla, C. *Science* **1998,** *282*, 63–68.
61. Baltz, R.; Brian, P.; Miao, V.; Wrigley, S. *J. Ind. Microbiol. Biotechnol.* **2006,** *33*, 66–74.
62. Unwin, J.; Standage, S.; Alexander, D.; Hosted, T., Jr.; Horan, A. C.; Wellington, E. M. *J. Antibiot. (Tokyo)* **2004,** *57*, 436–445.

2

THE BIOCHEMISTRY AND GENETICS OF AMINOGLYCOSIDE PRODUCERS

WOLFGANG PIEPERSBERG, KHALED M. ABOSHANAB, HEIKE SCHMIDT-BEIßNER, AND UDO F. WEHMEIER

BU Wuppertal, Chemical Microbiology, D-42097 Wuppertal, Germany

Aminoglycoside Antibiotics: From Chemical Biology to Drug Discovery, Second Edition, by Dev P. Arya

2.1. INTRODUCTION

Research on aminoglycoside antibiotics (AGAs; see Table 2.1 for AGA designations and abbreviations used throughout this chapter), since its start in the early 1940s, has gone through the phases of screening, elucidation of product patterns, and chemical structures, along with studies on the pharmacology. Still ongoing are studies on the mechanisms of action on the bacterial ribosome. Work on the biogenesis, biosynthesis, and physiology of AGAs in their producers by chemical and biochemical methods started later, which was paralleled by studying the biochemistry and genetics of resistance determinants in both producers and clinically relevant bacteria. Only recently the molecular biological analysis of the biosynthetic potential of AGA producers was started, also in order to find better conditions for their production or ways for the *in vivo* biocombinatorial production of engineered AGAs. The purpose of this chapter is to bring together the most recent progress in the genetics of aminoglycosides, especially during the last decade,[1,2] with the results of earlier work on the production profiles of their producers, biogenetic studies (e.g., *in vivo* labeling of aminoglycosides),

TABLE 2.1. Aminoglycoside Antibiotics (AGAs), Designations, and Abbreviations Used Throughout This Chapter

AGA (Complex, etc.)	Compound	Biosynthetic Genes	Biosynthetic Proteins
Apramycin	APR	*apr*	Apr
Bluensomycin	BLU	*blu*	Blu
Butirosin	BTR	*btr*	Btr
Dactimicin	DCT	(*for*)	(For)
Fortimicin	FOR	*for*	For
Gentamicin	GEN	*gen*	Gen
Hygromycin	HYG	*hyg*	Hyg
Istamycin	IST	*ist, iss*	Ist, Iss
Kasugamycin	KAS	*kas, ksg*	Kas, Ksg
Kanamycin	KAN	*kan*	Kan
Lividomycin	LIV	*liv*	Liv
Nebramycin	NEB	(*apr, tob*)	(Apr, Tob)
Neomycin	NEO	*neo*	Neo
Paromomycin	PAR	*par*	Par
Ribostamycin	RIB	*rib*	Rib
Sagamicin	SGM	(*gen*)	(Gen)
Streptomycin	STR	*str, sts*	Str, Sts
Spectinomycin	SPC	*spc, spe*	Spc, Spe
Sporaricin	SPO	n.a.	n.a.
Sisomicin	SSM	(*gen*)	(Gen)
Tobramycin	TOB	*tob*	Tob
Verdamicin	VDM	(*gen*)	(Gen)

n.a., not available; as yet no gene/protein sequences published.

mutant analysis of nonproducing mutants, biochemical studies on biosynthetic enzymes, and the like, in the aminoglycoside-producing bacterial strains used for their industrial production.[3–5] Our main goal in this chapter is to propose biosynthetic pathways for the major AGA classes as well as to discuss future possibilities with regard to understanding the biochemical and genetic evolution of AGAs and to use this knowledge for a pathway engineering leading to new end products with improved therapeutic properties.[6] Thus we will focus solely on the AGAs binding to the 16S rRNA component of the small subunit (30S) of the bacterial ribosome; we will leave out other aminoglycoside-related natural products, such as the glycosidic glucosidase inhibitors produced by actinomycetes or the major actinomycete thiol compound, mycothiol.[7–11] Recently we have isolated and analyzed the full-length biosynthetic gene clusters for most of the relevant AGAs in their genomic context from major producing strains and submitted the sequence data to the EMBL database; other groups also provided partial or complete sequence data for equal or further AGA clusters (see Table 2.2). Wherever differences exist in sequences of AGA gene clusters in doubly or triply submitted ones, we use our own data (without further notification), or this fact is stated without discussion of the nature of these differences themselves. The subdivision of the chapter in its core content will follow the genetic/biosynthetic families nature has formed during evolution.

2.1.1. The Genomes of Actinomycetes: Some General Remarks

Secondary metabolites (individualities) are *repellents* or *molecular weapons*, and an "antibiotic" would best fit into this category; they are metabolites of extra-organismal warfare directed against competitors, as well as against predators in the resident habitat. A special group in this category are the many extracellular enzyme inhibitors which do not have typical antimicrobial effects. Another group generally to be considered in this class are the many specific protein-based toxins and other macromolecules secreted by microorganisms that may be targeted against near relatives (e.g., bacteriocins) or distant taxa (general toxins), which may have a role in invasive attacks, quite apart from being protective.

Secondary metabolites are *chance products* of a steadily changing biosynthetic potential, based on a rapidly fluctuating genetic "playground of evolution." According to this view, the selective advantage would lie not in the product or the pathway of its synthesis, but rather in metabolic adaptability that allows rapid biochemical "invention" under changing environmental conditions. In this a larger group of organisms sharing a common exchangeable gene pool would participate to their advantage.

An attractive hypothesis is the independent evolution in bacteria of their diffusible individualites and the currently recognized secondary metabolic pathways, in parallel with their surface components and their biosynthesis. An indicator for this would be the use of the same gene pool.[12] The theory would include all substances that play a role in the build-up of glycan and other modified surface layers, lipids, murein, (glyco-) proteins (e.g., S-layers), polysaccharides, teichoic

TABLE 2.2. Partially or Fully Sequenced AGA Biosynthetic Gene Clusters Submitted to the International Databases[a,b]

Aminoglycoside (Complex) (Family)	Producer Strain	Accession Number	Boiling Point	ORFs in Cluster (Totally Sequenced)	Reference (Completely Sequenced)	Remarks
Streptomycin	*S. griseus* ssp. *griseus* DSM 40236	AJ862840	90,600	23 (126)	A	c
Streptomycin	*S. griseus* ssp. *griseus* DSM 40236	X53527	819	1 (1)	16	r
Hydroxystreptomycin	*S. glaucescens* GLA.0 (DSM 40716)	AJ006985	25,459	23 (23)	1	c
Bluensomycin	*S. bluensis* ISP5564	X78972	1,891	2	DS	
Bluensomycin	*S. bluensis* ATCC27420 (DSM40564)	AAD28516	4,094	3 (3)	DS	
Bluensomycin	*S. bluensis* ATCC27420 (DSM40564)	?	19,422	15 (15)	17	
Spectinomycin	*S. netropsis* DSM 40093	U70376	29,606	18 (19)	18	c
Spectinomycin	*S. spectabilis* DSM 40512	AF170704	6,682	6 (6)	DS	p
Spectinomycin	*S. spectabilis* DSM 40512	AF145039	5,635	2 (5)	DS	p
Kasugamycin	*S. kasugaensis* M338-M1 (DSM 40819)	AB120043	6,860	6 (6)	DS	p
Kasugamycin	*S. kasugaensis* M338-M1 (DSM 40819)	AB005901	7,581	8 (8)	19	p
Kasugamycin	*S. kasugaensis* M338-M1 (DSM 40819)	AB076838	3,755	3 (3)	20	p
Kasugamycin	*S. kasugaensis* M338-M1 (DSM 40819)	AB033992	4,236	4 (4)	21	p
Neomycin	*S. fradiae* DSM 40063	AJ629247	50,466	22 (38)	A	c
Neomycin	*S. fradiae* ATCC 10745	AJ786317	42,149	? (35)	DS	c
Neomycin	*S. fradiae* NCIMB 8233	AJ843080	30,837		22	
Neomycin	*S. fradiae*	AB211959	39,237		23	
Butirosin	*B. circulans* SANK 720703	AB097196	29,033	25 (27)	24,25	c

Butirosin	*B. circulans* ATCC 4513	X03364	1,320	1 (1)	26	r
Butirosin	*B. circulans* NRRL B3312	AJ494863	6,271		27	p
Butirosin	*B. circulans* NRRL B3312	AJ847918	4,422		DS	
Ribostamycin	*S. ribosidificus* NRRL B-11466	AJ744850	43,190	20 (41)	A	c
Ribostamycin	*S. ribosidificus* ATCC 21292	AJ748131	31,892		DS	
Paromomycin	*S. rimosus* ssp. *paromomycinus* NRRL 2455	AJ628955	31,940	22 (40)	A	c
Paromomycin	*S. rimosus* ssp. *paromomycinus* NRRL 2455	AJ749845	4,812	2 (7)	A	r
Lividomycin	*S.* sp. (*lividus*) CBS 844.73	AJ748832	40,579	23 (30)	A	c
Kanamycin	*S. kanamyceticus* DSM 40500	AJ628422	62,887	21 (28)	A	c
Kanamycin	*S. kanamyceticus* ATCC 12853 (DSM 40500)	AJ582817	47,050	20 (39)	28	c
Kanamycin	*S. kanamyceticus* M1164 (DSM 40500)	AB164642	25,467	24 (24)	29	p
Tobramycin	*S.* sp. (*tenebrarius*) DSM 40477	AJ810851	43,220	(37)	A	c
Tobramycin	*S. tenebrarius* ATCC 17920	AJ579650	13,802		30	
Gentamicin	*M. echinospora* DSM 43036	AY524043	38,146	(28)	31	p
Gentamicin	*M. echinospora* DSM 43036	AJ628149	84,222	(69)	A	c
Gentamicin	*M. purpurea*	AJ575934	32,668		32	p
Fortimicin	*M. olivasterospora* DSM 43868	AJ628421	42,807	(44)	A	c
Istamycin	*S. tenjimariensis* ATCC 31603	AJ845083	67,904	26 (67)	A	c
Apramycin	*S.* sp. (*tenebrarius*) DSM 40477	AJ629123	41,623	30 (35)	A	c
Apramycin	*Streptoall. hindustanus* DSM 44523	AJ875019	39,979	26 (31)	A	c(?)
Hygromycin B	*S. hygroscopicus* ssp. *hygroscopicus* DSM 40578	AJ628642	34,921	22 (33)	A	c

[a]Some database entries are included for AGA resistance genes that do not appear in the respective biosynthetic gene cluster from the same strain.
[b]Abbreviations: *S., Streptomyces; Streptoallo., Streptoalloteichus; M., Micromonospora; B., Bacillus;* A, authors; DS, direct submission to EMBL database; p, partially sequenced; r, resistance gene(s) outside biosynthetic clusters; sp., species; ssp., subspecies.

acids, and so on, which could have been precursors of antimicrobials and other extracellulary targeted secondary metabolites, or vice versa. The theoretical consideration would predict that all components that can form and variably modify the outer cell surfaces in a bound state (i.e., in polymer or membrane-integrated form) could be also used to produce diffusible low-molecular-weight compounds with cell–cell value for communication.

The highly adaptive extracellulary targeted metabolism of antibiotic producers requires a corresponding, highly flexible genetic basis. This seems to be the case in typical producers, as exemplified by the case of the streptomycetes and other differentiating, mycelial actinomycetes.[13] The two complete genomes of typical actinomycete producers of secondary metabolites, *Streptomyces (S.) coelicolor* and *S. avermitilis*, both contain most of the essential genes of central ("primary") metabolism centered around the genetically much more stable segment that contains the origin of vegetative DNA replication.[14,15] This is localized in about the middle of the single linear chromosome of $\approx$ 9 Mb in length, which is about double the size of the typical enterobacterial circular *Escherichia coli* (E.) genome. This gene pool (gene pool I) is ubiquitous and constant, at least in a given taxon, and the expression of most genes is constitutive or dependent on the essential needs of the stages of the cell cycle. Its purposes or targets are replication, transcription, translation, cell-cycle control (differentiation), central intermediary and energy-generating metabolism, and synthesis of the essential parts of the cell wall envelope. Gene pool II, in contrast, is strain-specific, highly variable, and only used under specific circumstances. Its individual genes or functional gene sets (e.g., production gene clusters) are highly regulated by environmental conditions and may be additionally dependent on cell differentiation.

Higher rates of horizontal gene exchange, based on the abundant presence of plasmid transfer (Tra), plasmid mobilization (Mob), transposons (Tn), insertion sequences (IS), and similar functions or elements, seem to be a rule in these variable gene pools.

One example might illustrate what we can expect from future analyses of microbial genomes, with respect to the gene pool used for the evolution of gene clusters for AGA production: Recently, an almost complete "AGA gene cluster" (*fra* cluster), which is mostly related to the *for* cluster from *Micromonospora olivasterospora* (cf. Tables 2.18 and 2.19, Section 2.2.3), came up in the just released genome sequence of *Frankia* sp. CcI3 (accession code CP000249; loci tags Francci3_3356 through Francci3_3376) which excellently fits into such a scenario. This gene set comprises "S"-, "E"-, "M"-, "Q"-, "L2"-, "L"-, and "O"-type and "HIJ"-related genes and a relative to typical AGA resistance genes encoding rRNA methyltransferase, a gene encoding a putative inositol dehydrogenase, and those encoding further members of the aminotransferase and radical SAM families of oxidoreductases-transferases, and so on. Close relatives (protein identity values higher than 60%) are especially found in the *for*- and *gen*- clusters. Thus, *Frankia* sp. CcI3 should be able to condense an amino sugar to a cyclitol moiety or two sugar units, attach amino groups, and modify the hypothetical pseudodisaccharide to form a FOR-like AGA metabolite. Above all, the presence of a possibly AGA-specific MFS efflux system ("HIJ"-like proteins, see Section 2.3.2) gives evidence for the ability to produce an excreted compound with AGA similarity in *Frankia* sp. It will be interesting to see whether such a compound is produced in the future.

2.1.2. Gene Clusters for AGAs

By now about 15 complete gene clusters for AGA production have been cloned and analyzed from their producers genomic DNA (see Tables 2.3–2.21). In some cases additional (mostly resistance) genes have been detected outside these clusters in a given strain. Also, some more incomplete data are available for additional AGA biosynthetic clusters. However, the genetic data are now sufficient to draw first conclusions on all aspects of their impact on different AGA production, such as questions of enzymology of the biosynthetic and export pathways, resistance mechanisms, cluster-specific regulation, genomic location of clusters and their putative ways of dissemination in bacterial, especially actinomycete, genomes, and last not least their evolution. Both the degrees of conservation/divergence of individual genes as well as of full-length clusters can now be quantitatively measured. Also, the studies of the evolutionary forces working on the biosynthetic potentials in secondary metabolic traits find here a very nice example, where the aspects of drift by functional selection and shift by mere DNA sequence alterations by mutation can be differentiated. Also, many details working by recombinational events on the DNA level can be pinpointed, such as rather recent recombination between gene clusters for quite different end products, which give additional examples of "natural biocombinatorial pathway engineering."

When we have a closer look at the gene clusters for AGAs described in this chapter, we will see that they are excellent model systems for studying the evolution of secondary metabolite gene clusters: First, all gene clusters analyzed in this study are obviously located in the variable regions of the chromosome. The only exception is the *str*-gene cluster from *S. griseus*, which lies in the center region. Second, clusters coding for closely related end products can be very similar as seen in the neomycin group. This indicates that they evolved from a common ancestor which had the majority of necessary genes already successfully assembled. The integration of additional genes and the independent evolution of the producer strains then led to the individual clusters. But we also find examples that the gene clusters for chemically related end products are differently organized—for example, the *for* and *ist* cluster. Here one has to speculate that these clusters have evolved independently and that the producers had acquired the necessary genes from an existing gene pool and that they were assembled in different ways. Third, the presence of *Tn* or *IS* elements within give evidence that some of the genes might have been transferred as groups. Fourth, gene clusters encoding end products which are chemically to be regarded as a mixture of two other end products (e.g., GEN, which shares subunits and modification patterns with KAN and FOR) may have evolved by fusions of whole clusters and subsequent recombinations and deletions. Fragments of *for* genes (e.g., the "erosion zone" in the *for* cluster, cf. Figure 2.22) which have close relatives in the *gen* cluster could be the results of such recent evolutionary events, where doubled or unnecessary genes had been deleted.

2.2. THE AGA FAMILIES

2.2.1. *Myo*-Inositol and 6-Deoxyhexose-Based AGAs: STR-Related AGAs

2.2.1.1. Producers, Genetics, and Biosynthesis of Streptomycins (STRs). Streptomycin (STR) became the first AGA to be detected and closely related AGAs were described later (Figure 2.1). STR had been originally described in 1944 and extensively investigated thereafter with respect to its occurrence, antibacterial scope, pharmacokinetics, and mode of action.[33–39] Later, quite extensive genetic and biochemical studies in the field of resistance development toward STR had been carried out.[40–43] The STRs also were the first class of AGAs for which attempts had been made throughout the 1980s and 1990s to elucidate the genetics and biochemistry of production in their producers. The results of these studies have been reviewed several times in the past.[1,2,7,41,44–47] Today, we know that *str* clusters are wide spread in natural populations of streptomycetes worldwide,[48] indicating that AGAs are a generally used and ecologically relevant metabolite in this group of organisms.[49]

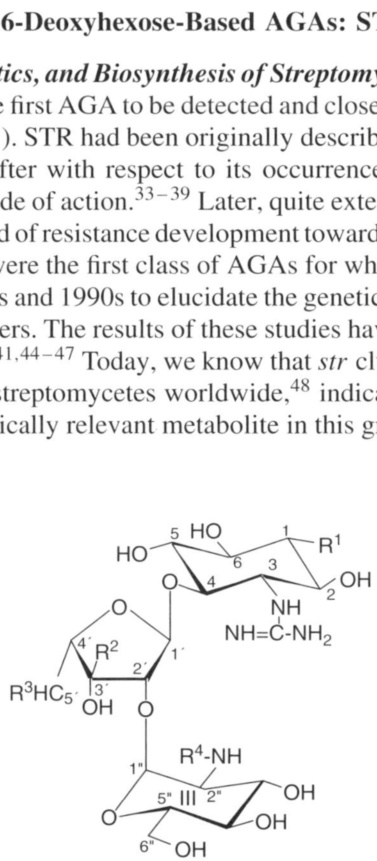

STR-related AGAs	R^1	R^2	R^3	R^4
Streptomycin	$NHCNHNH_2$	CHO	H	CH_3
Dihydrostreptomycin	$NHCNHNH_2$	CH_2OH	H	CH_3
5'-Hydroxystreptomycin	$NHCNHNH_2$	CHO	OH	CH_3
Bluensomycin	$OCONH_2$	CH_2OH	H	CH_3
N-Demethylstreptomycin	$NHCNHNH_2$	CHO	H	H

Figure 2.1. Structures of streptomycin (STR) family AGAs.

Sequence data for the *str*/*sts*-gene clusters for the production of STR from *S. griseus* DSM 40236 and of 5′-hydroxy-STR from *S. glaucescens* GLA.0 (DSM 40716, ETH 22794) had been collected in several laboratories and in a piece-by-piece approach; that is, sequences from small, individually subcloned genomic segments were determined and fused to larger contigs. Therefore, the completeness of the sequences from both gene clusters was uncertain until recently. Also, it turned out that a few genes present in the *S. glaucescens* cluster, *strPQ* and *strX*, could be absent from *S. griseus*. We, therefore, together with others set out to analyze the full-length sequence covering the *str*/*sts*-cluster and its genomic neighborhood in *S. griseus* DSM 40236 on a larger segment. The data obtained from a $\approx$ 90-kb segment (see accession code AJ862840) revealed that the *str*/*sts* cluster was integrated in a conserved genomic area in *S. griseus* DSM 40236 and that the *strU* and *strT* genes, incompletely analyzed before,[1] were really the genes at the respective ends of the cluster; an additional gene adjacent to *strT*, named *strZ*, could still be part of the cluster (see Table 2.3 and Figure 2.2). In comparison to the data from *S. glaucescens* GLA.0, no indications for the presence of additional genes related to the *strPQ* and *strX* genes could be found. For the *strQ* gene an earlier study had already suggested the absence of a counterpart from the *S. griseus* genome as a whole.[50,51] Such, clear and fundamental differences seem to exist in the gene composition between the otherwise closely related clusters and in the concomitant pathway design for STR and 5′-hydroxy-STR in *S. griseus* DSM 40236 and *S. glaucescens* GLA.0, respectively; this had already been suspected in earlier comparisons.[2,7] Also, it should be noted that not all the genes related to STR production are included in the *str*/*sts*-cluster, as the *aphE* resistance gene, encoding a STR-modifying APH(3″) enzyme, has not been found inside the *str*/*sts*-gene cluster or its immediate vicinity (cf. accession code AJ862840).[16,52,53] Thus, in summary the *str*/*sts*-cluster in *S. griseus* DSM 40236 contains 27 or 28 genes, dependent on whether or not the assigned *strZ* gene belongs to the cluster. The respective *str*/*sts*-cluster of *S. glaucescens* GLA.0 (DSM 40716) consists of 22 sequenced genes and some 7 or so genes mapped by hybridization only. It is unknown whether a gene equivalent to *strZ* is present or not; at least three of the genes in this cluster do not seem to occur in *S. griseus* DSM 40236 (see above; cf. Tables 2.3 and 2.4 and Figure 2.2).

The partial sequences and hybridization data available for the *blu*- (*blm*) gene cluster of the bluesomycin (BLU; cf. Figure 2.1 and Table 2.5) producer *S. bluensis* ATCC 27420 (DSM 40564) cover at least 15 ORFs, most of which seem to be homologous to respective genes in the known *str*/*spc* gene clusters.[2,17] The *blu*-cluster also contains a gene that hybridizes to *strQ* and, therefore, seems to be more closely related to the *str*/*sts*-cluster of *S. glaucescens* GLA.0.[2] Also, the complete *spc*-clusters for the biosynthesis of the related spectinomycins (SPCs) from *S. netropsis* (former "*S. flavopersicus*") NRRL B-2820 and from *S. spectabilis* NRRL 2792 had been reported.[54] Only rather incomplete sequence data have been reported on the *kas/ksg*-cluster for the production of kasugamycin (KAS) from *S. kasugaensis* M338-M1 (DSM 40819),[19–21] which is a third subclass of biosynthetically STR-related AGAs.[2] Proposals for partial or

TABLE 2.3. Proteins Encoded in the Genomic Area Covering the *str-/sts*-Cluster of *Streptomyces griseus* subsp. *griseus* DSM 40236 (Accession Code AJ862840)[a]

ORF/Locus Tag[a]	Gene Product[b] Symbol	aa	CDD Number[c]	Putative Function
SG7F10.3		176	25476	cHP
SG7F10.4c	StrU	427	26082	NAD(P)-dep. oxidoreductase
SG7F10.5	StrV	584	29340	ABC transporter; STR-phosphate
SG7F10.6	StrW	580	29340	ABC transporter; STR-phosphate
SG7F10.7	StsG	253	7925	L-Glutaminyl-2-*N*-MT
SG7F10.8	StsF	229	7651	GT
SG7F10.9	StsE	330	25797	ATP:*scyllo*-inosamine-4-PT
SG7F10.10	StsD	214	25595	Aminosugar-converting enzyme (KasE)
SG7F10.11c	StsC	424	15279	L-Glutamine:*scyllo*-inosose AT
SG7F10.12	StsB	507	15245	NADH:*N*-amidino-*scyllo*-inosamine oxidoreductase
SG7F10.13	StsA	410	15279	L-Alanine:*N*-amidino-3-keto-*scyllo*-inosamine AT
SG7F10.14c	StrO	259	30135	*N*-Amidino-*scyllo*-inosamine-4 phosphate phosphatase
SG7F10.15	StrN	319	25797	*N*-Amidino-streptamine-6-PT
SG7F10.16	StrB2	350	17123	*scyllo*-Inosamine-4-phosphate amidinotransferase
SG7F10.17	StrM	200	4377	dTDP-4-dehydrorhamnose 3,5 epimerase
SG7F10.18c	StrL	304	23531	dTDP-L-rhamnose synthase
SG7F10.19c	StrE	329	23188	dTDP-D-glucose 4,6-dehydratase
SG7F10.20c	StrD	355	16736	dTTP:alpha-D-glucose-1-phosphate thymidylyltransferase
SG7F10.21	StrR	350	26725	Transcriptional regulator
SG7F10.22	StrA	343	15952	APH(6), STR-resistance
SG7F10.23	StrB1	348	17123	L-Arginine:*scyllo*-inosamine-4 phosphate amidinotransferase
SG7F10.24	StrF	284	—	Unknown
SG7F10.25	StrG	199	—	Unknown
SG7F10.26	StrH	384	—	Phosphoglucosamine-mutase
SG7F10.27	StrI	348	6854	*myo*-Inositol:NAD$^+$2-oxidoreductase
SG7F10.28	StrK	449	28900	STR-6-phosphate-6-phosphohydrolase
SG7F10.29	StrS	378	15279	*N*-Methyl-L-glucosamine AT
SG7F10.30	StrT	317	25456	Oxidoreductase
SG7F10.31	StrZ	423	—	Membrane-spanning protein
SG7F10.32c		325	4793	3-Hydroxyacyl-CoA DH

[a]The coding sequences (ORFs) are given with their locus tag numbers.

[b]The list of gene products translated from the nucleotide sequence corresponds to the protein ID numbers. CAH94301–CAH94353.

[c]Refers to the CDD database.[255]

[d]Abbreviations are defined in the legend to Figure 2.3.

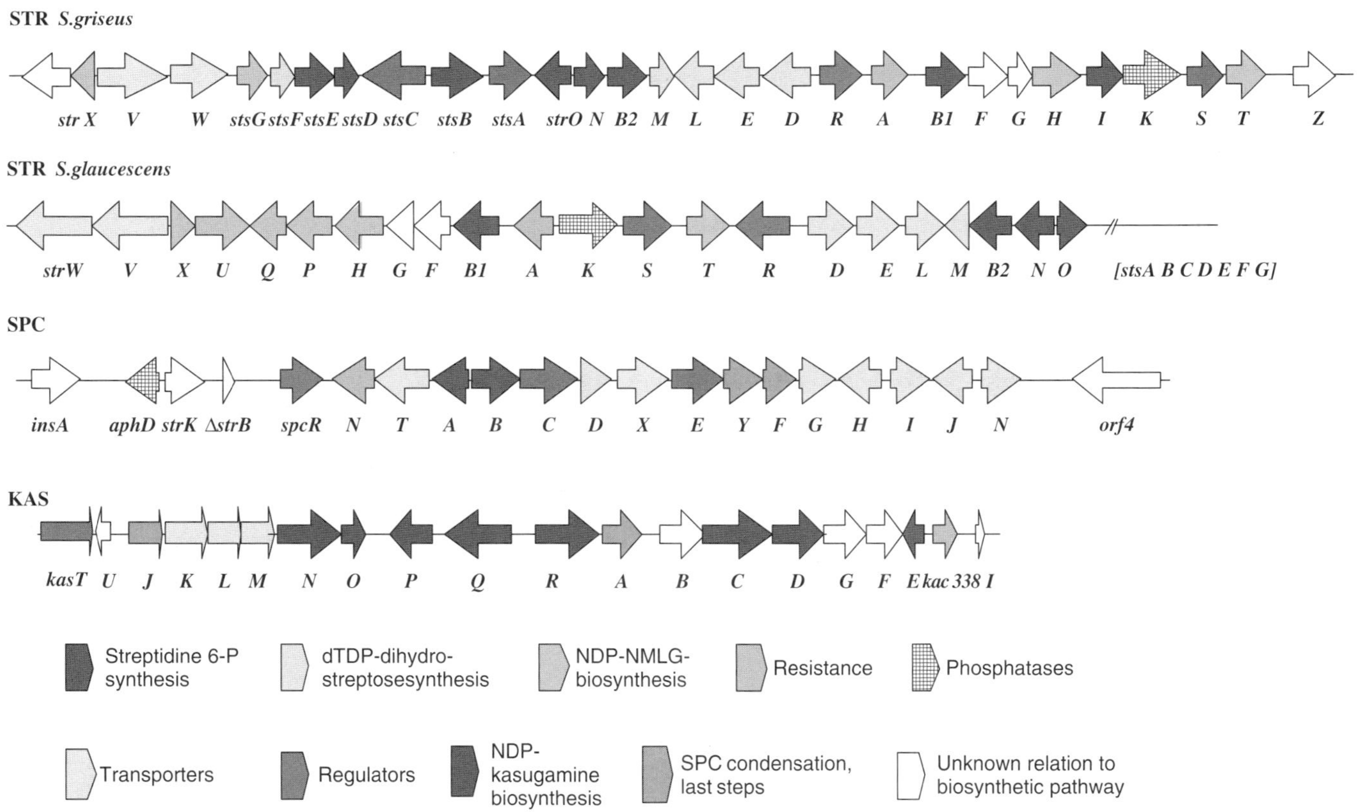

Figure 2.2. Biosynthetic gene clusters for STR-related AGAs. Functionally related groups of genes are differentiated by the color code given below. NMLG, *N*-methyl-L-glucosamine. Related groups of genes are differentiated by the color code given below. See color plates.

TABLE 2.4. Proteins Encoded in the Genomic Area Covering the *str*-Cluster of *Streptomyces glaucescens GLA.0* ATCC13032 (Accession Code AJ006985)

Gene Product[a] Symbol	aa	CDD Number[b]	Putative Function
StrW	591	29340	ABC transporter; 5′-hydroxydihydro-STR-phosphate
StrV	584	29340	ABC transporter; 5′-hydroxydihydro-STR-phosphate
StrX	182	4377	dTDP-4-ketohexose 3,5-epimerase
StrU	428	26082	NAD(P)-dep. oxidoreductase
StrQ	299	16736	Glucose-1-phosphate cytidylyl-transferase
StrP	358	23188	Glucose 4,6-dehydratase
StrH	385	—	Phosphoglucosamine mutase
StrG	201	—	Unknown
StrF	281	—	Unknown
StrB1	348	17123	L-Arginine:*scyllo*-inosamine-4 phosphate amidinotransferase
StrA	307	15952	APH(6), STR-resistance
StrK	460	28900	(5′-hydroxy)-STR-phosphate-phosphatase
StrS	378	15279	*N*-Methyl-L-glucosamine AT
StrT	319	25456	Oxidoreductase
StrR	424	26725	Transcriptional regulator
StrD	356	16736	dTTP:alpha-D-glucose-1-phosphate thymidylyltransferase
StrE	329	23188	dTDP-D-glucose 4,6-dehydratase
StrL	305	23531	dTDP-L-rhamnose synthase
StrM	200	4377	dTDP-4-dehydrorhamnose 3,5-epimerase
StrB2	319	17123	*scyllo*-Inosamine-4-phosphate amidinotransferase
StrN	315	25797	*N*-Amidino-streptamine-6-PT
StrO	256	30135	*N*-Amidino-*scyllo*-inosamine-4 phosphate phosphatase

[a]The list of gene products translated from the nucleotide sequence corresponds to the protein ID numbers. CAA07371–CAA07392.
[b]Refers to the CDD database.[255]

preliminarily complete biosynthetic pathways have been published for all three of the STR-related subclasses.[1,20,47,54] We combine all currently available data to actualize these proposals in the postulated pathways (Figures 2.3, 2.5, and 2.7). However, none of these biosynthetic pathways has been analyzed to completeness on the biochemical/enzymological level so far.

2.2.1.2. The STR Pathway. Characteristic common features of the STR, SPC, and KAS pathways (Figures 2.3, 2.5, and 2.7) are the involvement of both a *myo*-inositol-based cyclitol branch and a dTDP-6-deoxyhexose branch.[1,2] Most efforts

TABLE 2.5. Proteins Encoded in the Genomic Region Covering Part of the *blu*-cluster of *Streptomycetes bluensis* ISP5564 (ATCC27420) (Accession Codes X78972, AF126354)

Gene Product Symbol	aa	CDD Number[a]	Putative Function
BlmW		29340	ABC transporter
BlmV		29340	ABC transporter
BlmC		23354	*O*-Carbamoyltransferase
BlmY		—	Unknown
BlmA		25797	BLU-6-phosphotransferase, resistance
BlmX		7925	MT
BlmU		26082	NDP-hexose oxidoreductase
blmH		10312	GT
BlmG		—	NDP-hexose epimerase
StrF/BlmF	(267)	—	NDP-hexose epimerase
StrB/BlmB	(355)	17123	Amidinotransferase
StrS/BlmS	378	15279	AT
StrT/BlmT	318	25456	Aldo/keto reductase
StrD/BlmD	355	16736	dTDP-D-glucose synthase/nucleotidyl-transferase
BlmE		23188	dTDP-D-glucose dehydratase

[a]Refers to the CDD database.[255]

in this field so far have been invested into the elucidation of the STR pathway, starting with work of the groups of S. A. Waksman,[36,37,55,56] J. B. Walker,[3,46,57,58] and W. H. Horner[59–62] and later continued into the genetic area.[1,2,7,47,52]

Not much has been added to the biochemistry of STRs in the recent years.[1,7] Therefore, we will only briefly treat the biosynthetic pathway here. Biosynthesis of the aminocyclitol (streptidine) and 6-deoxyhexose (streptose) moieties are well understood, with some remaining gaps in the enzymology.[1,3,7] Much less is known about the residual biosynthetic phases, the formation of *N*-methyl-L-glucosamine, the condensation of subunits, and the late modification steps potentially coupled with export.[7,63]

2.2.1.2.1. The General myo-Inositol and scyllo-Inosamine Pathway. Two distinct cyclitol pathways, which we earlier called Ca (started by *myo*-inositolphosphate synthase) and Cb (started by 2-deoxy-*scyllo*-inosose synthase), are used in the initiation of AGA pathways.[2] Here we will describe the Ca route and the Cb pathway in the paragraph on NEOs (see Section 2.2.2.1.3). The cyclitol pathway in the biosynthetic pathways for STRs, but also for other AGAs such as SPCs, KAS, and FORs and also the mixed type antibiotic HYG-A (see Section 2.2.4.3.1), starts with the formation of *myo*-inositol in a two-step pathway (Ca) that is not encoded in the *str/sts*-clusters. The first step in streptidine biosynthesis (and any other *myo*-inositol utilising pathway) is the formation of a *myo*-inositol monophosphate (D-*myo*-inositol-3-phosphate or L-*myo*-inositol-1-phosphate) via L-*myo*-inositol-1-phosphate synthase, which in actinomycetes

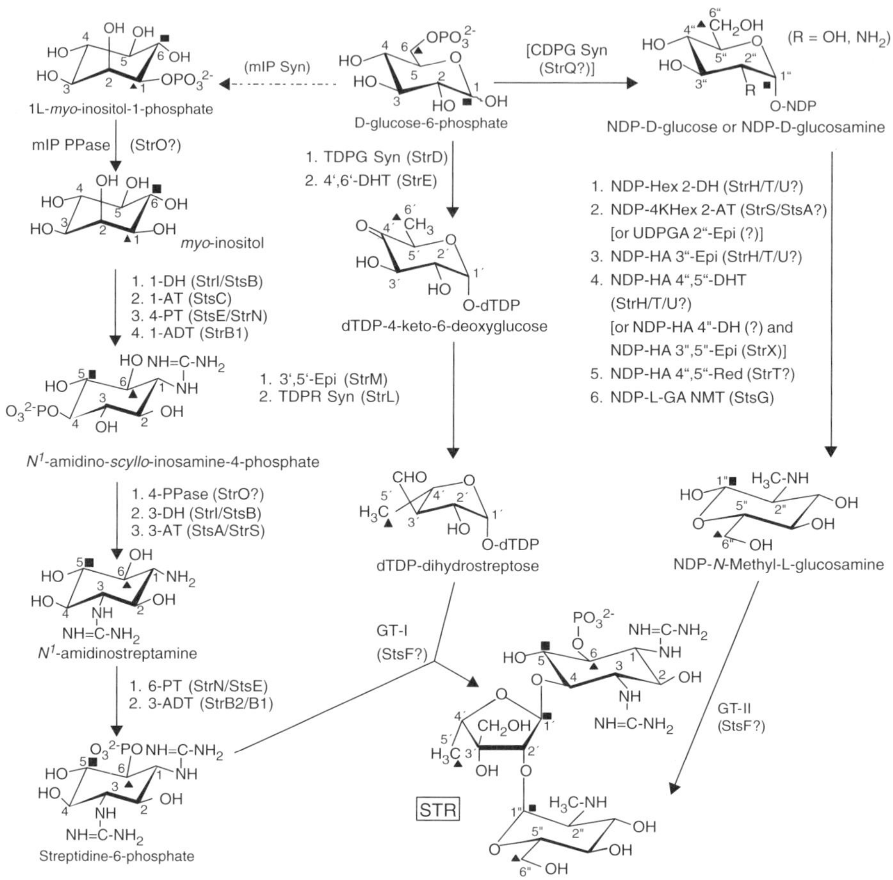

Figure 2.3. The proposed biosynthetic pathway for STRs. The enzymes known or postulated (indicated by a question mark) to be involved in the pathway are given in abbreviated form (cf. Tables. 2.3–2.23 for explanation). The squares and triangles mark the C-atoms derived from the C-1 and C-6 of *D*-glucose, respectively. Abbreviations generally used in figure legends on biosynthetic pathways: ADT, (N-)amidinotransferase; AT, aminotransferase; Cyc, Cyclase; DeAc, deacetylase (amidase); DH, dehydrogenase; DHT, dehydratase; Epi, epimerase; FIT, formimidoyltransferase; GlyT, glycyltransferase; GT, glycosyltransferase; GA, glucosamine; HA, hexosamine; Hex, hexose; mIP, *myo*-inositol-1-phosphate; (C,N,O)MT, (*C*-,*N*-,*O*-) methyltransferase; OR, oxidoreductase; PL, phospholyase; PPase, phosphatase; PT, phosphotransferase; rad SAM, radical SAM OR family; Red, reductase; Syn, synthase/synthetase; major intermediates and/or end products proven experimentally to be produced in either wildtype or mutant strains are boxed. Dashed arrows indicate housekeeping pathways or steps catalyzed by enzymes that are not encoded in the respective production gene cluster.

is an essential metabolic step. Earlier it was believed to be a sole property of eukaryotes. Now we know that some bacterial groups, especially all actinobacteria, produce and use two types of essential metabolites containing *myo*-inositol, namely mycothiol (the major cellular thiol and redox co-catalyst)[10,11] and *myo*-inositol-containing lipids, such as phosphoinositides and lipoglycans.[64] From whole genome data of actinobacteria (*Actinomycetales*)—for example, streptomycetes, corynebacteria, and mycobacteria—it is now known that they all contain at least one gene, *ino1*, encoding a *myo*-inositol monophosphate synthase of a specifically eubacterial type.[65] For instance, the genomes of *S. coelicolor* A3(2) and *S. avermitilis* MA-4680 contain two genes (SCO3899 and SCO6573) and one gene (SAV4296), respectively, with significant similarity to eukaryotic *myo*-inositolphosphate synthases.[15,65,66] In another actinobacterium, *Mycobacterium (M.) tuberculosis*, the *ino1* gene is essential for growth and virulence and the encoded enzyme has been intensively studied biochemically.[66,67] As a second step in *myo*-inositol biosynthesis, the dephosphorylation of L-*myo*-inositol-1-phosphate via an inositol monophosphatase activity has to follow. This function might be provided by the *strO* gene, which encodes a protein with monophosphatase signature; however, since the streptidine pathway involves two dephosphorylation steps, it is not clear at present whether or not StrO is catalyzing phosphorolysis from L-*myo*-inositol-1-phosphate (cf. Tables 2.3 and 2.4). Also the other clusters for *myo*-inositol-derived AGAs encode such enzymes—for example, the *spcA* and *forA* genes in the respective clusters for SPC and FOR production (cf. Tables 2.6 and 2.17, Figures 2.4 and 2.23a). Alternatively, this step could be also provided from primary metabolism. The actinobacterial genomes harbor at least one gene (sometimes several) with inositol monophosphatase signature; for instance, *M. tuberculosis* has four possible monophosphatases: Rv2701c (*suhB*), Rv1604 (*impA*), Rv2131c (*cysQ)*, and Rv3137. One of the four gene products of this protein family, SuhB*tb*, having highest similarity to the human inositolphosphate monophosphatase protein, was shown to have the expected properties and to be involved in inositol biosynthesis.[68] The ForA protein has also significant similarity to SuhB*tb* ($\approx$ 37% sequence identity), ImpA*tb* ($\approx$ 35% sequence identity), and Rv3137 ($\approx$ 31% sequence identity). Therefore, we assume that in the STR producers an L-*myo*-inositol-1-phosphate phosphatase is provided by general housekeeping metabolism, such that the pool of *myo*-inositol from *myo*-inositol-1-phosphate is high enough for the initial steps of STR biosynthesis during the production phase. In other AGA pathways—for example, those for SPC and FOR productions—this activity obviously must be coordinately made or co-regulated to promote optimal supply of cyclitol precursor during the AGA production phase. In the general *myo*-inositol-dependent aminocyclitol pathway the cyclitol is postulated to be first converted via two enzymes, third step in Ca pathway *myo*-inositol 3-dehydrogenase (ketocyclitol synthase I; StrI, ForG, or SpcB/SpcH [?] in the pathways regarded); these three proteins all belong to the GFO/IDH/MocA family of oxidoreductases, and the fourth step L-glutamine:*scyllo*-3-inosose 3-aminotransferase (ketocyclitol aminotransferase I; StsC, ForS, SpcE/C[?]), to *scyllo*-inosamine (3-deoxy-3-amino-*scyllo*-inositol;

cf. Figures 2.3, 2.5, and 2.23, first part). This first transamination step is quite unusual in using the alpha-amino group of L-glutamine as an amino donor group and a special subfamily of a new class (class V) of secondary metabolic aminotransferases (SMATs).[3,6,58,69] From this intermediate the three diaminocyclitol-forming pathways split into quite different routes: (i) in the case of STRs to the highly modified streptidine (cf. Figure 2.3); (ii) in the formation of SPCs to the more simple streptamine and further on to actinamine (cf. Figure 2.5); and (iii) in the FOR-pathway it becomes directly involved in a glycosylation step and only modified later on pseudodisaccharidic intermediates (cf. Figure 2.23, first part).

2.2.1.2.2. Biosynthesis of Streptidine. Streptidine is postulated to be further made from *scyllo*-inosamine via the following route (cf. Figure 2.3): Rephosphorylation in the 6-position could be catalyzed by either one of the two putative kinases StsE (favoured candidate) or less likely StrN, which both show the typical signature of aminoglycoside and protein kinases.[43] On the resultant intermediate, *scyllo*-inosamine-4-phosphate, a first amidino transfer step from the guanidino group of L-arginine is catalyzed by the StrB1 enzyme, and the product is subsequently dephosphorylated again, probably by StrO (see above) and yielding *N*-guanidino-*scyllo*-inosamine.[3,70,71] The next step involves another dehydrogenase reaction for the introduction of a second keto group in position 3 of the cyclitol, for which candidate enzymes are either of the two dehydrogenases StrB (preferred candidate) or StrI again. The 3-keto-*N*-guanidino-*scyllo*-inosamine again becomes transaminated, probably via transfer of the alpha-amino group of L-alanine and catalyzed by either of the two class V aminotransferases StsA (preferably) or StrS (less likely candidate). Next the diaminocyclitol intermediate becomes again phosphorylated, this time on the 6-hydroxyl group, for which again the putative kinases StrN (preferred candidate) and StsE could account. To complete the streptidine pathway, also the second amino group becomes transamidinated by the guanidino group of L-arginine and a second amidinotransferase enzyme, possibly StrB2, which yields streptidine-6-phosphate. This metabolite is the likely aglycone receptor for the two glycosyltransfer steps to come.

2.2.1.2.3. Biosynthesis of the (Dihydro-)streptose Moiety. The 6-deoxyhexose-derived streptose unit of STR is clearly made from D-glucose via a dTDP-glucose pathway and under retention of the hexose C-chain with the C-3 atom becoming branched out during an enzyme-catalyzed intramolecular cleavage between C-2 and C-3 and rejoining C-2 with C-4.[2,63,72–79] The StrD,E,M,L proteins were heterologously expressed in strains of *E. coli* K-12 and *S. lividans* 66 and tested for their four individual activities: Activation of D-glucose is catalyzed by the StrD enzyme, a member of the wide spread dTDP-D-glucose synthases (pyrophosphorylases), the product of which becomes dehydrated by the dTDP-D-glucose 4,6-dehydratase StrE; the product is epimerized by the 3,5-epimerase StrM to the dTDP-4-keto-L-rhamnose intermediate; the StrL protein clearly is a 4-ketoreductase and its product was shown to be dTDP-L-rhamnose by NMR. This is in contradiction to earlier results that suggested that dTDP-L-dihydrostreptose is the product of an oxidoreductase analogous to a dTDP-L-rhamnose synthase.[80–83] We could not confirm these data and demonstrate conversion to a dTDP-activated L-dihydrostreptose in the crude extracts of *S. griseus* DSM 40236 under any condition given in the literature (S. Verseck, J. Distler, and W. Piepersberg, unpublished). Thus, the enzymology and mechanism of the last step in dihydrostreptose formation, the branching out of the C-3 hydroxymethyl group, remains mysterious.

2.2.1.2.4. Biosynthesis of the N-Methyl-L-glucosamine (NMLGA) Moiety. The biosynthesis of the NMLGA unit of STRs is not understood at all, though this problem had attracted several early biogenetic studies with isotope-labeled hexoses/hexosamines which indicated that the C-chain of D-glucose is retained unchanged in C-1 to C-6 of NMLGA and that D-glucosamine is likely to be the direct precursor of this unusual sugar.[72,84–87] Therefore, multiple epimerization—for example, via successive oxidation/reduction steps catalyzing the inversion of all the asymmetric C-atoms of the D-glucose precursor and/or isomerization steps, for which no similar example is known so far—seemed to be the most likely mechanisms involved.[4,76] Also, work on isolation and analysis of NDP-activated sugar metabolites in wildtype and mutant strains of *S. griseus*, such as UDP-*N*-methyl-D-glucosamine-phosphate,[88] or on the identification of a gene that complemented the deficiency in STR production in a mutant strain, SD1, of *S. bikiniensis* and was suggested to be involved in NMLGA biosynthesis, was not continued and its interpretations remained speculative.[89] The genetic record laid down in the *str*/*sts*-clusters is uneasy to interpret with respect to the NMLGA pathway (cf. Tables 2.3 and 3.4). To complicate the situation, two distinctly different NDP-hexose/hexosamine pathways could be involved in the NMLGA branches of the STR pathways of *S. griseus* DSM 40236 and *S. glaucescens* DSM 40716 (ETH 22794) since the production gene cluster in the latter strain contains some additional genes (*strP,Q,X*) pointing to this possibility (see above; cf. Figure 2.2; Tables 2.3 and 2.4). The pathway outlined in Figure 2.3 includes a number of candidate gene products/enzymatic steps, which could be involved in the NMLGA formation and account for several alternative routs: First, activation of D-glucose-1-phosphate could be catalyzed by the CDP-glucose synthase StrQ (in *S. glaucescens*); alternatively, in *S. griseus*, an already NDP-activated hexose, such as UDP-D-glucose or UDP-D-glucosamine, could be recruited as the equivalent from primary metabolism. Several factors could be involved in the second step. A 2-oxidation (possible candidates include StrH, StrT, StrU) and a 2-amino transfer (StrS or StrA) could occur. Also possible is a 2-epimerization of a 2-aminohexose such as D-glucosamine by an enzyme related to enzymes epimerizing the NDP-*N*-acetylhexosamine precursors of the *N*-acetyl-mannosamine, *N*-acetyl-mannosaminuronic acid. A third factor is the possible presence of 4-acetamido-4,6-dideoxy-D-galactose moieties occurring in many bacterial cell wall components. Extracellular capsular polysaccharides [the rffE gene encoding UDP-GlcNaC-2-epimerase involved in synthesizing the enterobacterial common antigen (ECA) and "Wec" proteins] could also result. Third, 3-epimerization of the NMLGA precursor could be achieved by various mechanisms, such as direct epimerization by a reversible oxidation/reduction—for example, initiated by an enzyme of the GFO/IDH/MocA family, such as StrU, or by a 3,5-epimerase, such as StrX, after an involvement of a 4-oxidase, creating a 4-keto intermediate as the prerequisite for this kind of enzyme (cf. dTDP-L-rhamnose pathway). Fourth, the latter mechanism would have the advantage that the 4-position also could be stereochemically inverted in this reaction sequence by a 4-ketoreductase—for example, the StrT protein. Alternatively, the 4,5-epimerization could be also the result of a 4,5-dehydratation and subsequent rehydroxylation under stereoinversion (candidate proteins again are StrH/StrT/StrU). The postulated enzyme for the putatively fifth and final *N*-methylation step is the StsG protein that shows significant similarity to other (*N*-)methyltransferases. It is not known whether the methylation step occurs before or after the glycosyltransfer of the (NM)LGA unit to the presumed dihydrostreptosylstreptidine-6-phosphate acceptor substrate; the first is more likely since NMLGA was detectable in the sugar nucleotide fraction of the cell lysate of *S. griseus*.[82]

2.2.1.2.5. Condensation of the Pseudotrisaccharide, Late Modification, and Export. The two glycosylation steps on streptidine-6-phosphate as the acceptor substrate and assumed to be involved in the condensation of the pseudotrisaccharidic cytoplasmatic precursors of STRs also remain enigmatic. The dTDP-L-dihydrostreptose:streptidine-6-phosphate dihydrostreptosyltransferase described in enzymatic studies by Kniep and Grisebach[82] has not been identified so far among the gene products encoded by the *str*/*sts*-clusters, nor have any other clear cut members of one of the major glycosyltransferase groups been identified among these (cf. Tables 2.3 and 2.4). The StsF protein has distant similarity to a putative glycosyltransferase (27% identity in 117 aa; ZP_00050674.1) from *Magnetospirillum magnetotacticum*, but also to hydrolases, such as peptidases and Nudix hydrolases. There is also weak evidence, from a study on the mutual complementation by cloned DNA segments, indicating that either of the StrF,G,H proteins may be involved in glycosyltransfer. A fragment covering the *strFGH* genes (the locus was provisionally called *strC*) complemented mutant SD245 of *S. griseus* ATCC 1037. It was discovered that mutant SD245 had a phenotype consistent with a block in the glycolysation step between dihydrostreptose and streptidine-6-phosphate.[91,92] The StrF protein is the only one out of these candidate proteins that shows a distant similarity to a domain in a multidomain protein of *Carboxydothermus hydrogenoformans* Z-2901 (accession code ABB14724; locus-tag CHY_0980; 27% identity in 136 aa) having similarity to group 2 glycosyltransferases and, therefore, could be the dihydrostreptosyltransferase or a subunit thereof. Intriguingly, this gene lies in a putative operon also containing the four conserved genes for a dTDP-L-rhamnose pathway (related to *strD,E,M,L*) in this organism. Also, the L-hexosaminyltransferase transferring the NMLGA moiety to the postulated precursor dihydrostreptosylstreptidine-6-phosphate[80,82] is not yet known.

Grisebach and co-workers also demonstrated that crude membrane preparations of *S. griseus* N-2-3-11 converted dihydro-STR (DH-STR) to STR and suggested that this is a step occurring during, or directly after, export of DH-STR to the environment and that DH-STR-6-phosphate is the even better substrate for this oxidoreductase.[83]

2.2.1.3. Producers, Genetics, and Biosynthesis of Spectinomycins (SPCs). Spectinomycin (SPC; actinospectacin; Figure 2.4) was detected independently by researchers of the Upjohn Company (from *S. spectabilis* ATCC 27465 (DSM 40512)) and Abbot Laboratories (from *S. netropsis* [former "*S. flavopersicus*"] ATCC 19756 [DSM 40093]) in 1961.[93,94] 3′-dihydro-SPC can be observed as a second natural component in the culture broth of fermentations with *S. spectabilis*.[95] SPC became the first member of a second group of AGAs related to the STRs in terms of structural features and pathway design.[2,96,97] Spenolimycin, another spectinomycin-type AGA, was isolated from *S. gilvospiralis* and first described in 1984.[98–100] SPC is composed of an aminocyclitol, designated actinamine, and actinospectose, a 4,6-dideoxy-2,3-diketohexose (Figure 2.2.). The SPCs resemble the STRs in terms of structural features and in pathway design based on branches starting from *myo*-inositol and a dTDP-activated 6-deoxyhexose, dTDP-D-glucose.[1,2,63] Labeling studies *in vivo*, carried out in both *S. netropsis* and *S. spectabilis*, proved that both subunits are derived from D-glucose without reorganization the C-backbone and that the cyclitol unit comes directly from *myo*-inositol; also these investigations resulted in first proposals for biosynthetic routes to SPC.[101–103]

The gene clusters for SPC biosynthesis from two different streptomycete strains, *S. netropis* DSM 40093 (former "*flavopersicus*") and *S. spectabilis* ATCC 27465 (DSM 40512), have been reported (see Table 2.6 and Figure 2.2); also, a pathway suggesting the involvement of the Spc-gene products has been proposed recently.[18,54] Therefore, we here sketch only briefly the pathway design for the formation of SPC under use of the proposals of Altenbuchner and co-workers.[54]

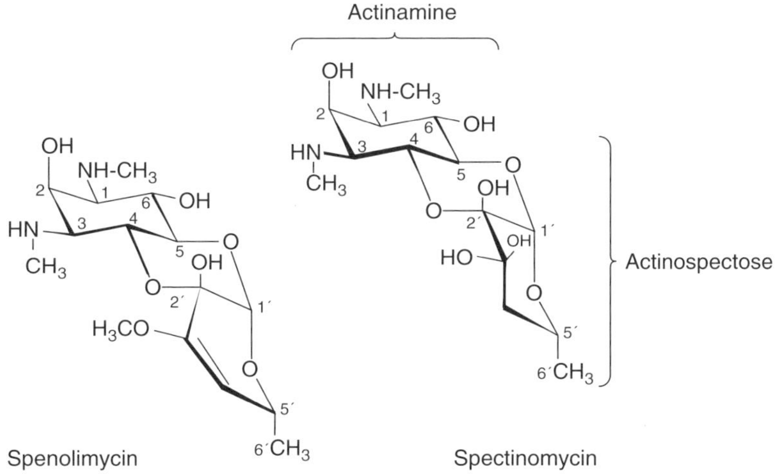

Figure 2.4. Structures of spectinomycin (SPC) family AGAs.

TABLE 2.6. Proteins Encoded in the Genomic Region Covering the *spc*-Cluster of *Stromyces netropsis* NRRL 2820 (Accession Code U70376)

Gene Product Symbol	aa	CDD Number[a]	Putative Function
InsA	390	14743	Transposase (IS1374-like)
Orf1	256	28900	6-Phosphate phosphatase (fragment)
Aph(6)	307	15952	Aminoglycoside PT, SPC resistance
Orf2	104	—	Amidinotransferase (fragment)
Orf3	90	—	Transposase (IS112-like)
SpcR	330	26725	Transcriptional regulator
SpcN	330	12511	SPC-PT; SPC resistance
SpcT	433	24850	Transport; export
SpcA	266	30137	*myo*-Inositol monophosphatase
SpcB	392	10542	*myo*-Inositol DH
SpcC	441	25439	AT
SpcD	262	10804	Epimerase
SpcX	397	11890	Sugar-binding protein, transporter
SpcE	427	15279	AT
SpcY	302	26181	Molybdene cofactor synthesis; rad SAM
SpcF	270	25539	GT
SpcG	272	8054	MT
SpcH	346	10542	DH
SpcI	312	23188	dNDP-glucose-4,6-dehydratase
SpcJ	328	23188	Dehydratase
SpcK	303	16736	Thymylyltransferase
Orf4	679	25582	Integrase

[a]Refers to the CDD database.[255]

The first steps in cyclitol formation are obviously very similar as in the STR pathway (see Section 2.2.1.2.2): (i) *myo*-inositol is provided from the cells pool of L-*myo*-inositol-1-phosphate and becomes dephosphorylated by the inositol-monophosphatase SpcA; (ii) a *myo*-inositol dehydrogenase, putatively either of the two proteins SpcB or SpcH, yields the *scyllo*-inosose intermediate; (iii) either of the two aminotransferases SpcC or SpcE is proposed to catalyze the first transamination step, though only SpcE is a member of the class V aminotransferases and related to the "S"-type L-glutamine-dependent cyclitol aminotransferases (we suggest that this is the bifunctional cyclitol aminotransferase I and II). A repetition of the last two steps that introduces the second amino group in the 3-position of the cyclitol is thought to reoccur with the alternative use of the same candidate enzymes, SpcB/H and SpcC/E, thus yielding the streptamine intermediate.

For shaping the 1,3-diaminocyclitol into its final configuration and into the actinamine structure, first an epimerization step has to invert the stereochemical placement of the 2-hydroxyl group, which is proposed to be catalyzed by the SpcD enzyme, being related to sugar epimerases; second, the actinamine branch of the SPC pathway has to be extended by a rather unspecific aminocyclitol *N*-methyltransferase (NMT), methylating both amino groups and putatively represented by the SpcG protein (see Figure 2.5). In his study on the aminocyclitol NMTs from two actinomycete AGA producers, those for HYG-B and SPC, Walker[104] has described an activity in crude extracts of *S. netropsis* (former "*S. flavopersicus*") ATCC 19756 that seems to result from the expression of this enzyme. Surprisingly, the extracts of both of the these AGA producers methylated the N^1- and N^3-amino groups of 2-deoxystreptamine, streptamine, and 2-epi-streptamine; however, no proof for the presence of a single NMT enzyme

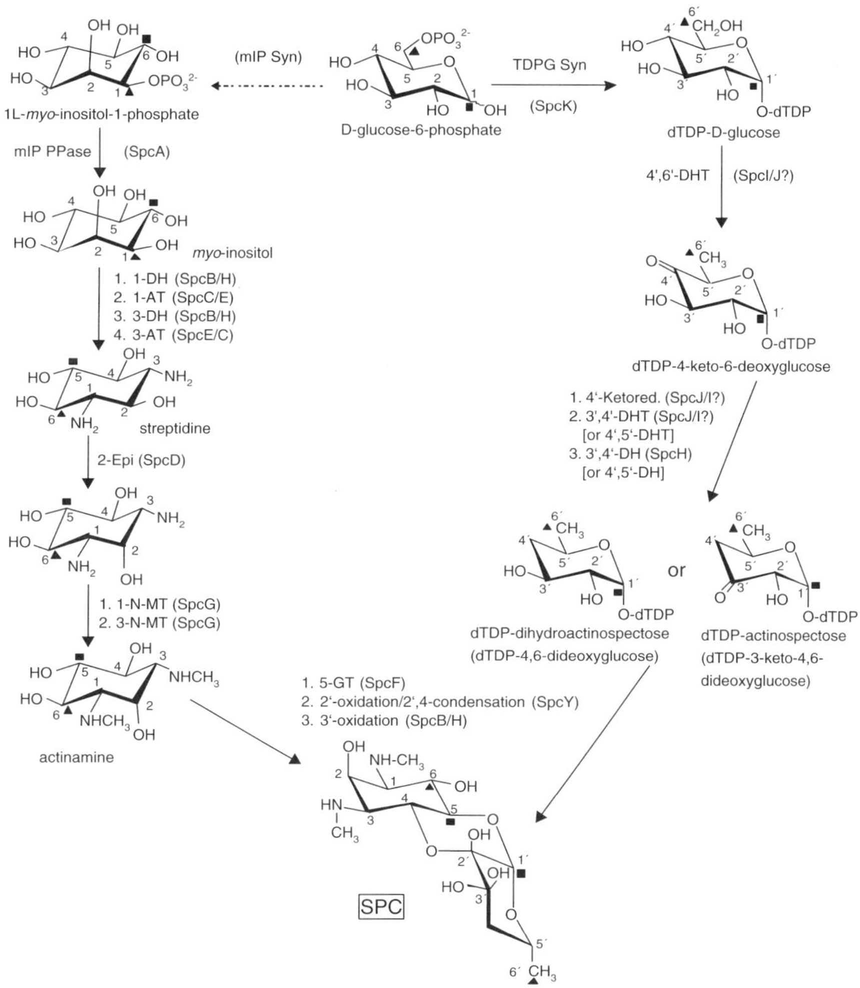

Figure 2.5. Biosynthetic pathway for SPCs. Other explanations and abbreviations are as given in text and in the legend to Figure 2.3.

was presented; also it is unknown whether the NMT enzymes tested are those encoded in the respective *spc*- and *hyg*-clusters.

Next, as in the STR pathway, the biosynthesis of the 6-deoxyhexose derivative actinospectose is likely to first occur in the form of a dTDP-activated hexose unit: (i) SpcK is a member of the dTDP-D-glucose synthases and would provide the nucleotide sugar precusor; (ii) the dTDP-D-glucose-4,6-dehydratase SpcJ and the dTDP-4-keto-6-deoxyglucose isomerase SpcI would synthesize a dTDP-3-keto-6-deoxyhexose intermediate; and (iii) finally the 4-dehydroxylation of the latter compound in a pyridoxalphosphate-dependent dehydratase reaction catalyzed presumably by the SpcE enzyme, which is related to class III aminotransferases and could function according to the ketosugar-forming mechanism described by Liu and Thorson,[105] would yield the nucleotide-activated actinospectose precursor for the glycosyltransfer reaction.

The condensation step is catalyzed by the glycosyltransferase SpcF. Further enzyme-catalyzed oxidation is probably needed for the introduction of the hemiketal linkage between cyclitol and sugar units. The SpcY enzyme, which has a similar counterpart, HygY, among the *hyg*-cluster encoded proteins (see Section 2.2.4.3.1), is a candidate enzyme for this reaction. SpcY is a member of the "radical SAM" superfamily of proteins and relatives of SpcY have been found before all in connection with molybdenum-cofactor biosynthesis; but to our knowledge no details of the mechanism involved is known for those.

2.2.1.4. Producers, Genetics, and Biosynthesis of Kasugamycins (KASs). Kasugamycins (KASs, Figure 2.6), being produced by *S. kasugaensis* and other *Streptomyces* sp. and first described in 1965 represent another pseudodisaccharidic AGA group biosynthetically related to the STRs.[106] Another member of this AGA subfamily is the minosaminomycin described in 1974.[107–109] The mode of action of these compounds is distinctly different from most other AGAs, which induce miscoding and bind to the decoding site of 16S rRNA in the 30S subunit of bacterial ribosomes, in that it inhibits the initiation of protein synthesis by preventing EF-Tu-dependent binding of aminoacyl-tRNA to ribosomes.[110] Chemically, KAS [= 1L-1,3,4/2,5,6-1-deoxy-2,3,4,5,6-pentahydroxycyclohexyl 2-amino-2,3,4,6-tetradeoxy-4-(α-iminoglycino)-α-D-*arabino*-hexopyranoside] is

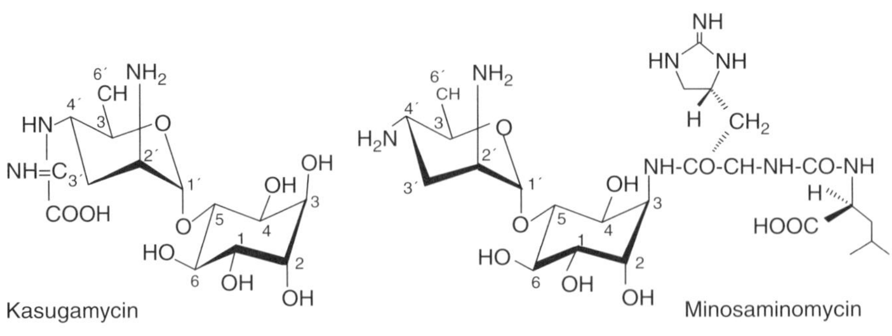

Figure 2.6. Chemical structures of natural kasugamycins (KASs).

a pseudodisaccharide composed of a D-*chiro*-inositol and a substituted 6-deoxyaminosugar, called kasugamine. Biogenetic studies were carried out extensively[63,111–115] and suggested that (i) glycine provides the carboxyformidoyl side chain; (ii) the C-chain of D-glucose migrates unsplit into the two sugar-derived subunits; (iii) *myo*-inositol (and not D-*chiro*-inositol) is the cyclitol precursor; (iv) D-glucosamine is a much better precursor for the kasugamine moiety than D-glucose; and (v) *N*-acetyl-D-glucosamine is directly incorporated into this subunit. From these data a provisional biosynthetic pathway for kasugamine could be proposed, which is in accord with what is described below and shown in Figure 2.7.

The *kas*-cluster for KAS production from *S. kasugaensis* M 338-M1 [DSM 40819] has been partially sequenced on four DNA fragments (see Table 2.7), which sum up to a length of 22.4 kb altogether.[19–21] From these data a preliminary map of the *kas*-cluster can be drawn as given in Figure 2.2. Also, a tentative biosynthetic pathway was proposed by these authors,[20] which we extend to the one outlined in Figure 2.7. The pathway involves three phases, (i) the cyclitol pathway, (ii) the 6-deoxyhexosamine pathway which is postulated to form the

TABLE 2.7. Proteins Encoded in the Genomic Region Covering Part of the *kas*-Cluster of *Stromyces kasugaensis* M 338-M1 [DSM 40819] (Accession Codes AB076838, AB033992, AB120043, AB005901)

Gene Product Symbol	aa	CDD Number[a]	Putative Function
KasT	346	26725	StrR-like regulator
KasU	103	—	Unknown
KasJ	445	11983	Cyclitol DH, oxidoreductase
KasK	329	10851	ABC exporter, ATPase component
KasL	257	10709	ABC exporter, permease component
KasM	240	10709	ABC exporter, permease component
KasN	383	7951	Gly oxidase, oxidoreductase
KasO	149	13750	Decarboxylase?
KasP	281	29429	Monoxygenase, oxidoreductase, 3′-deoxygenation
KasQ	413	25900	UDP-NAcGA 2-epimerase
KasR	399	15279	NDP-4-keto-6-deoxyglucose-3-dehydratase or AT
KasA	391	7651	GT, NDP-hexosamine
KasB	253	—	Unknown
KasC	436	25439	AT
KasD	329	23188	dNDP-hexose 4,6-dehydratase/epimerase
KasG	122	—	Unknown
KasF	283	25558	Acetyltransferase
KasE	265	25595	Hydrolase, PHase
Kac338	141	25558	AAC(2′), KSG resistance
KasI	(78)	17417	Monoxygenase

[a]Refers to the CDD database.[255]

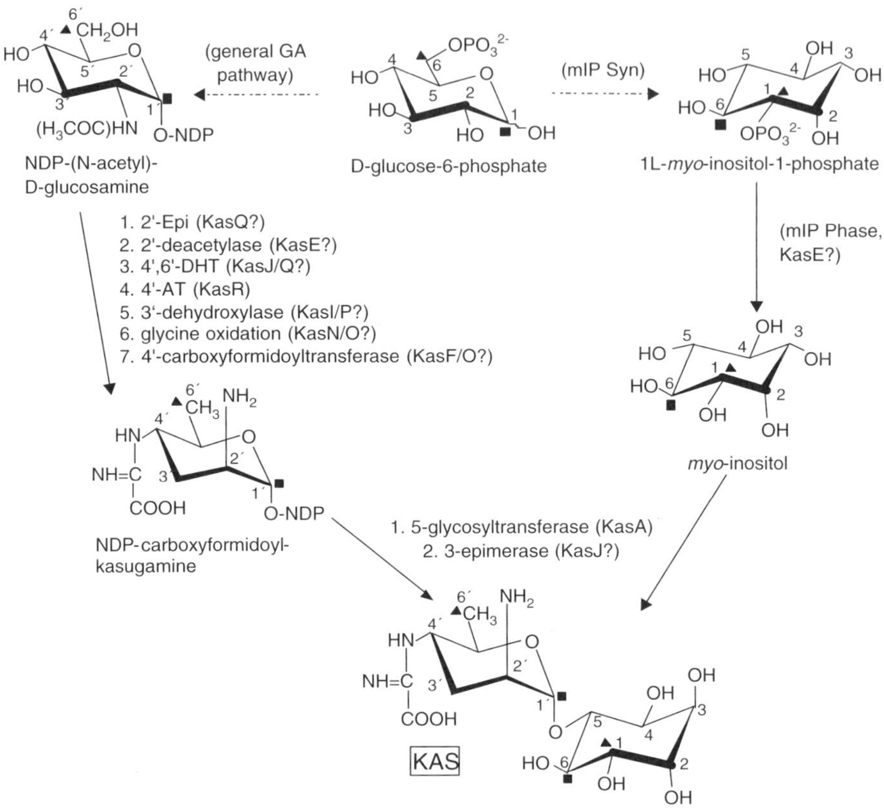

Figure 2.7. A hypothetical biosynthetic pathway for KAS (according to Ikeno et al. 2002)[20], modified. Abbreviations: GA, glucosamine; for further explanations see legend to Figure 2.3.

nucleotide-activated sugar intermediate NDP-carboxyformidoylkasugamine, and (iii) the condensation and late modification pathway leading to a pseudodisaccharidic end product.

The first two steps providing *myo*-inositol are the same as in the STR pathway and are already described there (see paragraph on STRs). The possibility that a hydrolase, KasE, could be a possible *myo*-inositolmonophosphate monophosphatase is provided by the expression of the respective gene in the *kas*-cluster. The *myo*-inositol is assumed to be directly incorporated into the first pseudodisaccharidic intermediate.

We propose that the first step in NDP-kasugamine biosynthesis is 2-epimerization of the postulated UDP-*N*-acetyl-D-glucosamine precursor, which is suggested by the similarity of the KasQ protein with known UDP-(*N*-acetyl-)D-glucosamine 2-epimerases and catalyzes the conversion to UDP-*N*-acetyl-D-mannosamine. To provide this route with enough precursor, the general bacterial pathway(s) for filling up UDP-(*N*-acetyl-)D-glucosamine in its primary metabolic pool is probably used; it obviously does not need any KAS-specific biosynthetic functions. Also, no typical amidase-encoding gene, for expressing the deacetylase removing the acetyl residue from the 2-amino group, is observed in the *kas*-cluster; the suggestion that KasE (a member of a hydrolase family with wide scope) could alternatively catalyze this reaction is highly speculative.

The 6-deoxygenation and the introduction of a 4-keto group to prepare this site for the later 4-transamination (by the aminotransferase KasR) would require an NDP-mannosamine 4,6-dehydratase, for which the oxidoreductase KasD, which has similarity to the dTDP-glucose 4,6-dehydratases, is the candidate protein of choice.[19] The aminotransferase KasC, which belongs to the class III aminotransferases, probably catalyzes the 4-transamination. The 3-deoxygenation step is possibly following an oxidation–reduction mechanism, which inverts the chirality at C-3. The putative gene products KasR and KasI or KasP represent a pyridoxalphosphate-dependent aminotransferase/dehydratase and two oxidoreductases, respectively, which could account for this biosynthetic step by catalyzing a dehydration under dehydroxylation at C-3 and subsequent reduction of the double bond created. The carboxyformidoyl side chain could be preformed from glycine by oxidation and activation, in which the putative oxidoreductases KasI and KasN, as well as the possible protein KasU, whose function is unknown, could be involved.

KasF is related to acyltransferases and, therefore, could be the possible carboxyformidoyltransferase. The condensation of the modified NDP-2,4-diamino-2,3,4,6-tetradeoxyhexose (NDP-kasugamine) and the *myo*-inositol precursors by glycosyltransfer is probably catalyzed by the KasA protein, which is similar to *N*-acetylglucosaminyltransferases. The last step, 3-epimerization of the cyclitol unit to the D-*chiro*-inositol, is postulated to be catalyzed by the KasJ protein belonging to the choline dehydrogenase family of flavoproteins and yields probably the *N*-acetylated form of KAS, due to the activity of the AAC(2′) resistance protein encoded by the *kac* gene.[20]

2.2.2. 2-Deoxystreptamine (2DOS) and D-Glucosamine-Based AGAs Forming Paromamine as an Intermediate: Neomycin- and Kanamycin-Related AGAs

2.2.2.1. The Molecular Genetics and Biosynthesis of the Neomycin-Related 2DOS-AGAs. A core compound and regarded as an evolutionary ancestor in the neomycin (NEO) family of 2DOS-AGAs is ribostamycin (RIB) as a common precursor of all other family members and, at the same time an end product of some of the producers (e.g., *S. ribosidificus* NRRL B-11466), and the NEOs themselves (examples of producers: *S. fradiae* DSM 40063 or *M. chalcea* 69–683 [Schering-Pl.]) (see Figure 2.8). Other family members are modified RIBs, such as the N^1-acylated butirosins (BTRs; producer, e.g., *Bacillus circulans* ATCC

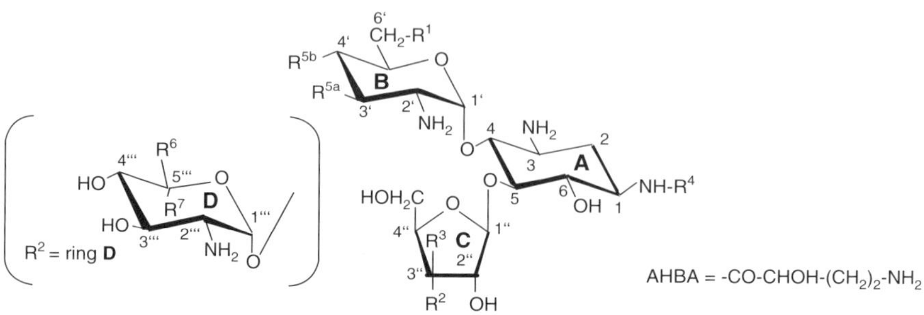

Aminoglycoside	R^1	R^2	R^3	R^4	R^{5a}/R^{5b}	R^6	R^7
6'-OH-Ribostamycin	OH	OH	H	H	OH/OH	-	-
Ribostamycin	NH_2	OH	H	H	OH/OH	-	-
Butirosin A	NH_2	H	OH	AHBA	OH/OH	-	-
Butirosin B	NH_2	OH	H	AHBA	OH/OH	-	-
Butirosin E_1	OH	H	OH	AHBA	OH/OH	-	-
Butirosin E_2	OH	OH	H	AHBA	OH/OH	-	-
Butirosin C_1	NH_2	H	OH	AHBA	OH/H	-	-
Butirosin C_2	NH_2	OH	H	AHBA	OH/H	-	-
(Butirosin D_1)	NH_2	H	OH	AHBA	H/H	-	-
(Butirosin D_2)	NH_2	OH	H	AHBA	H/H	-	-
(Butirosin F_1)	$NHCH_3$	H	OH	AHBA	H/H	-	-
(Butirosin F_2)	$NHCH_3$	OH	H	AHBA	H/H	-	-
6'''-OH-Neomycin C	NH_2	ring **D**	H	H	OH/OH	CH_2OH	H
Neomycin C	NH_2	ring **D**	H	H	OH/OH	CH_2NH_2	H
Neomycin B	NH_2	ring **D**	H	H	OH/OH	H	CH_2NH_2
Paromomycin I	OH	ring **D**	H	H	OH/OH	H	CH_2NH_2
Paromomycin II	OH	ring **D**	H	H	OH/OH	CH_2NH_2	H
Lividomycin B	OH	ring **D**	H	H	H/OH	H	CH_2NH_2

(Lividomycin A = 4'''-alpha-D-Mannosyl-LIV-B)
(Inosamycins A-E = 2-hydroxy analogs of NEOs, PARs, RIB)

Figure 2.8. Structures of neomycin (NEO)-family AGAs.

21557), or are structurally related to NEOs directly, such as the paromomycins (PARs; 6′-hydroxy-NEOs; producer, e.g., *S. rimosus* ssp. *paromomycinus* DSM 41429) and lividomycins (LIVs; 6′-hydroxy-3′-deoxy-NEOs; producer, e.g., *S. lividus* CBS 844.73). NEOs were first described in 1949;[116] several other *Streptomyces* sp. and *Micromonospora* sp. have been found to produce NEOs and PARs thereafter in industrial screening programs (e.g., *S. albogriseolus* DSM 4003, *S. roseoflavus* DSM 40536, *S. chrestomyceticus* DSM 40545, *S. albus* var. *metamycinus*, *M. chalcea* 69–683, and many others).[117,118] Neomycins D, E, and F were found to be identical with paromamine, paromomycin I, and paromomycin II.[119] Interestingly, NEO producers more in general seem to be able to coproduce PARs: the typical bulk preparations of NEO from industrial fermentations with descendants of the original Waksman and Lechevalier isolate[117] of *S. fradiae* contain, besides NEO-B as main component and NEOs C and A (neamine = 6′-aminoparomamine; cf. Figs. 2.8 and 2.11) as minor components, traces (below 1%) of NEOs D (= paromamine), E (= PAR-I), and F (= PAR-II).[119] In a strain of *S. hygroscopicus* (ATCC 39150) a complex of 2DOS-AGAs closely related to NEOs and called inosamycins (components A to E; analogs of

NEOs, PARs, RIB) was detected that had a streptamine in place of the 2DOS moiety.[120,121] The antibacterial activity of inosamycin A, the major component of the complex, turned out to be comparable to that of NEOs, but its acute toxicity was significantly lower. Later the NEO family AGAs were also studied in many other respects relevant to production, application, and an understanding of its molecular biology—for instance, its mode of action, therapeutic application and toxicology, and resistance in pathogens and producers, as well as including an investigation of the biogenesis and biosynthesis in the producer *S. fradiae*.[4,122]

In the early phase of molecular genetics in streptomycetes there was indirect evidence—such as spontaneous phenotypic instability, sensitivity toward curing agents of antibiotic associated traits such as production, resistance, incorporation of precursors, and so on, for a participation of mobile genetic elements, plasmids, transposons, and the like—in the inheritance of genes needed for the production of antibiotics. The producers of NEOs became a model system in investigations on this topic.[63,123] However, it turned out later that the actinomycete chromosome is the usual location of the productive gene clusters, as is emphasized by the data presented in this report; this is also the case for the *neo*-cluster of *S. fradiae* DSM 40063 in particular, since there is overwhelming evidence for its chromosomal location: (i) The mapped cosmid clones isolated from the region harboring the *neo*-cluster are contiguous and do not indicate presence of a defined genomic end structure; (ii) the stretches of about 10 kb DNA each sequenced upstream and downstream of the *neo*-cluster (cf. accession code AJ629247) do not contain any genes typically localized on plasmids; (iii) rather these genes represent typically chromosomal (and "housekeeping") ones, since they are all well-conserved in the two fully sequenced streptomycete chromosomes of *S. coelicolor* A3(2) and *S. avermitilis* MA-4680 (accession codes AL645882 and BA000030, respectively), but do not come from a particular and conserved chromosomal region (see Section 2.3.1).

2.2.2.1.1. Gene Clusters for the Neomycin-Related 2DOS-AGAs. For an elucidation of the genetics and the different pathways for the production of the NEO-family of 2DOS-AGAs, we set out to analyze the respective gene clusters and their immediate genomic environment. Genomic cosmid libraries were constructed and the insert sequences of single or overlapping cosmid clones were determined and annotated; the sequences were submitted to the EMBL database and received the following accession codes AJ629247 (*neo*-cluster; 31940 bp), AJ744850 (*rib*-cluster; 43190 bp), AJ628955 (*par*-cluster; 48169 bp), AJ748832 (*liv*-cluster; 40579 bp), and AJ781030 (*btr*-cluster; 19248 bp). The producers of NEOs, RIB, PARs, and LIVs share the possession of a 3-acetyltransferase encoding resistance gene (aacC).[1,254] Surprisingly, the *aacC7* gene of *S. rimosus* was not found in the *par*-cluster, obviously being an exception in the gene clusters of the NEO family. Therefore, a partial insert sequence (4812 bp) around the *aacC7* resistance gene was determined from cosmid SriA13 and also submitted to the database under accession code AJ749845. The ORFs encoded on these contiguous genomic segments are listed in Tables 2.8 to 2.12.

TABLE 2.8. Proteins Encoded in the Genomic Area Covering the *neo*-Cluster of *Streptomyces fradiae* DSM 40063 (Accession Code AJ629247)[a]

ORF /Locus Tag[a]	Gene Product[b] Symbol	aa	CDD Number[c]	Putative Function
SfrA10.4		82	9271	Pectin esterase (fragment)
SfrA10.5		180	—	Membrane protein (fragment)
SfrA10.6c		98	—	HP (fragment)
SfrA10.7c	AphA	268	25797	APH(3′), NM-resistance
SfrA10.8c	NeoG	431	—	Component of sensor/response regulator
SfrA10.9c	NeoH	173	—	Component of sensor/response regulator
SfrA10.10c	NeoI	175	28977	Component of sensor/response regulator
SfrA10.11	NeoE	340	25397	Aminocyclitol 1-DH
SfrA10.12	NeoS	424	15279	L-Glutamine:ketocyclitol AT I + II
SfrF04.2	NeoC	430	17039	2-Deoxy-*scyllo*-inosose synthase (cyclase)
SfrF04.3	NeoM	421	7651	2DOS 4-glucosaminyltransferase, GT I
SfrF04.4	NeoT	666	10852	ABC transporter
SfrF04.5	NeoU	594	10852	ABC transporter
SfrF04.6	NeoQ	541	16494	6′-(6‴-) DH
SfrF04.7	NeoN	299	26181	5′-Epimerase or 6‴-DH
SfrF04.8	NeoP	233	16692	Sugar phosphate phosphatase
SfrF04.9	NeoX	83	—	cHP, regulation
SfrF04.10	NeoF	366	10312	5‴-GT, GT-III
SfrF04.11	NeoD	279	11828	*N*-Acetyl-hexosaminyl deacetylase or amidase
SfrF04.12c	NeoL	660	—	Unknown, ribosyltransfer
SfrF04.13	NeoB	416	25439	6′-/6‴-AT
SfrF04.14	NeoA	1293	—	Unknown, ribosyltransfer
SfrF04.15c	AacC8	287	3032	AAC(3), NM resistance
SfrF04.16c	NeoR	886	13212	Regulator
SfrF04.17	(NeoY)	73	—	Unknown
SfrF04.18		173	—	Unknown
SfrF04.19c		140	—	Unknown
SfrF04.20c		454	8453	Serine protease

[a]The coding sequences (ORFs) are given with their locus tag numbers as found on the respective cosmids analyzed.

[b]The list of gene products translated from the nucleotide sequence corresponds to the protein ID numbers CAF33300–CAF33330.

[c]Refers to the CDD database.[255]

The biosynthetic pathways for the neomycin-related 2DOS-AGAs and their corresponding gene clusters seem to represent an ancient metabolic invention. The basic *neo*-gene cluster has been well-conserved during evolution, with considerable deviation, both in gene organization and sequence identity—for example, seen in the *btr*-cluster of *B. circulans* (cf. Figure 2.9). It could have evolved before other families of paromamine-forming and 2DOS-containing AGAs [e.g., those for the KANs and GENs, including the FORs (see the paragraph on these

TABLE 2.9. Proteins Encoded in the Genomic Area Covering the *rib*-Cluster of *Streptomyces ribosidificus* NRRL-B-11466 (Accession Code AJ744850)[a]

ORF /Locus Tag[a]	Gene Product[b] Symbol	aa	CDD Number[c]	Putative Function
SribP10.12c		253	—	Unknown
SribP10.13c		149	17345	Helicase-like protein
SribP10.14c	AacC	287	3032	AAC; RM-resistance
SribP10.15c	RibA	1242	—	Unknown; ribosyltransfer
SribP10.16c	RibB	416	25439	6′-AT
SribP10.17	RibL	652	—	Unknown; ribosyltransfer
SribP10.18c	RibD	278	11828	*N*-Acetyl-hexosaminyl deacetylase or amidase
SribP10.19c	RibF	352	10312	5‴-GT, GT III
SribP10.20c	RibX	82	—	cHP, regulation
SribP10.21c	RibP	223	10312	Sugarphosphate phosphatase
SribP10.22c	RibN	299	16692	5‴-Epimerase or 66‴-DH
SribP10.23c	RibQ	541	26181	6′- (and 6‴-) DH
SribP10.24c	RibU	594	10852	ABC transporter
SribP10.25c	RibT	617	10852	ABC transporter
SribP10.26c	RibM	419	7651	2DOS 4-glucosaminyltransferase, GT I
SribP10.27c	RibC	391	17039	2-Deoxy-*scyllo*-inosose synthase (cyclase)
SribP10.28c	RibS	424	15279	L-Glutamine:ketocyclitol AT I+ II
SribP10.29c	RibE	340	25397	Aminocyclitol 1-DH
SribP10.30	RibI	166	28977	Component of sensor/response regulator
SribP10.31	RibH	177	—	Component of sensor/response regulator
SribP10.32	RibG	390	—	Component of sensor/response regulator
SribL03.8	AphA	468	25797	APH(3′), RM-resistance
SribL03.9		81	—	Polyketide synthase (fragment)
SribL03.10c		107	17139	Transposase (partial)

[a]The coding sequences (ORFs) are given with their locus tag number as found on the respective cosmids analyzed.
[b]The list of gene products translated from the nucleotide sequence corresponds to the protein ID number CAG34022–CAG34045.
[c]Refers to the CDD database.[255]

compound families)] evolved. Evidence for this comes from the following facts: (i) The gene clusters for NM-related compounds are found in taxonomically more distant bacterial producers such as in members of the families *Streptomycetaceae, Micromonosporaceae* (both actinobacteria) and *Bacillaceae* (firmicutes); (ii) the number of genes necessary for the basic 2DOS/paromamine pathway are less than in other pathways (in the NM-pathway as compared to or in the KAN and GEN pathways, respectively); (iii) conserved and equivalent genes show a higher degree of divergence; (iv) no gene duplications have been introduced among the typically conserved basic set of 2DOS-AGAs genes; (v) they use *aphA* genes encoding aminoglycoside 3′-phosphotransferases as resistance genes,

TABLE 2.10. Proteins Encoded in the Genomic Area Covering the *par*-Cluster of *Streptomyces rimosus* subsp. *paromomycinus* NRRL 2455 (Accession Code AJ628955)[a]

ORF /Locus Tag[a]	Gene Product[b] Symbol	aa	CDD Number[c]	Putative Function
SriD03.19		271	—	Unknown
SriD03.20	ParE	339	25397	Aminocyclitol 1-DH
SriD03.21	ParS	424	15279	L-Glutamine:ketocyclitol AT I + II
SriD03.22	ParC	386	17039	2-Deoxy-*scyllo*-inosose-synthase (cyclase)
SriD03.23	ParM	417	7651	2DOS 4-glycosyltransferase; GT-I
SriD03.24	ParT	604	10852	ABC transporter
SriD03.25	ParU	628	10852	ABC transporter
SriD03.26	ParQ	546	16494	6‴-DH
SriD03.27	ParN	298	26181	5‴-Epimerase or 6‴-DH
SriD03.28	ParP	231	16692	Sugarphosphate phosphatase
SriD03.29	ParX	90	—	cHP, regulation
SriD03.30	ParF	367	10312	5‴-GT, GT III
SriD03.31	ParD	253	11828	*N*-Acetyl hexosaminyl deacetylase or amidase
SriD03.32	ParL	637	—	Unknown; ribosyltransfer
SriD03.33	ParB	417	25439	6‴-AT
SriD03.34	ParA	1302	—	Unknown; ribosyltransfer
SriL03.20c	ParZ	455	—	PM-phosphate phosphatase
SriL03.19	ParI	163	28977	Component of sensor/response regulator
SriL03.18	AphA	262	25797	Aph(3′), PM-resistance
SriL03.17c	ParY	229	—	cHP
SriL03.16c	ParG	400	—	Component of sensor/response regulator
SriL03.15c	ParH	184	—	Component of sensor/response regulator
SriL03.14c	(PasA)	76	—	UDP-*N*-acetylglucosamine synthase
SriL03.13	(PasB)	775	—	Chitinase
SriL03.12	(PasC)	284	—	NDP-sugar epimerase
SriL03.11	(ParR1)	209	—	Transcriptional regulator
SriL03.10	(ParR2)	292	—	Transcriptional regulator
SriL03.9	(PasD)	285	—	Oxidoreductase, DH

[a]The coding sequences (ORFs) are given with their locus tag number as found on the respective cosmids analyzed.
[b]The list of gene products translated from the nucleotide sequence corresponds to the protein ID number CAF32369–CAF32386 and CAG44621–CAG44632.
[c]Refers to the CDD database.[255]

instead of having them recruited for encoding biosynthetic functions involved in dehydroxylation subpathways.

The original postulate that RIB is the antecedent to all NEO-like AGAs led us to expect that the *rib*-cluster would lack a glycosyltransferase gene for the third glycosylation (2nd hexosaminyltransfer step). This postulate also established the expectation that this *rib*-cluster would perhaps lack other genes needed to complete the subsequent steps in the modification pathway, which in

TABLE 2.11. Proteins Encoded in the Genomic Area Covering the *liv*-Cluster of *Streptomyces lividus* CBS 844.73 (Accession Code AJ748832)[a]

ORF /Locus Tag[a]	Gene Product[b] Symbol	aa	CDD Number[c]	Putative Function
SliD01.5		692	9264	Membrane transport protein
SliD01.6	LivE	339	25397	Aminocyclitol 1-DH
SliD01.7	LivS	424	15279	L-Glutamine:ketocyclitol AT I+ II
SliD01.8	LivC	384	17039	2-Deoxy-*scyllo*-inosose synthase (cyclase)
SliD01.9	LivM	414	7651	2DOS 4-glucosaminyltransferase, GT I
SliD01.10	LivT	601	10852	ABC transporter
SliD01.11	LivU	621	10852	ABC transporter
SliD01.12	LivQ	546	16494	6′ (and 6‴-) DH
SliD01.13	LivN	299	26181	5‴-Epimerase or 6‴-DH
SliD01.14	LivP	229	16692	Sugarphosphate phosphatase
SliD01.15	LivF	357	10312	5‴-GT, GT III
SliD01.16	LivD	252	11828	*N*-Acetyl-hexosaminyl deacetylase or amidase
SliD01.17c	LivL	617	—	Unknown; ribosyltransfer
SliD01.18	LivB	416	25439	6‴-AT
SliD01.19c	LivZ	456	28900	LM-phosphate phosphatase
SliD01.20	LivI	166	28977	Component of sensor/response regulator
SliD01.21	LivH	170	—	Component of sensor/response regulator
SliD01.22	LivG	407	—	Component of sensor/response regulator
SliD01.23c	LivA	1355	—	Unknown; ribosyltransfer
SliD01.24c	LivX	80	—	cHP, regulation
SliD01.25c	LivO	62	—	Unknown, (fragment)
SliD01.26	LivV	430	—	cHP
SliD01.27	LivW	458	22645	Fe–S oxidoreductase; 3′-dehydroxyl. (?)
SliD01.28	LivY	238	25396	Oxidoreductase; 3′-dehydroxyl. (?)
SliD01.29c	LiwA	407	26181	Coenzyme PQQ synthesis protein
SliD01.30		152	29446	Nitrate reductase alpha chain

[a]The coding sequences (ORFs) are given with their locus tag number as found on the respective cosmids analyzed.
[b]The list of gene products translated from the nucleotide sequence corresponds to the protein ID number CAG38693-CAG38719.
[c]Refers to the CDD database.[255]

turn would lead to the pseudotetrasacchardic members of the neomycin family. Therefore, it was surprising to find that the *neo*- and *rib*-gene clusters are practically identical in gene number and order (see Figure 2.9). Both gene clusters comprise 20 genes and are flanked by the two resistance genes encoding AGA-modifying enzymes AphA or APH(3′) (*aph, pph*) and AAC(3) (*aacC8*). The residual 18 genes are believed to be necessary and sufficient to express the functional NEO pathway, in which RIB is an intermediate (cf. Figure 2.11). Genes of type "C,S,E,M,D" are known or postulated to be involved in paromamine

TABLE 2.12. Proteins Encoded in the Genomic Area Covering Part of the *btr*-Cluster of *Bacillus circulans* ATCC 21558 (Accession Code AJ781030) Supplemented with Data from *Bacillus circulans* SANK 720703 (Accession Code AB097196)[a]

ORF /Locus Tag[a]	Gene product[b] Symbol	aa	CDD Number[c]	Putative Function
70720832	ORF1	171	—	Unknown
70720833	BtrR1	217	11025	Transcriptional Regulator
70720834	BtrU	191	8331	Oxidoreductase
70720835	BtrR2	380	16371	Transcriptional Regulator
70720836	BtrT	77	—	Unknown
70720837	BtrS	418	15279	L-Glutamine:ketocyclitol AT I + II
70720838	BtrN	250	26181	Oxidoreductase, rad. SAM
70720839	BtrM	389	7651	2DOS 4-glucosaminyltransferase
BciH11.2	BtrL	604	—	Unknown, ribosyltransfer
BciH11.3	BtrA	1225	—	Unknown, ribosyltransfer
BciH11.4	BtrB	432	25439	6′-AT
BciH11.5	BtrC	368	17039	2-Deoxy-*scyllo*-inosose synthase (cyclase)
BciH11.6	BtrD	275	11828	*N*-Acetyl-paromanine deacetylase or amidase
BciH11.7	BtrE	349	25397	Aminocyclitol 1-DH
BciH11.8	BtrF	232	25396	Oxidoreductase; hexose-DH
BciH11.9	BtrG	156	5716	cHP in bacilli; AHBA-transfer (?)
BciH11.10	BtrH	302	—	HP; AHBA-transfer (?)
BciH11.11	BtrI	87	25546	ACP, AHBA-synthesis
BciH11.12	BtrJ	419	3293	Carboxylase; AHBA synthesis
BciH11.13	BtrK	428	25963	Glu-decarboxylase; AHBA synthesis
BciH11.14	BtrO	341	11849	FMN monoxygenase, AHBA synthesis
BciH11.15	BtrP	213	16692	Phosphatase or mutase
70720854	BtrV	82	—	cHP, regulation
BciH11.16	BtrQ	504	16494	6′-DH
70720856	BtrW	418	15279	ABC-transporter
70720857	BtrX	250	26181	ABC-transporter

[a]The coding sequences (ORFs) are given with their locus tag numbers created on analyzing the respective cosmid clones; alternatively the GI numbers are given (see accession code AB097196).
[b]The list of gene products translated from the nucleotide sequence corresponds to the protein ID numbers CAG77417-CAG77432.
[c]Refers to the CDD database.[255]

biosynthesis (cf. Figure 2.10) and are, therefore also conserved in the gene clusters for other 2DOS/paromamine-AGAs, such as the KAN/GEN family and even in the FOR/IST family, since most of the basic pseudodisaccharide pathway is congruent there (see paragraph on these AGAs). Further biosynthetic enzymes shared with pathways for other, non-NEO, AGAs are the "Q,B"-type proteins probably connected with the 6′- (or 6‴-) transamination of hexosamine units in the pathways of the same three AGA families. Of the remaining 11 *neo*-genes,

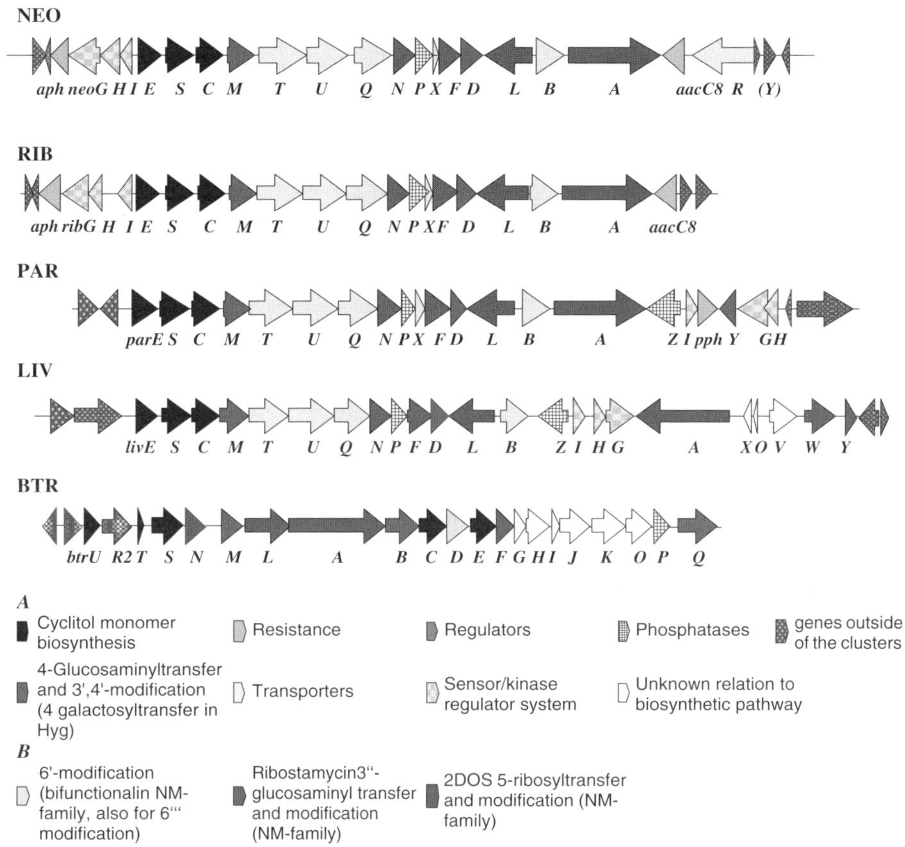

Figure 2.9. Biosynthetic gene clusters for NEO-family AGAs. Functionally related groups of genes are differentiated by the color code given below: (A) color code used in all genetic maps for 2DOS and 2DOS-related AGAs; (B) color code for pathway-specific functions in the NEO family. See color plates.

those conserved in all five clusters for NEO-type AGAs (including BTR) should be likely involved in the residual steps all pathways have in common, namely the 5-ribosyltransfer to the 2DOS unit. These are the "A,L"-type proteins, which both show indications for having transmembrane domains. Both these proteins do not belong to protein families with a known functional scope. The "G,H,I"-type genes are postulated to encode a sensor/response regulator system, and the "T,U"-type genes are postulated to encode a drug exporter belonging to the ABC transporter superfamily. Thus, only four genes remain for a differentiation among the close relatives of the NEO-type pathways, including those for PAR and LIV (in addition to those for NEO/RIB); these are the *neo-/rib*-genes of type "F,N,P,X." The encoded "F" protein belonging to the glycosyltransferases of group 1 is the only

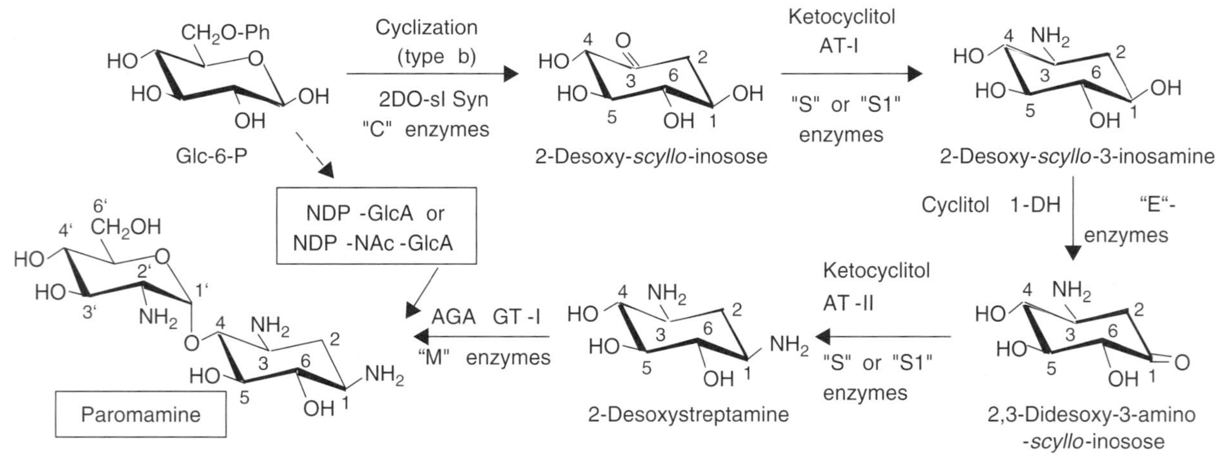

Figure 2.10. The general paromamine pathway. The enzymes involved are named only by the enzyme specific letters "C," "S," or "S1," "E," and "M," according to the nomenclature given in Tables 2.1–2.16. Other explanations and abbreviations are as given in the legend to Figure 3.

candidate from the glycosyltransferase (GT-III) that can transfer the second hexosaminyl residue (ring D in Figure 2.8) to the 3″-position of the ribose moiety in the RIB intermediate (cf. Figure 2.11); surprisingly, the *ribF* gene does not show any indication for being nonfunctional or not being expressed. Therefore, the reason for the absence of the fourth sugar moiety from RIB has to be sought elsewhere. The "P"-type (member of the phosphoglycerate mutase/phosphatase family) and the "X"-type (possibly a regulatory protein) proteins do not show any evidence for being different between the NEO- and RIB-biosynthetic tools. Therefore, only the "N"-type proteins, being members of the Fe–S cluster-containing "radical SAM" oxidoreductases (suggesting a biosynthetic function; e.g., for 5‴-epimerization), remain for differentiating the NEO- and RIB-pathways from each other. The RibN protein shows some interesting difference in its primary structure relative to those of the NeoN, ParN, and LivN proteins: A stretch of nonconserved amino acid sequence is seen in positions from 185 to 261 in the RibN amino acid sequence. Moreover, this deviating sequence is created by two compensating frame-shifts (+1/−1) in the nucleotide sequence of *ribN* (cf. accession code AJ744850) between which the sequence shows the same degree of conservation as that of the *neoN* gene as outside this region. Since frame-shifts of this kind easily can occur as artifacts either via reading errors during automatic analysis of DNA sequences or via mutational alteration in heterologous genetic backgrounds after cloning of the DNA, we analyzed the particular DNA sequence twice again from independently isolated fragments obtained directly by PCR-amplification of the genomic DNA of *S. ribosidificus* NRRL B-11466, which confirmed the same result.[124] This rare structural feature in a naturally occurring gene can be interpreted as follows: Either no RibN protein is made at all or a nonfunctional, but glycosyltransfer-blocking, RibN* polypeptide is made. How an altered and nonfunctional enzyme protein could exert such an effect cannot be explained at present.

2.2.2.1.2. The Biosynthetic Pathways of NEO-Related 2DOS-AGAs.

2.2.2.1.3. The General 2DOS Pathway. The general pathway for making the 2DOS and paromamine intermediates, common to the biosyntheses of several 2DOS-AGAs, is shown in Figure 2.10. Most of the attempts to elucidate the biosynthetic pathway for 2DOS by *in vivo* labeling and enzymological methods in the crude extracts of their producing strains have been made using the producers of NEO-type AGAs.[4,125,126] Similar observations were made in GEN and BTR producers, in which 2-deoxy-*scyllo*-inosose and 2-deoxy-*scyllo*-inosamine could be demonstrated to be precursors of 2DOS.[4,125,126] Only for the BtrC-related enzymes KanC/GenC (2-deoxy-*scyllo*-inosose synthases or cyclases) and for the StsC-related aminotransferase enzymes KanS1/GenS1/ForS (L-glutamine:ketocyclitol aminotransferases I and II) a clear assessment of their functions can be postulated on the basis of published enzymological data.[27,58,69,129–132]

The BtrC-related cyclases are specific for the 2DOS pathway and represent dihydroquinic acid synthase-related enzymes as is the 2,5-*epi*-valiolone

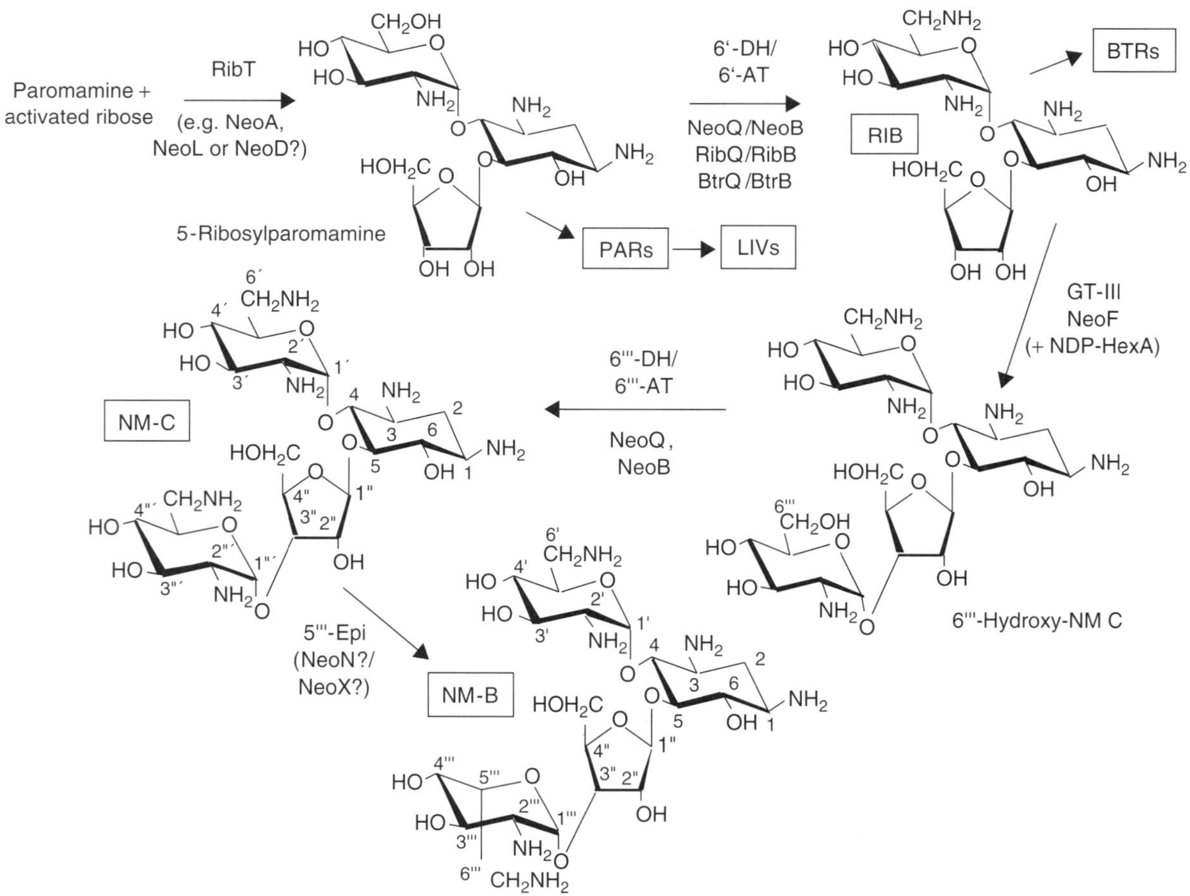

Figure 2.11. Biosynthetic pathway for NEOs. Other explanations and abbreviations are as given in the legend to Figure 2.3.

synthase of the acarbose pathway in *Actinoplanes* sp.[133,134] In contrast, the L-glutamine:ketocyclitol aminotransferases are probably the only highly conserved link among the enzyme sets for the biosyntheses of most aminocyclitol-containing AGAs and have first been described for the streptomycin pathway.[3,69] Therefore, several probes specific for the *stsC*-related genes could be designed for heterologous detection of other AGA gene clusters.[54,124] Despite these few firmly known functions, a relatively safe prediction on the involvement in the 2DOS pathway of the oxidoreductases belonging to the "E"-type gene products (putative cyclitol [1-]dehydrogenases) could already be made on in the basis of their distribution and level of conservation among the clusters for 2DOS-AGAs. We have confirmed this hypothesis by cloning and heterologous expression of the KanC, KanS, KanE, and NeoE proteins in both *E. coli* K-12 and *S. lividans* 66, which could reconstitute the formation of 2DOS from glucose-6-phosphate and L-glutamine, in presence of NAD_+ (K. M. Aboshanab, H. Schmidt-Beissner, U. F. Wehmeier, and W. Piepersberg, unpublished). Also, the ForE enzyme from the biosynthetic toolkit for FOR/IST family of pseudodisaccharidic AGAs belongs to this enzyme family (cf. Table 2.17 and Figure 2.23). However, it seems likely that ForE does not act on a monomeric substrate, but acts instead on a pseudodisaccharide (cf. Section 2.2.3).

2.2.2.1.4. The Formation of Paromamine and Neamine. Researchers engaged in a lengthy debate as to whether paromamine was an intermediate, or if neamine (6′-amino-6′-deoxyparomamine) was synthesized directly from 2DOS and a preformed and NDP-activated sugar unit neosamine C (6′-amino-6′-deoxy-D-glucosamine). Rinehart and co-workers could not detect any conversion of paromamine and neamine into antibiotic by 2DOS-negative mutants of *S. fradiae*, *S. kanamyceticus*, and *S. rimosus* ssp. *paromomycinus*.[135] In contrast, experiments by Pearce et al.,[136,137] who measured the incorporation of isotope-labeled precursors (^{3}H- and ^{14}C-labeled neamines) by idiotrophic mutants of *S. fradiae* and *S. rimosus* ssp. *paromomycinus*, suggested that the second route toward neamine was realized. The first route, formation of the pseudodisaccharidic intermediate paromamine by hexosaminylation and further modification to neamine, is now the accepted route (see below). A series of blocked mutants of *B. circulans* suggested the route via paromamine in studies on the utilization and bioconversion to active end products of various precursors (2DOS, paromamine, neamine, and 1-AHBA derivatives thereof; ribostamycin and xylostasin, 6′-hydroxy-derivatives of xylostasin, BTR-B, and BTR-A).[63,128,138–143] However, the pathway to paromamine (and further to neamine) has not been demonstrated by enzyme assays so far. We postulate here that the "M"-type genes conserved in most of the gene clusters for 2DOS-AGA and related AGAs express paromamine synthases (cf. Figure 2.10). The nucleotide sugar as a cosubstrate for this glycosyltransfer could be both UDP-D-glucosamine or UDP-*N*-acetylglucosamine. At present we favor the second possibility, since this would use a substrate abundantly provided by intermediary metabolism. However, an *N*-acetylated intermediate would require a later deacetylation step catalyzed by an amidase. There is a putative amidase encoded by the "D"-genes/enzymes conserved in most of the 2DOS-AGA clusters and being related to the MshB/MitE/LmbE family of possible *N*-acetylhexosamine deacetylases (accession code NP_{215686}).[144] Nevertheless, this notion is weak and the "D"-type amidases also could be involved in other steps of crucial importance for the biosynthetic pathway, such as the specific hydrolytic removal of 2-ketoglutaramate (converted into 2-ketoglutarate), which is an unavoidable by-product of the two L-glutamine-dependent transamination steps of the 2DOS pathway (see above) and is otherwise prone to spontaneously cyclize to 2-hydroxy-5-ketoproline, a likely toxic nonmetabolic compound.[58] Surprisingly, the equivalent "D"-type enzyme, BtrD, from *B. circulans* had been claimed to be a UDP-glucosamine synthase.[145] This puzzling notion is in contrast to any other knowledge on the NDP-sugar nucleotidyltransferase enzyme family, which shows uniformly conserved primary structure motifs throughout (e.g., the nucleotidyltransferase motifs summarized under CDD40572) and which does not fit to sequences in the "D"-type enzymes encoded by genes conserved in many AGA and the lincomycin (*lmbE*) gene clusters. Future enzymological studies will have to solve this conflict.

The two additional enzymatic steps needed, an oxidation for 6′-dehydrogenation and 6′-aminotransfer, for the 6′ transamination of paromamime to yield neamine are probably in strong competition with the successive second glycosyltransfer and other following reactions, since most NEO-type AGAs can be constructed as 6′-desamino-6′-hydroxy variants (such as the PARs and LIVs). The gene products needed for creating these two additional steps are the "Q"-type oxidoreductases belonging to the choline dehydrogenase family and the "B"-type aminotransferases; members of the gamma-aminobutyric acid semialdehyde aminotransferases [Figure 2.11; cf. Tables. 2.8–2.14 and 2.15–2.18]). At least, it seems likely at present that the formation of 5-ribosylparomamine is a preferred route over that of RIB biosynthesis via neamine. The placement of 6′-transamination, therefore, seems to be possible at various stages of the pathway, either before or after pseudotrisaccharide formation.

2.2.2.1.5. Biosynthesis of the Ribosylated Pseudotrisaccharides 5-Ribosylparomamine and RIB. The NEO-type pathway starts with a glycosyltransfer reaction coupling a D-glucosamine unit glycosidically to 2DOS the aminocyclitol precursor, thus forming a pseudodisaccharide, paromamine, as an intermediate (cf. Figure 2.10). The respective putative glycosyltransferases, NeoM, RibM, ParM, LivM, and BtrM, are highly conserved in many of the producers for other pathways that do or do not (e.g., the FOR/ISTs) produce a paromamine intermediate, despite the difference in substrate requirements. However, no clear candidate gene for the second glycosylation step introducing the ribose unit becomes evident from analysis for the *neo*-like gene clusters.

Since its detection, RIB was believed to be a common intermediate and evolutionary precursor of all the NEO-family 2DOS-AGAs. The precursor function was proven, since a mutant of *S. fradiae* blocked in NEO production was found to produce RIB and to yield single-step revertants to the wild-type phenotype.[146] However, until today we do not have any indication on how the ribose moiety is activated and introduced into the paromamine precursor, and the current data set rather suggests that PARs and LIVs are derivatives of 5-ribosylparomamine (cf. Figure 2.12). This in turn is a realistic candidate for an intermediate only, if neamine (NEO-A) is not formed before ribosylation. However, the occurrence of neamine as a regular by-product in all NEO fermentations[147] and the fact that neamine is readily incorporated and converted into NEOs or BTRs when fed to 2DOS-negative mutants of *S. fradiae* and *B. circulans*[136,138–140] rather suggests the opposite; an alternative explanation for these puzzling facts could be that the occurrence of neamine as a minor component in fermentation broths results from degradation of NEOs and that its conversion in mutants blocked in the cyclitol

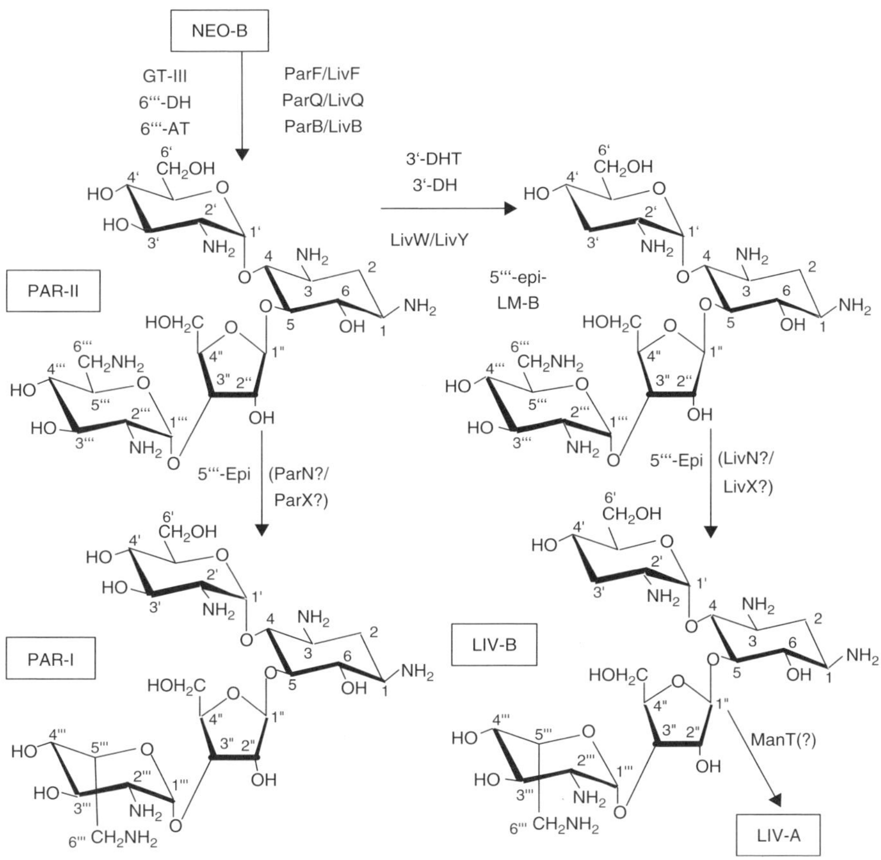

Figure 2.12. Biosynthetic pathway for LIV- and PAR-AGAs. Other explanations and abbreviations are as given in the legend to Figure 2.3.

pathway also results from prior hydrolysis into 2DOS and an amino sugar. The latter route was excluded by feeding studies with the labeled pseudodisaccharides, neamine and 5-ribosyl-2DOS, which are either incorporated directly (neamine) or not (5-ribosyl-2DOS) and do not show indications for prior hydrolysis and reuse of the monomers;[136,137] neosamine C (6′-amino-6′-deoxy-D-glucosamine) is not incorporated at all.[63,122]

2.2.2.1.6. The Biosynthesis of the Pseudotetrasaccharides, NEOs, PARs, and LIVs. The isolation from NEO and PAR producers of 6‴-desamino-6‴-hydroxy derivatives, suggests that these are intermediates in the biosynthesis of the pseudotetrasaccharidic AGAs of the NEO-family as are the 6′-desamino-6′-hydroxy pseudotrisaccharides.[148] The transfer and modification of a second glucosamine moiety, probably provided again from the pool of UDP-(*N*-acetyl)-D-glucosamine, involves the following steps: glycosyltransfer to the 3″-position of the RIB intermediate by the NEO-specific glycosyltransferase GT-III, for which the sole candidate is the "F"-type gene product, only encoded in the gene clusters for the pseudotetrasaccharidic NEO-related AGAs (see Tables 2.8–2.12; Figure 2.9).

Surprisingly, there are two cases of unexpected occurrences in the *rib*- and the *par*-clusters of genes which at first glance appear to be unnecessary for the production of the individual end products: (1) the *ribF* gene, the equivalents of which putatively encode the second hexosaminyltransferase in the NEO, PAR, and LIV pathways, and (2) the *parQ/livQ* and *parB/livB* genes, the counterparts of which encode the dehydrogenase/aminotransferase couple necessary for the introduction of the 6′-amino group into the paromamine subunit in the NEO and BTR pathways. These genes obviously nevertheless co-evolved in their respective gene clusters with other genes. Whereas the ParQ/ParB set of enzymes does not show any indication for nonfunctionality in their primary structures (as compared to the other pairs of 6′-dehydrogenases/6′-aminotransferases), there is a clear stretch of deviation in the primary structure of the RibN protein (see above). All these observations can be interpreted as follows: (i) The non-occurrences of the second hexosamine residue in RIB and of a 6′-amino group in PARs are effects of cluster/pathway degeneration rather than representing primordial states relative to the ones in the NEOs and LIVs. (ii) The RibF protein has lost its function; its production cannot be excluded because there are no features to be seen in the DNA sequence which would indicate its exclusion from transcription or translation. (iii) The phenomenon of the production of 6′-deamino derivatives of neomycins (called paromomycins) in *S. rimosus* subsp. *paromomycinus* is even more mysterious at present, because nonfunctionality of the respective enzymes necessary for 6′-transamination is not evident. One possible explanation would be that 6′-transamination is the last step in biosynthesis of neomycin-type aminoglycosides and that export of the 6′-hydroxylated precursor became more efficient in the PM producer and, therefore, competes with transamination. This would suggest that 6′-transaminated product(s) would be likely to arise as minor components in the culture broth of *S. rimosus* subsp. *paromomycinus* and that the specific exporter system (probably the ABC-transporter complex ParT/ParU) has slightly altered its substrate specificity relative to the one used for export of NEOs, RIB, or LIVs.

2.2.2.1.7. The Late Modification Pathway in BTR Biosynthesis. The acquisition of the late modification pathway leading to the transformation of the NEO pathway into that for the BTRs, which is characteristic for the sole version of an AGA production pathway in a member of the firmicute (low G+C gram-positive bacteria) branch of the bacterial tree (in contrast to *Actinobacteria* [high G+C gram-positive bacteria], which are the typical AGA producers), seems to be a typically bacillar biosynthetic solution: Secondary metabolic traits in bacilli are known to be especially strong in the production of nonribosomal peptides (and less in polyketides, etc., including secondary sugar derivatives). Therefore, it seems consequent that instead of the fourth (aminated!) sugar unit in the NEOs, an amino acid derivative ([2S]-4-amino-2-hydroxybutyric acid = AHBA; Figure 2.13) is coupled via an amide bond to the core pseudotrisaccharide RIB and thereby replaces some of the lost sugar functionality in form of a hydroxyl and an amino group. In the evolution of the *btr*-cluster, loss of biosynthetic capacity occurred possibly during or after the horizontal transfer of a *neo*-type gene cluster from an actinomycete to a firmicute organization, where the genetic material needed for the pseudotrissacharide function was firmly established. The genetic material for this biosynthesis and transfer of the AHBA moiety was possibly added later to the "core *btr*-cluster."

The AHBA unit of BTRs is specifically labelled from L-glutamate and gamma-aminobutyric acid (GABA).[149] Several genes have been implicated to be involved in the AHBA pathway, including the *btrG, H, I, J, K, O*, and *btrV* genes.[24,25,150] Recently, an integrated genetic and biochemical study revealed the mechanism of most of the steps involved in formation of the AHBA side chain, which occur on acylcarrier protein (ACP)-bound intermediates (cf. Figure 2.13)[150]: (i) ATP-dependent transfer of the gamma-carboxyl group of L-glutamate to the ACP BtrI by the ligase BtrJ; (ii) decarboxylation of the alpha-C atom by the pyridoxalphosphate-dependent decarboxylase BtrK performs the BtrI-bound GABA unit; (iii) a second gamma-glutamyl transfer, again catalyzed by BtrJ, to synthesise an *N*-acyl derivative of GABA; (iv) stereospecific hydroxylation of the alpha-C atom (S configuration) of the GABA precursor by the two-component monooxygenase BtrO/BtrV, which is $FMNH_2$-dependent and in which the BtrV protein component recycles FMN by catalyzing its NAD(P)H-dependent reduction to $FMNH_2$, yields the final precursor unit BtrI-S-AHBA-γ-L-Glu; (v) the transfer of the AHBA moiety from the precursor under release of L-glutamate and BtrI is proposed to be catalyzed by the postulated acyltransferase BtrG/BtrH, which, however, is not yet tested so far.[150] All proteins postulated to be involved in the AHBA pathway are unique gene products of the *btr*-cluster, except for the *btrV* encoded protein. In view of these data the occurrence of *btrV*-related genes, *neoX, ribX, parX*, and *livX*, in the other gene clusters for NEO-related AGAs remains enigmatic. Since there is no evident enzyme counterpart related to the monooxygenases of

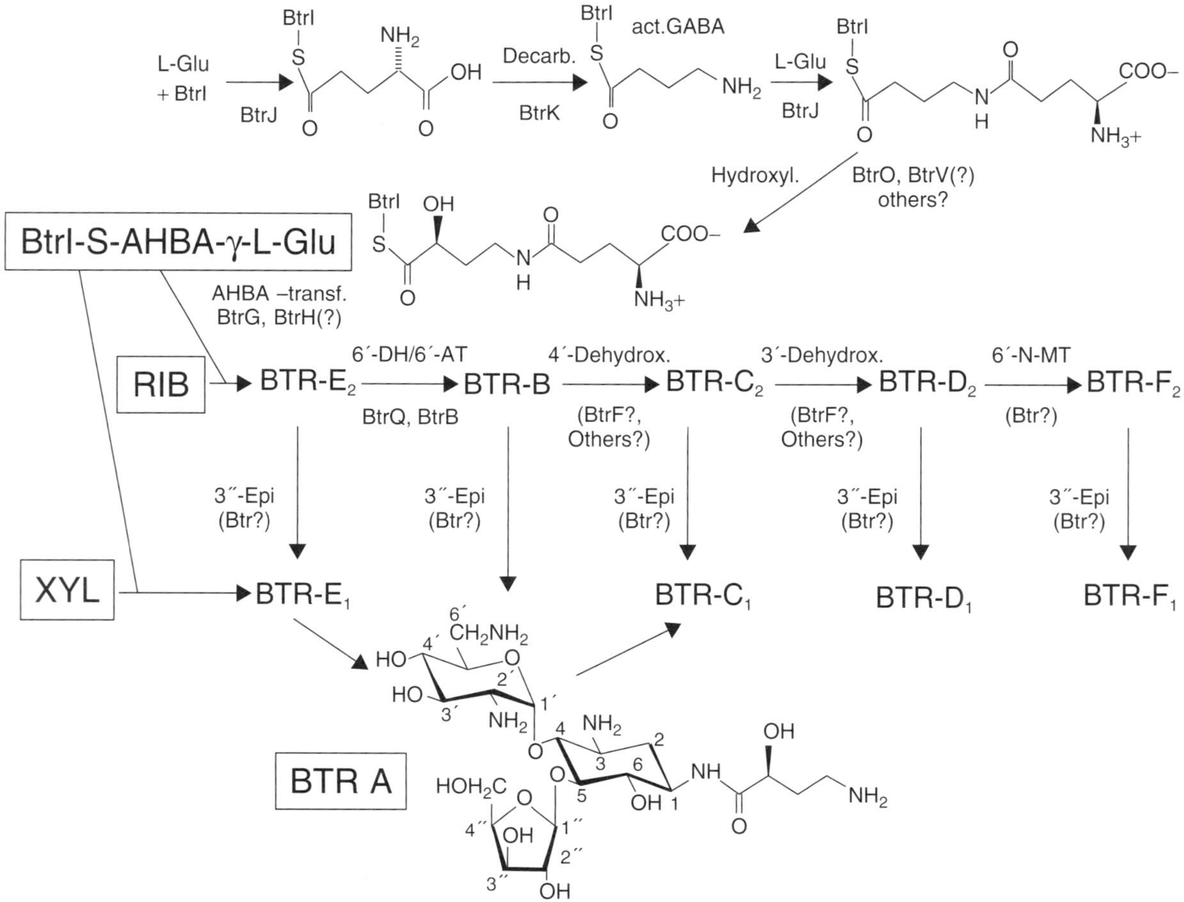

Figure 2.13. Biosynthetic pathway for BTR AGAs. The predicted route for the side chain was postulated,[150] and all other steps were postulated from the published sequences (cf. Table 2.12). Other explanations and abbreviations are as given in the legend to Figure 2.3.

BtrO-type encoded in these clusters, it will be interesting to see which step and enzyme system in the biosynthesis of NEO will be necessary: either a regeneration of $FNMH_2$ or a regeneration of the NeoX enzyme.

2.2.2.2. 4,6-Disubstituted 2DOS-AGAs.

2.2.2.2.1. The Genetics and Biochemistry of the Biosynthetic Pathway for KAN-Related 2DOS-AGAs. Kanamycin (KAN; Figure 2.14) was first described as a product of *S. kanamyceticus* K-2J (DSM 40500) in 1957.[151,152] The complex comprises KAN-A as the main component, with the two minor components KAN-B and KAN-C.[153,154] The trace components Paromamine and 6-O-(3-amino-3-deoxy-D-glucopryanosyl)-2DOS were also subsequently found during kanamycin preparations.[155] Biogenetic studies with D-[1-^{14}C]glucose and D-[1-^{14}C]glucosamine fed to cultures of *S. kanamyceticus* showed that D-glucose is an equally well incorporating precursor for all three moieties of KAN-A, but that D-glucosamine is a preferred precursor of the 6-amino-6-deoxyglucose unit of KAN-A.[156] This is a surprise, since it implicates that the glucosamine moiety of paromamine becomes deaminated during KAN-A biosynthesis (likely from KAN-B as an immediate precursor, see below).[63] For a long time it was believed

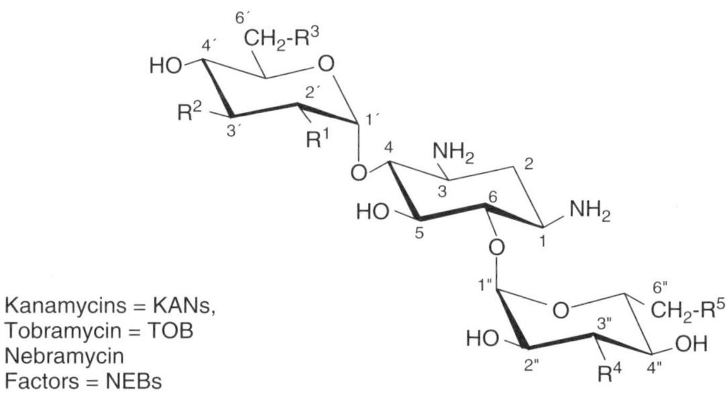

KANs	R^1	R^2	R^3	R^4	R^5
KAN-C	NH_2	OH	OH	NH_2	OH
KAN-B	NH_2	OH	NH_2	NH_2	OH
KAN-A	OH	OH	NH_2	NH_2	OH
NK-1001	OH	OH	NH_2	OH	OH
NK-1012-1	NH_2	OH	NH_2	OH	OH
TOB	NH_2	H	NH_2	NH_2	OH
NEB-4	NH_2	OH	NH_2	NH_2	$OCONH_2$
NEB-5‘	NH_2	H	NH_2	NH_2	$OCONH_2$

Figure 2.14. Structures of kanamycin (KAN) family AGAs.

that the subunits of KAN were synthesized independently and assembled then directly to the 2DOS moiety.[157]

However, studies in other 2DOS-AGA producers, such as the producers for NEO, BTR, and GEN (see the respective paragraphs), clearly proved the general paromamine pathway (cf. Figure 2.10) and subsequent addition of further sugar units to this pseudodisaccharide.

To explain the various monomers, pseudisaccharides, and pseudotrisaccharides, found to be produced by a mutant of *S. kanamyceticus* (2DOS, paromamine, neamine, kanamine, 6-*O*-glucosylneamine, 4,6-diglucosyl-2DOS, and KAN-A, -B, -C),[154] Okuda and Ito proposed a "metabolic grid" for interconversion of KANs and their precursors with a preferred route,[63] which is in accord with the biosynthetic pathway for KANs described below (see Figure 2.16a). The TOB pathway (Figure 2.16b) should be largely congruent with that for the KANs, but complicated by the fact that NEB-factors and APR are made in the same strains (cf. Table 2.2).

The *kan*-gene cluster of *S. kanamyceticus* DSM 40500 was identified on the cosmid clones SkaJ19 and SkaJ15. A 62887 bp contiguous genomic DNA sequence from *S. kanamyceticus* DSM 40500 was obtained comprising 65 ORFs (accession code AJ628422), the characteristics and nomenclature of which are outlined in Table 2.13 and the map locations and orientations of which are shown in Figure 2.15. The absolute limits of the functional *kan*-gene cluster cannot be exactly predicted at present due to lacking functional studies and/or analyses of closely related gene clusters; the analysis of a cluster for the 2DOS AGA TOB, the closest relative of KANs, from *S.* sp. "*tenebrarius*"[28,30] (our own published sequence analyses; cf. Tables 2.2 and 2.14), helps to determine their structural limits and to interpret their informational content (cf. Figure 2.15). Therefore, currently we suggest these limits to lie somewhere between ORFs SkaJ19.7c and SkaJ19.11c at the left-hand side and between ORFs SkaJ19.31 and SkaJ15.17 (genes designated *kanL* through *kanU*) at the distal side (right-hand side; cf. Table 2.13 and Figure 2.15). Therefore, the total potential of kanamycin production genes is in the range between 19 (core cluster) and 29 genes in total. From the genes present in the *kan*-cluster, only the two kanamycin resistance genes *kmr* (CAA75800.1; encoding a 16S rRNA methyltransferase) of *S. kanamyceticus* ISP1375 and *kac* (AB164230.1; encoding aminoglycoside 6′-acetyltransferase) were known from earlier cloning and sequencing studies;[158,159] both genes found in our data set turned out to be identical in sequence with those submitted to the database before. Several of the other protein sequences deduced from the core *kan*-genes show clear relationships to genes already known from other treated AGA clusters, such as KanR related to StrR, and the proteins of "C," "S," "E," and "M" type involved in the general paromamine branch of the KAN/TOB pathways (see above, Section 2.2.2.1.3).

The *tob*-Gene Cluster of *S. sp.* DSM 40477. For the sequence analysis of the tobramycin (*tob*) gene cluster of *S. sp.* DSM 40477, a persistent DNA stretch of 43,220 bp was analyzed, encoding 37 ORFs from which 20 genes are potentially attributed to the *tob*-cluster. Eleven genes are dedicated properly to encode

TABLE 2.13. Proteins Encoded in the Genomic Area Covering the *kan*-Cluster of *Streptomyces kanamyceticus* DSM 40500 (Accession Code AJ628422)[a]

ORF /Locus Tag[a]	Gene Product[b] Symbol	aa	CDD Number[c]	Putative Function
SkaJ19.7c	KaoA	832	11915	cHP, regulator
SkaJ19.8c	KanZ	811	25388	Efflux protein
SkaJ19.9c	KanY	147	—	Unknown
SkaJ19.10c	KanX	420	29728	Efflux protein
SkaJ19.11c	KanW	64	—	Unknown
SkaJ19.12	KanA	184	25558	AAC(6′), KM-resistance
SkaJ19.13c	KanG	382	—	Component of sensor/response regulator
SkaJ19.14c	KanH	195	—	Component of sensor/response regulator
SkaJ19.15c	KanI	160	—	Component of sensor/response regulator
SkaJ19.16	KanR	369	26725	KM-specific pathway regulator
SkaJ19.17	KanE	343	25397	Aminocyclitol 1-DH
SkaJ19.18	KanT	418	—	Exporter (MFS family)
SkaJ19.19	KanB	392	25439	6′-AT
SkaJ19.20	KanQ	508	16494	6′-DH
SkaJ19.21	KanS1	427	15279	Cyclitol AT, AT I
SkaJ19.22	KanC	390	17039	2-Deoxy-*scyllo*-inosose synthase (cyclase)
SkaJ19.23	KanD2	369	25397	3″-DH
SkaJ19.24	KanM2	388	7651	6-Glucosyl-GT, GT II
SkaJ19.25	KanD	249	11828	*N*-Acetyl-hexosaminyl deacetylase or amidase
SkaJ19.26	KanS2	419	6445	3″-AT; AT II
SkaJ19.27	KanJ	322	24783	Unknown, 2′ deamination?
SkaJ19.28	KanK	329	—	Unknown, 2′ deamination?
SkaJ19.29	Kmr	277	27198	16S rRNA MT; KM-resistance
SkaJ19.30	KanM1	414	7651	2DOS 4-glucosaminyltransferase, GT I
SkaJ19.31	KanL	335	10881	Unknown, GTPase?
SkaJ19.32	KanO	132	30229	cHP
SkaJ19.33c	KanP	223	—	Unknown
SkaJ15.16	KanD3	593	4793	DH
SkaJ15.17	KanU	316	7613	Transcript. repressor (TetR fam)

[a]The coding sequences (ORFs) are given with their locus tag numbers as found on the respective cosmids analyzed.
[b]The list of gene products translated from the nucleotide sequence corresponds to the protein ID numbers CAF31574-CAF31602.
[c]Refers to the CDD database.[255]

biosynthetic proteins, and two gene products could be involved in transport. However, neither genes encoding the proteins needed for the 3′-dehydroxylation nor a gene encoding a resistance protein could be detected (Table 2.14, Figure 2.15). We assume that the lacking genes are all located within the *apr*-gene cluster, which coexists in the producer strains (see Table 2.2 and Section 2.2.4.1). The lack of these genes in the *tob*-cluster might explain the observation that

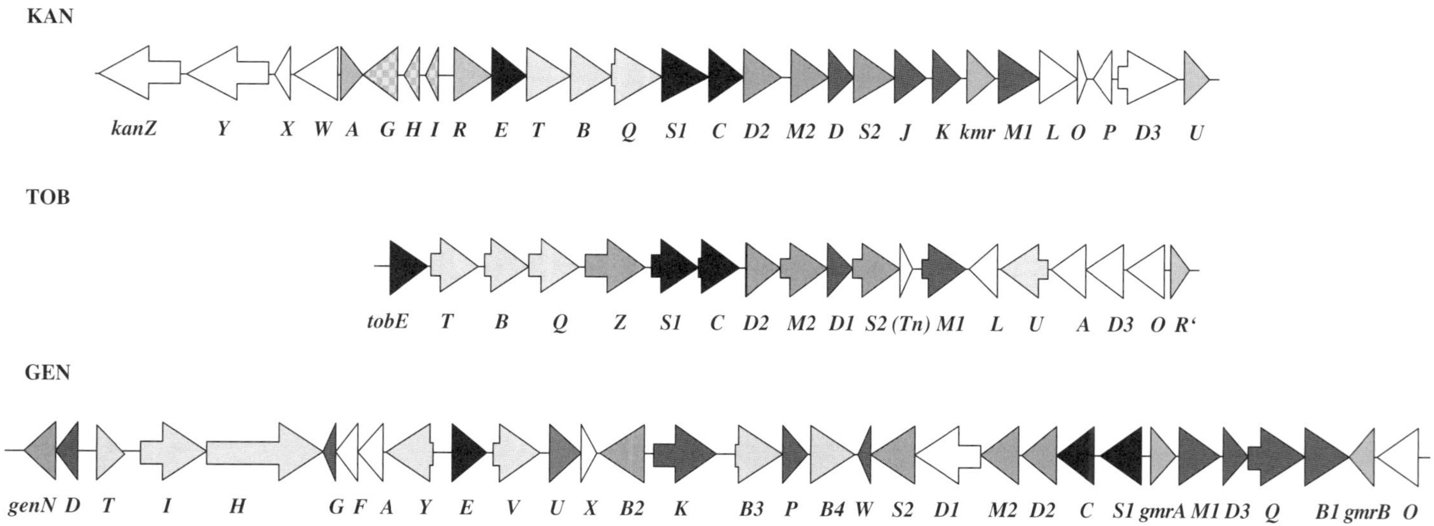

Figure 2.15. Biosynthetic gene clusters for KAN/GEN-related AGAs. 2DOS 6-glycosyltransfer and modification-specific genes, all other colors see Figure 2.9.

TABLE 2.14. Proteins Encoded in the Genomic Area Covering the *tob*-Cluster of *Streptomyces* sp. DSM 40477 (Accession Code AJ810851)[a]

ORF /Locus Tag[a]	Gene Product[b] Symbol	aa	CDD Number[c]	Putative Function
SteK17.16	TobY	165	—	Unknown
SteK17.17	TobX	217	—	Oxygenase superfamily
SteK17.18	TobE	339	25397	Aminocyclitol 1-DH
SteK17.19	TobT	436	—	Transporter (MFS family)
SteK17.20	TobB	395	25439	6′-AT
SteK17.21	TobQ	508	16494	6′-DH
SteK17.22	TobZ	579	23354	6″-Carbamoyltransferase
SteK17.23	TobS1	424	15279	Cyclitol AT, AT I
SteK17.24	TobC	386	17039	2-Deoxy-*scyllo*-inosose synthase (cyclase)
SteK17.25	TobD2	347	6445	3″-DH
SteK17.26	TobM2	420	7651	6-Glucosyl-GT (GT II)
SteK17.27	TobD1	261	11828	*N*-Acetyl-hexosaminyl deacetylase or amidase
SteK17.28	TobS2	416	15279	3″-AT
SteK17.29		123	—	Retrotransposon (fragment)
SteK17.30	TobM1	416	7651	2DOS 4-glucosaminyltransferase, GT I
SteK17.31c	TobL	410	10331	carbamoyl-phosphate synthase
SteK17.32c	TobU	325	—	Transporter (DMT-superfamily)
SteK17.33c	TobA	372	16647	AT
SteK17.34c	TobD3	358	7803	DH (FMN-dep.)
SteK17.35c	TobO	327	29325	Oxygenase
SteK17.36	TobR	163	1588	Transcription regulator (AsnC family)

[a]The coding sequences (ORFs) are given with their locus tag numbers as found on the respective cosmids analyzed.
[b]The list of gene products translated from the nucleotide sequence corresponds to the protein ID numbers CAH18548–CAH18568.
[c]Refers to the CDD database.[255]

the production gene clusters for TOB and APR always coexist in the producer strains.

2.2.2.2.2. The Biosynthetic Pathways for KANs. The general pathway for making the 2DOS and paromamine intermediates, common to the biosyntheses of the NEO family has been described in details above (see Section 2.2.2.1.4; Figure 2.10). The postulated KAN/TOB pathways starting from the paromamine intermediate are outlined in Figures 2.16a and 2.16b. At first, it seems that the glycosyltransferase for the introduction of a glucose moiety into the 6-position of the paromamine precursor has evolved after a gene duplication event involving an ancestral "M"-type gene and further shifting its properties by mutation into a *kanM2*/KanM2 gene/enzyme (which is much more distant from the "M1"-type genes/enzymes than these are, for example, to NEO-type "M" genes/enzymes)

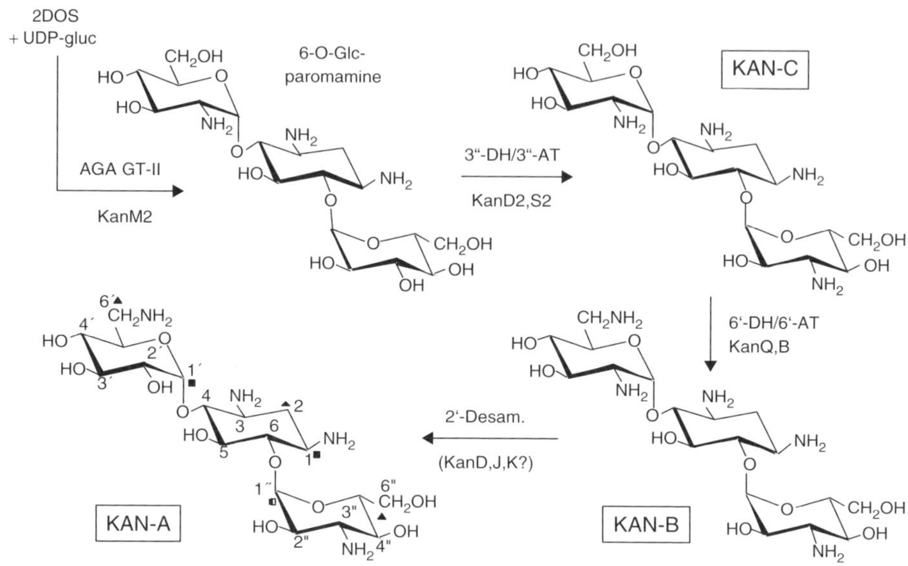

Figure 2.16A. Biosynthetic pathway for KAN AGAs. Other explanations and abbreviations are as given in the legend to Figure 2.3.

and that this step is the preceding one relative to first converting the pseudo-disaccharide into neamine as an alternative intermediate. This, in turn, also could mean that the nucleotide-activated sugar as co-substrates should be very similar for both types of "M" proteins; therefore, we postulate that UDP-D-glucose should be the sugar donor.

Second, the introduction of the third amino group into the 3″-position of the glucose unit could be, at least in part, the result of a gene duplication event involving the "S(1)"-type ancestor and evolving a ketopyranose aminotransferase of "S2"-type genes/enzymes (again, in phylograms the KanS2/TobS2 proteins are in a distant branch when compared to the "S1"-type proteins, which, in turn, form a common branch with the NeoS-related sequences; see Section 2.3.2). The KanS2-type aminotransferases putatively again use the alpha-amino group of L-glutamine as donor group, which is characteristic for the "S"-type enzymes (e.g., StsC in the STR pathway).[3,58,69] This suggests that a partial duplication of a primordial *kan*-cluster or the fusion of two similar clusters—for example, each providing a paromamine pathway—was the starting point for the evolution of the KAN-related pathways, later also evolving into that for the GEN-related 2DOS-AGAs (see Section 2.2.2.2.4). The sugar 3″-dehydrogenase required for introducing the 3″-keto group could be represented by the "D2"-type oxidoreductases, being members of the GFO-IDH-MocA family of dehydrogenases, to which also the *myo*-inositol DHs belong (e.g., StrI; see Section 2.2.1.2).

Third, whether 3″-transamination precedes or follows 6′-transamination cannot be decided from the presently available data, but both seem realistic from the

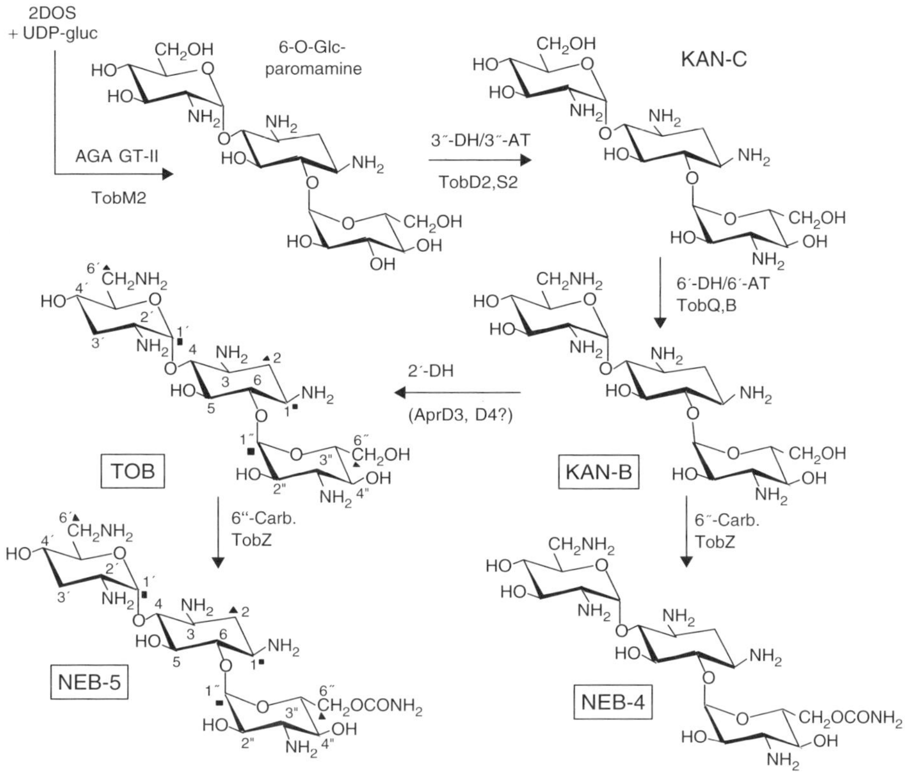

Figure 2.16B. Biosynthetic pathway for TOB and NEB. 6″-Carb. stands for carbamoyltransfer. Other explanations and abbreviations are as given in the legend to Figure 2.3.

metabolites encountered in fermentations of *S. kanamyceticus* and its mutants.[63] However, again the "Q"/"B" pair of conserved dehydrogenases/aminotransferases are clearly the likely candidates for the 6′-transamination step as already described (see Section 2.2.2.1.4). Finally, more specific and mysterious is the 2′-desamination-hydroxylation step converting KAN-B into the main end product KAN-A, which obviously does not occur in the TOB producers. The enzymatic mechanism for this "hydrolytic" removal of an amino group from an amino sugar could follow different routes, but has no well-studied counterpart in any known pathway. A conservative view would suggest that the reverse reactions of a typical hydroxyl oxidation/amino transfer could be involved—that is, introduction of a keto group by an aminotransferase transferring the amino group to (i) pyridoxalphosphate (thereby synthesizing the pyridoxamin phosphate form of the coenzyme and leaving a keto group) and then (ii) a keto co-substrate under regeneration of the pyridoxalphosphate; (iii) would be the action of a (likely NAD[P]H-dependent) reductase which reduces the 2′-keto group stereospecifically into the 2′-hydroxyl of gluco-configuration. Candidate genes/enzymes for this step are the *kanJ*/*kanK* (KanJ/KanK) pairs which during evolution of the *tob*-cluster from an ancestral *kan*-cluster might have been knocked out by a transposition event (see below, Section 2.3). Neither of the two proteins show any sequence similarity to protein families expected to be involved in such a mechanism; KanJ is related to phytanoyl-CoA dioxygenases, which could not be easily be harmonized with a function in the desamination/hydroxylation step. Alternatively, the KanD enzyme (putative 2′-*N*-actyl-hexosamine deacetylase, could be involved in hydrolytic desamination of this site if it had changed its enzyme mechanism accordingly and the KAN-B type intermediate still would be *N*-acetylated.

2.2.2.2.3. The Biosynthetic Pathways for TOB/NEB. The biosynthetic routes for TOB/NEB are quite similar to the biosynthesis of KAN and differ only in the final steps (see above and Figures 2.16a and 2.16b). There are three differences in the TOB/NEB pathway compared to the KAN pathway to be discussed here. The *tob*-cluster does not encode equivalents to the genes needed for deamination at position 2′. This finding corresponds to the production pattern of the TOB-producing strains, where we find only end products in the 2′-aminated form. There are two additional steps necessary in the TOB biosynthesis: the 3′-dehydroxylation and the carbamoylation found at position 6″ in the NEBs.

As mentioned above, we could not find any genes in the *tob*-cluster which could encode the enzymes for the 3′-dehydroxylation. The postulated dehydratase/oxidoreductase proteins involved are likely to be the AprD3/AprD4 enzymes encoded by the *apr*-cluster in the same strain, which have conserved counterparts only in the complement of gene products encoded by the *liv*-cluster (LivW/LivY; see above). We assume that the 3′-dehydroxylation is the last step in the biosynthesis of TOB for the following reasons: the producers produce a mixture of TOB, KAN-B, NEB-4 and NEB-5 and if the 3′-dehydroxylation would occur in earlier steps the strains should not be able to produce KAN-B and NEB-4.

At least two gene products, the putative carbamoylsynthase TobL and the carbamoyltransferase TobZ are needed for the carbamoylation found in the NEBs. The respective enzymes are unique among the AGA gene clusters analyzed so far. Because NEBs are mainly found as a complex of NEB-4 and NEB-5, the TobZ protein must recognize both KAN-B and TOB as substrates. Therefore the carbamoylation must be the final step in the NEB biosynthesis, as illustrated in Figure 2.16b.

2.2.2.2.4. The Genetics and Biochemistry of the Biosynthetic Pathway for GEN-Related 2DOS-AGAs. The GENs were first isolated from cultures of *M.* "*purpurea*," which later became renamed to *M. echinospora*, in 1964 and its structure analysed later[160–166] (see Figure 2.17). Later also in the cultures of other species of *Micromonospora sp.* very similar compounds have been described. The nomenclature for several of these isolates has been changed since then, such that reading the original literature became confusing; therefore, we here introduce

TABLE 2.15. *Micromonospora* sp. Producing GEN-Type AGAs and Some Derived Strains

Species	Original /NRRL	Strain Strain Number, ATCC	DSM	Major Product(s)[a], AGAs (cf. Figure 2.17)	Remarks (ref.)[b]
M. echinospora (former *purpurea*)	—/2953	15837	43816	GEN C_1, C_{1a}, C_2	(169,178)
M. echinospora ssp. *echinospora*	—/2985	15837	43816	GEN C_1, C_{1a}, C_2	(169)
M. echinospora (former *purpurea*)	JI-20			JI-20A, JI-20B	Mutant 2953 (169)
M. echinospora (former *purpurea*)	JI-33			GEN C_{2b}	Mutant 2953(169)
M. echinospora ssp. *ferruginea*	2995	15836	43141	GEN C_1, C_{1a}, C_2	(169)
M. echinospora ssp. *pallida*	—/2996	15838	43817	GEN C_1, C_{1a}, C_2	(169)
M. echinospora (former *rhodorangea*)	—/5326	27932	43822	G418	(169)
M. inyoensis	—/3292	27600	46123	SSM	(165,169)
M. inyoensis	—/5742			1-*N*-ethyl SSM	3929 (169)
M. zionensis	—/5466			G-52	(169,179)
M. sp. (former *grisea*)	—/3800	35853	1043	VDM	(169)
M. sagamiensis	MK-65, —/11334	21803	44886	SGEN (GEN C_{2b})	(169)
M. purpureochromo genes	—/B-16094	27007		GEN (?)	

[a]Abbreviations: see Table 2.1.
[b]NRRL strain numbers are given.

a list of the mostly used strains among the producers of GEN-related AGAs (see Table 2.15). Since soon after its introduction into the antibiotic market GENs became the major commercial AGA, it is not surprising that a lot of efforts have been invested into its development and research. The commercial preparations of GEN, the so-called GEN-C complex, obtained from fermentations with descendants of *M. echinospora*, are composed of GEN-C_1, GEN-C_2, and GEN-C_{1a}. A series of minor components or major components produced in some of the other producing strains, such as SGM (GEN-C_{2b}), VDM, or SSM, have later been described, too.[165–174] Even in a strain of *Streptomyces* a complex of AGAs related to GENs, the seldomycins, have been detected.[99,175–177]

2.2.2.2.5. The Genetics of GEN-Related AGAs. **The Gentamicin (*gen*)-Gene Cluster of *M. echinospora* DSM 43036**. From cosmids MecP21, MecE04, MecG05, and MecO02 84222 bp contiguous genomic DNA sequence from *M. echinospora* DSM 43036 were obtained comprising 43 ORFs (accession code AJ628149; Table 2.16, Figure 2.15). A subsegment of the sequenced genomic

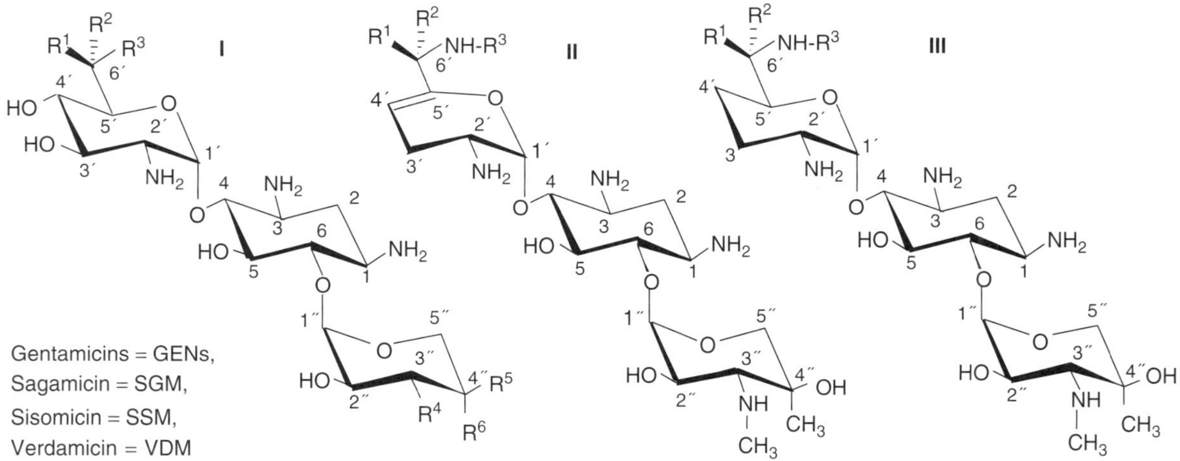

Gentamicins = GENs,
Sagamicin = SGM,
Sisomicin = SSM,
Verdamicin = VDM

GENs-I	R^1	R^2	R^3	R^4	R^5	R^6
GEN-A2	H	H	OH	OH	H	OH
UAA-41	H	H	OH	NH_2	H	OH
GEN-A	H	H	OH	NH_2CH_3	H	OH
GEN-X2	H	H	OH	NH_2CH_3	OH	CH_3
JI-20A	H	H	NH_2	NH_2CH_3	OH	CH_3
G-418	CH_3	H	OH	NH_2CH_3	OH	CH_3
JI-20B	H	CH_3	NH_2	NH_2CH_3	OH	CH_3
TPJ-B	CH_3	H	NH_2	NH_2CH_3	OH	CH_3

GENs-II	R^1	R^2	R^3
SSM	H	H	H
VDM	H	CH_3	H
G-52	H	H	CH_3
GENs-III			
GEN-C_{1a}	H	H	H
SGM	H	H	CH_3
GEN-C_{2a}	H	CH_3	H
GEN-C_2	CH_3	H	H
GEN-C_1	CH_3	H	CH_3

Figure 2.17. Structures of gentamicin (GEN) family AGAs.

area reported herein was also analyzed by two other groups from *M. echinospora* strain ATCC 15835 and submitted to the database (accession codes AY524043.1 and AJ575934.2). The overlapping sequences are practically identical: Part of an rRNA operon is also comprised in the published sequence, which seems to flank the GEN-production cluster at one side. This *rrn* operon [called *rrnE* because its closest relationship is to the *rrnE* operon of *S. coelicolor* A3(2)] is completely enclosed in our sequence and located between ORFs MecP21.23c and MecE04.1. A total of 32 ORFs have been assigned to the *gen*-gene cluster for the reasons discussed below. Of these, 19–25 could encode biosynthetic enzyme functions, two could be involved in resistance formation (*gmrA, B*; encoding 16S rRNA methyltransferases), four in transport (e.g., export of gentamicins; *genH,J,T,Y*), and one in regulation (*genU*). Of these, only the *gmrA* gene was already known from earlier genetic studies and database submissions.[180]

According to our original postulate, the *gen*-cluster could have basically originated from a fusion of each a *kan*- and a *for*-like precursor gene cluster. To test for this hypothesis, we both checked the similarity of DNA and protein sequences in dot plots and FASTA comparisons, respectively, for all pairwise comparisons. The results can be summarized as follows: (i) Though the *kan*- and *for*-cluster DNAs showed no similarities, the *gen*-cluster DNA did show significant stretches of similarity to both other DNA types in mutually exclusive regions (cf. Figure 2.15). (ii) The distribution of similarity regions turned out to be quite complicated in that generally the two flanking segments of the *gen*-cluster comprised segments of strong similarity toward the *for*-cluster, whereas the central part of the *gen*-cluster essentially contained the similarity regions toward the *kan*-cluster, however interrupted by another stretch of *for*-similarity. Within these segments, many rearrangements such as displacements, deletions, inversions, insertions of nonconserved genetic material, and so on, could be detected. (iii) The pairs of significantly conserved segments between the *kan*- and *gen*-clusters covered the following genes: *kanS1CD2M2/genS1CD2M2, kanE/genE, kanS2/genS2, kmr, kanM1/gmrA, genM1*. (iv) Segments of conservation between the *for*- and *gen*-clusters covered the following sets of equivalent gene sets: *forHIJ/genHJ, forD2/genQ, forPBK/genKB3PB4* (genes *genB3* and *genB4* seem to have originated from a recent gene duplication), *forT/genI, fosDEFG/genYAFG, fosC/genW, fmrB, fosA/gmrB, genP*, as well as several truncated (obviously nonfunctional) reading frames (cf. Figure 2.22; "erosion zone"). (v) Generally the level of conservation between the *gen/for* pairs of equivalent genes/proteins is much higher (55–95% sequence identity), and most of these genes do not occur in any other of the currently analyzed AGA gene clusters, as judged from the data on the gene clusters for the production of NEO, PAR, RIB, LIV, TOB, APR, and HYG-B; in turn, the *kan/gen* pairs of significantly related genes are generally also present in other 2DOS-AGA gene clusters (cf. the NEO-family clusters[24]) as well as in the *for*-cluster, with the exception of a *kanC/genC*-related gene which is lacking in the *for*-cluster. (vi) Most of the gene/protein families basically conserved in the gene clusters for the majority of the 2DOS-AGAs, which form a paromamine precursor in their biosynthetic pathways (see Figure 2.10),[1,63] are more strongly conserved between the *kan*- and *gen*-clusters, except for the *genB1*/*forB* and *genQ*/*forD2* encoded pairs of equivalent proteins (see Tables 2.16 and 2.17). (vii) The surrounding genomic regions obviously are not conserved at all; however, these regions are enriched with ORFs that occur highly and are conserved in both completely analyzed streptomycete genomes, those of *S. coelicolor* A3(2) and *S. avermitilis*,[14,15] whereas the clusters themselves do not contain such genes. (viii) The three clusters are not located in a commonly conserved chromosomal region. The *kan*-cluster lies in either the central core or at one of its borders to the variable arms. The *gen*- and *for*-clusters rather are located in one of the variable arms of their host strains; the presence of an rRNA operon (*rrnE*) in close vicinity to the *gen*-cluster in *M. echinospora* does not contradict this suggestion, since also in the two completely analyzed streptomycete genomes two to three of the six rRNA operons lie outside the core regions.[14,15] Also, it became apparent that the two *Micromonospora* strains have a higher proportion of genes having closest ortho- (or para-) logs in non-streptomycete genomes outside their AGA clusters than in *S. kanamyceticus*.

2.2.2.2.6. The Pathway for GEN-Related AGAs. First proposals for a biosynthetic route toward GEN stem from bioconversion studies, in a paromamine-producing mutant of *M. echinospora* (former "*purpurea*"), involving the feeding and transformation of GEN minor components into the GEN-C complex.[181] There exist well-developed postulates of the GEN pathway from work on a set of mutants induced and obtained by screening from a SGM-producing strain of *M. sagamiensis* KY11525,[182] which underlie our hypothetical pathway given below (cf. Figures 2.18a,b). From these attempts a branched pathway has been suggested to split into two different routes from the last common intermediate GEN-X_2 (cf. Figure 2.18).

To bring together the *gen*-cluster encoded gene products with the postulated steps in the GEN pathway, we suggest the enzymes/gene products to be involved in the following individual steps: GenC (2-deoxy-*scyllo*-inosose synthase; cyclase), GenS1 (2-deoxy-*scyllo*-3-inosamine synthase; ketocyclitol aminotransferase I), GenE (2-deoxy-*scyllo*-3-inosamine 1-dehydrogenase); and again GenS1 (L-glutamine:1-keto-2-deoxy-*scyllo*-3-inosamine 1-aminotransferase; ketocyclitol aminotransferase II or 2-deoxystreptamine synthase) catalyzes the well-known 2DOS pathway shared with all other producers of 2DOS-AGAs (Figure 2.10; Section 2.2.2.1.2). The next steps also are those following in the general paromamine pathway at this stage (see above): They involve the action of the putative UDP-(*N*-acetyl)-D-glucosamine:2DOS 4-glycosyltransferase ([2′-*N*-acetyl]paromamine synthase) GenM1 and the postulated 2′-amidase GenD 2′-*N*-acetyl-paromamine deacetylase. Similar to the KAN pathway, the glycosyltransferase GenM2 (UDP-xylose:paromamine 6-glycosyltransferase or GEN-A_2 synthase introduces a neutral second sugar unit, a pentose, to form the pseudodisacchridic core structure. This neutral sugar unit is also modified by transamination via the hypothetical GEN-A_2 3″-dehydrogenase GenD2 (xylosyl 3″-DH) and the 3″-keto-GEN-A_2 3″-aminotransferase GenS2 (3″-AT; demethyl-GEN-A

TABLE 2.16. Proteins Encoded in the Genomic Area Covering the *gen*-Cluster of *Micromonospora echinospora* DSM 43036 (Accession Code AJ628149)[a]

ORF /Locus Tag[a]	Gene product[b] Symbol	aa	CDD Number[c]	Putative Function
MecE04.1	ArgRS	228	28809	Arg-tRNA synthetase-related
MecE04.2	TrpRS	359	28814	Trp-tRNA synthetase-related
MecE04.3	GenO	385	41735	TGT, tRNA queosine ribosyl-transferase
MecE04.4	GmrB	272	27198	16S rRNA MT; GEN resistance
MecE04.5c	GenB1	417	25439	6′-AT
MecE04.6c	GenQ	557	16494	6′-DH
MecE04.7c	GenD3	269	25396	Oxidoreductase, DH
MecE04.8c	GenM1	415	7651	2DOS 4-glucosaminyltransferase, GT-I
MecE04.9c	GmrA	274	27198	16S rRNA MT, GEN-resistance
MecE04.10	GenS1	420	15279	L-Glutamine:ketocyclitol AT I/II
MecE04.11	GenC	397	17039	2-deoxy-*scyllo*-inosose synthase (cyclase)
MecE04.12	GenD2	341	6445	Pentose-3″-DH
MecE04.13	GenM2	390	7651	Xylosyltransferase (GEN-A-synthase), GT-II
MecE04.14	GenD1	659	26181	(Fe-S) oxidoreductase, rad. SAM
MecE04.15	GenS2	418	15279	3″-AT
MecE04.16	GenW	145	31123	NADPH-dep. nitrile reductase, queuosine biosyn. pr. QueF
MecE04.17c	GenB4	445	25439	6′-AT or -Epi
MecE04.18c	GenP	268	25797	APH(3′), GEN resistance
MecE04.19c	GenB3	490	25439	6′-AT or -Epi
MecE04.20c	GenK	638	26181	6′-*C*-MT (and 4″-*C*-MT?)
MecE04.21	GenB2	414	25439	6′-AT or -Epi
MecE04.22c	GenX	170	14112	Unknown
MecE04.23c	GenU	311	29257	WD-repeat protein, regulation
MecG05.15c	GenT	468	10350	Efflux protein
MecG05.16c	GenE	336	25397	Aminocyclitol 1-DH
MecG05.17	GenY	504	25668	Antiporter/exporter
MecG05.18	GenA	245	10473	Queuosine biosyn. pr. QueC
MecG05.19	GenF	212	10472	Rad. SAM oxidored. QueE
MecG05.20	GenG	117	29761	Queuosine biosyn. pr. QueD
MecG05.21	GenH	1138	11977	Efflux protein complex
MecG05.22	GenI	645	11977	Efflux protein complex
MecG05.23	GenJ	312	16776	Unknown
MecG05.24	GenD	206	11828	*N*-Acetyl-hexosaminyl deacetylase or amidase
MecG05.25	GenN	321	25197	(*N*-) MT
MecG05.26c		167		HP

[a]The coding sequences (ORFs) are given with their locus tag numbers as found on the respective cosmids analyzed.

[b]The list of gene products translated from the nucleotide sequence corresponds to the protein ID numbers CAF31423–CAF31455.

[c]Refers to the CDD database.[255]

TABLE 2.17. Proteins Encoded in the Genomic Area Covering the *for*-Cluster of *Micromonospora olivasterospora* DSM 43868 (Accession Code AJ628421)[a]

ORF /Locus Tag[a]	Gene Product[b] Symbol	aa	CDD Number[c]	Putative Function
MolI14.2c		306	12746	Transposase
MolI14.3	ForY	440	25668	Cation antiporter
MolI14.4	ForH	668	11977	Transporter system, FTM-efflux
MolI14.5	ForI	503	11977	Transporter system, FTM-efflux
MolI14.6	ForJ	620	12168	Transporter system, FTM-efflux
MolI14.7	ForQ	559	16494	FTM-AO 6′-DH
MolI14.8c	ForW	287	—	Oxidoreductase complex
MolI14.9c	ForV	516	12168	Fms13(?); exporter
MolI14.10c	ForM	427	7651	Fms3(?); 4-GT, FTM-FU-10 synthase
MolI14.11c	ForS	423	15279	Fms4(?); 3- and 6-AT
MolI14.12c	ForE	353	25397	Fms5(?); (3-) or 6-DH
MolI14.13c	ForO	244	23935	Fms10(?); FTM-AP 1-O-MT
MolI14.14c	ForL2	379	26181	Fms12(?); 6′-epim. or 3′, 4′-reductase
MolI14.15c	ForD	302	11828	Fms2(?); *N*-acetyl hexosaminyl deacetylase or amidase
MolI14.16c	ForP	274	25797	Fms8; FTM-KK1 3′-PT
MolI14.17c	ForB	442	25439	6′-AT
MolI14.18c	ForK	553	26181	Fms7; FTM-KL1 6′-C-MT
MolI14.19c	ForZ	482	7951	Fms14; glycyl-N-formimidoyltransferase FI-FTM-A synthase
MolI14.20	ForX	155	25931	cHP, decarboxylase?
MolI14.21	ForG	353	6445	*myo*-inositol 3-DH
MolI14.22	ForN	280	25711	6-*N*-MT
MolI14.23	ForA	281	30137	Fms1(?); 3’-phosphate phosphatase
MolI14.24c	ForL	262	26181	Fms11(?); 1-OMe-epimerase
MolI14.25	FmrR	99	27198	16S rRNA MT (fragmentary)
MolI14.26	FmrO	293	27198	16S rRNA MT, FTM-resistance
MolI14.27	ForT	313	16776	MT
MolI14.28c	FosG	118	29761	3′-Phospholyase or queuosine biosyn. pr. QueD
MolI14.29c	FosF	212	26181	radical SAM oxidored. QueE
MolI14.30c	FosE	233	10473	queuosine biosyn. pr. QueC
MolI14.31c	FosD	137	25668	Unknown, (fragmentary)
MolI14.32c	FosC	123	25715	NADPH-dep. nitrile reductase QueF
PORF1		40	15279	AT (fragmentary)
MolI14.33	FmrP	158	27198	16S rRNA MT (fragmentary)
PORF2		103	—	Oxidoreductase (fragmentary)
MolI14.34c	FmrB	269	27198	16S rRNA-MT, FTM resistance
MolI14.35c	FosA	385	2235	TGT, tRNA queosine ribosyl-transferase

[a]The coding sequences (ORFs) are given with their locus tag numbers as found on the respective cosmids analyzed.

[b]The list of gene products translated from the nucleotide sequence corresponds to the protein ID numbers CAF31531–CAF31564.

[c]Refers to the CDD database.[255]

synthase); the last three proteins highly resemble the equivalents encoded by the *kan-/tob*-clusters, suggesting common evolutionary origin of this subpathway in the GEN/KAN relationship. The 3-aminopentose moiety is modified in two steps by methylating the 3″-amino group likely by use of the SAM:demethyl-GEN-A 3″-*N*-methyltransferase (3″-NMT; GEN-A synthase), which is possibly represented by either one the GenJ or the GenN proteins, and by *C*-methylation via the SAM:GEN-A 4″-*C*-methyltransferase GenK (GEN-X_2 synthase). The GenK enzyme is a close relative of the 6′-*C*-methyltransferase ForK from used in the FOR pathway (see below). The ForK enzyme is a methylcobalamine-dependent enzyme[183] and explains the strict Co^{2+} requirement of this step *in vivo*,[182,184] and we postulate that it can catalyze two different reactions on successive GEN precursors, on GEN-A and GEN-X_0. From here the pathway branches and follows two distinct routes by alternate modification at the C-6′ group of Gen-X_2, whose reactions lead to either G418 or IJ-20A intermediates (cf. Figure 2.18).

2.2.2.2.7. The First Branch in GEN Biosynthesis: The GEN-C_1 Pathway. The first branching step is probably catalyzed again by the putatively bifunctional *C*-methyltransferase GenK (SAM:GEN-X_2 6′-*C*-methyltransferase or G418 synthase). This step is possibly kinetically rate limiting and competed out by the enzyme, GenQ, which converts part of the GEN-X_2 pool intermediate. During the next step, this intermediate acts on the same C-atom and thereby irreversibly withdraws part of the substrate (see below). The G418 6′-dehydrogenase GenQ belongs to the "Q" protein oxidoreductases which are highly conserved among most of the pathways of 2DOS-AGAs and related AGAs. This same notion also holds for the next step enzyme, an "B"-type aminotransferase, probably represented by GenB1 (or GenB4; these two "B"-proteins are closest to the "B"-type proteins of the NEO-family AGA producers), one of which should have 6′-dehydro-G418 6′-aminotransferase activity (TPJ-B synthase). The TPJ-B 6′-epimerase (JI-20B synthase) needed in the next step in this branch could be identical to the GenB2 protein (or GenB3; these two proteins are much closer related to each other than to any other "B"-protein and their next relative is ForB), which is another member of the "B"-type protein family of 6′-aminotransferases; we here postulate that the occurence of four "*genB*/GenB"-type genes/enzymes in GEN producers indicates that the additional GenB2 and GenB3 proteins serve as "amino group epimerases" in having acquired an altered substrate recognition with respect to the stereospecificity of their attack on the carbonyl substrate. This type of reaction sequence is unique and might require the action of the preceding or an additional aminotransferase, GenB1 (or GenB4, GenB2/GenB3), for reintroduction of the 6′-aldehyde group. The next steps are devoted to 3′,4′-dehydroxylation/deoxygenation and occurs in an equivalent manner in the FOR/IST pathways (see the comments in Section 2.2.3.2). We postulate this joint removal of two adjacent hydroxyl groups to occur via a five-step reaction sequence, involving 3′-phosphorylation, 3′,4′-phospholysis,

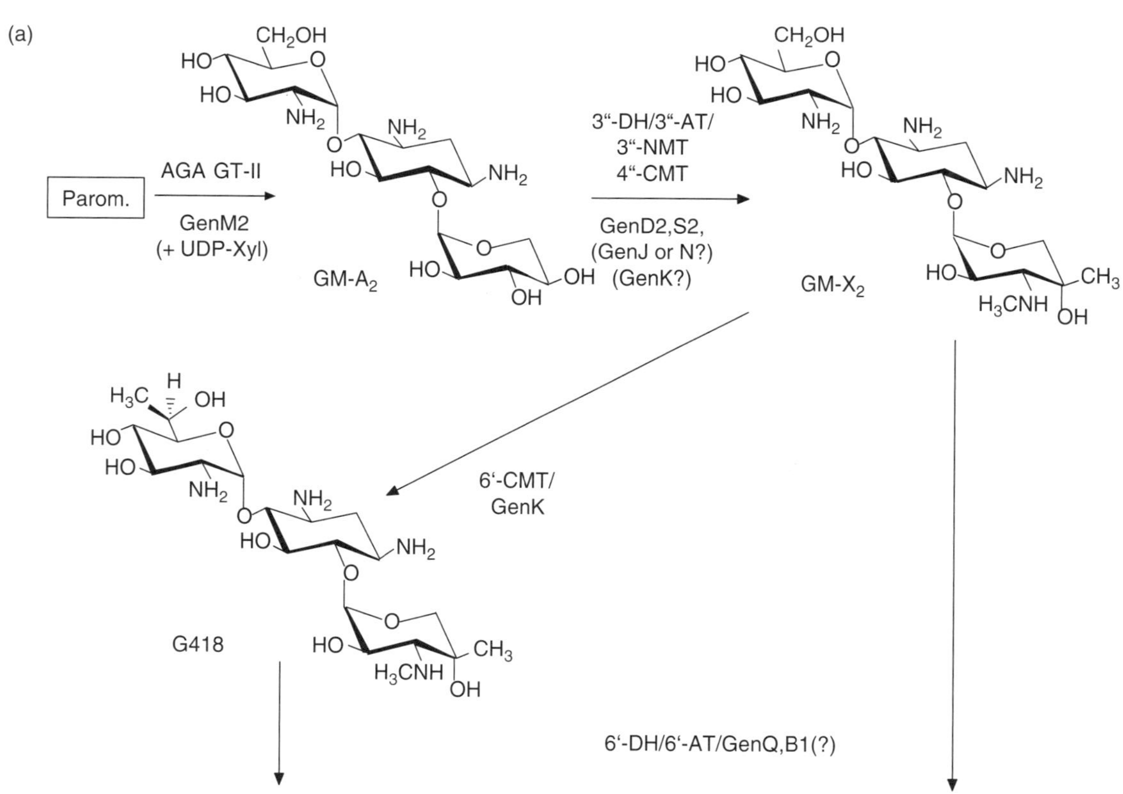

Figure 2.18. The branched biosynthetic pathway for GEN-related AGAs. Other explanations and abbreviations are as given in the legend to Figure 2.3.

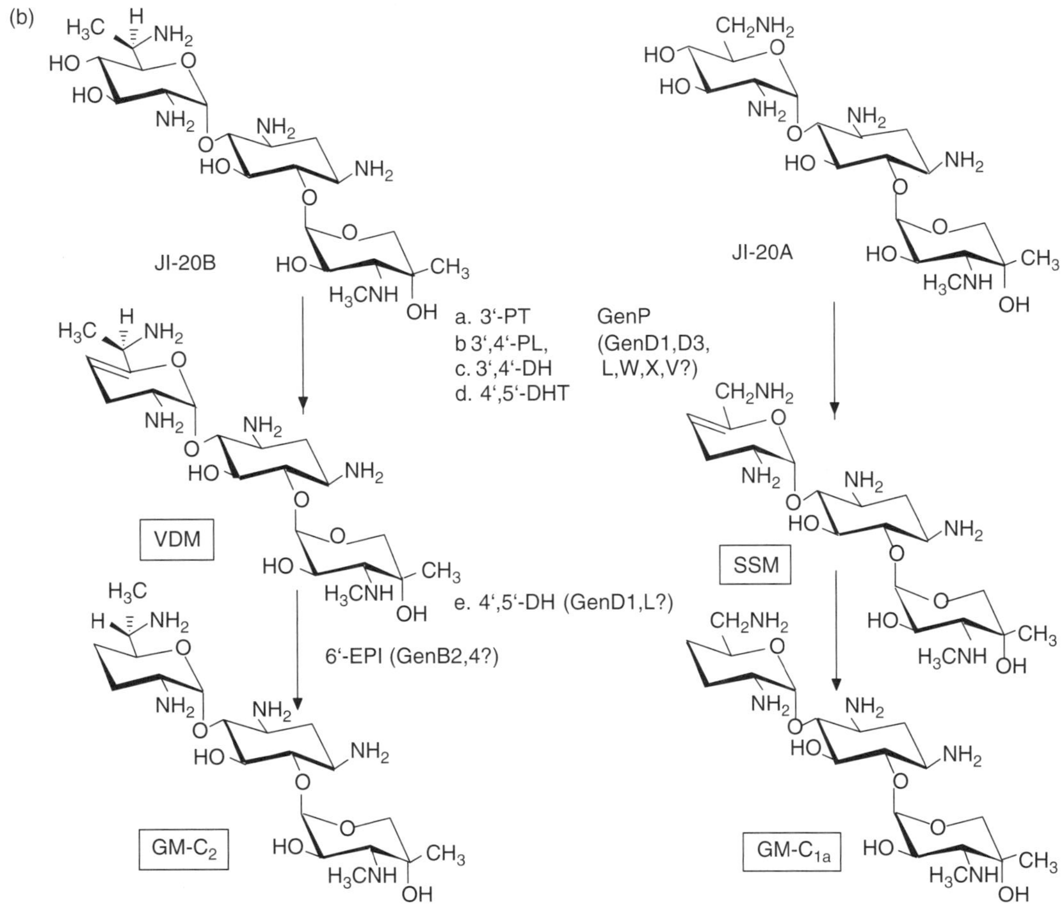

Figure 2.18. *(continued)*

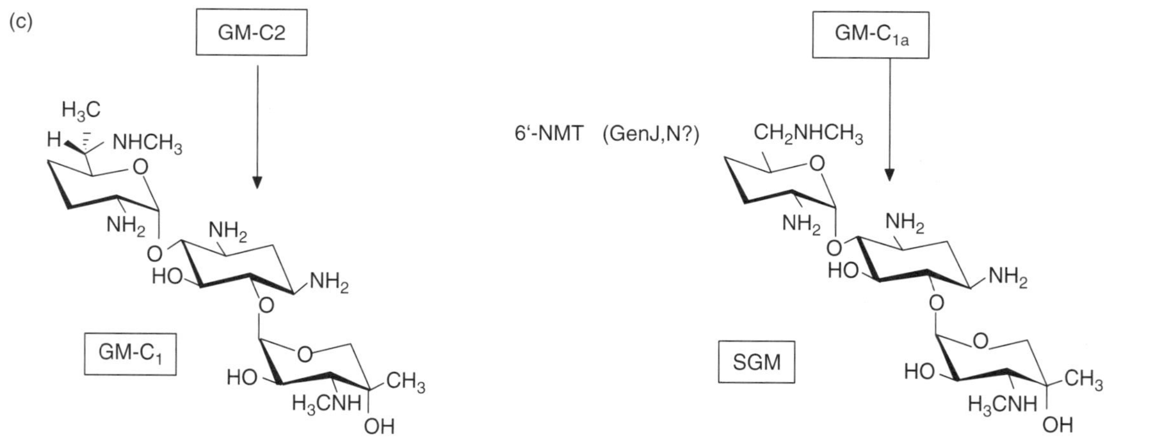

Figure 2.18. *(continued)*

3′,4′-reduction, 4′,5′-dehydratation, and again 4′,5′-reduction, where some intermediates are further converted by final steps and excreted under omission of the last step, being the source of the GEN-related compounds with a 4′,5′-double bond, such as SSM and VDM (cf. Figure 2.17). The proteins/enzyme activities postulated to be involved in this process are first the 3′-phosphotransferase GenP, an APH(3′)-related enzyme that phosphorylates IJ-20B; IJ-20B-3′-phosphate is dephosphorylated by the hypothetical 3′,4′-phospholyase GenG; the next step is catalyzed by a 3′-desoxy-3′,4′-dehydro-IJ-20B 3′,4′-reductase, which could be one of the oxidoreductases and other proteins with unkown specificity encoded in the *gen*-cluster (e.g., the GenD1, GenD3, GenL, GenW, GenV, GenX proteins); among these could also be the candidate proteins for the next two steps, which according to our postulate should be a 3′-desoxy-IJ-20B 4′,5′-dehydratase (VDM synthase) and a VDM 4′,5′-reductase (GEN C_{2a} synthase). The following step would require again a GEN-C_{2a} 6′-epimerase (GEN-C_2 synthase), which we propose to be searched for among the GenB1 through GenB4 enzymes again. Finally, we require a GEN-C_2 6′-*N*-methyltransferase (GEN-C_1 synthase), probably identical to the GenN protein (alternatively GenJ), which belongs to the methyltransferase superfamily.

2.2.2.2.8. The Second Branch in GEN Biosynthesis: The SGM Pathway. In the second branch of the GEN pathway a largely parallel sequence of enzyme-catalyzed conversions yield the other portion of the GEN complex: In the branching phase a GEN-X_2 6′-dehydrogenase activity, probably as a result of an alternative substrate recognition of the GenQ dehydrogenase (see above) on non-6′-C-methylated intermediates, initiates pool separation among the the GEN-C1 and SGM precursors. The following 6′-dehydro-GEN-X_2 6′-aminotransferase (IJ-20A synthase) could be again either GenB1 or GenB4; this requirement might even be the reason for the gene duplication that leads to these two "B"-type proteins. Probably, this conversion completes the separation of the intermediate pools by withdrawing part of the precursors from 6′-*C*-methylation. Next the reaction chain already described above to eliminate the 3′,4′-hydroxyls are used to synthesize SSM via a IJ-20A 3′-phosphotransferase (GenP), a IJ-20A-3′-phosphate 3′,4′-phospholyase (GenG), a 3′-deoxy-3′,4′-dehydro-IJ-20A 3′,4′-reductase (GenD1, D3, L, W, V, X [?]), and a 3′-deoxy-IJ-20A 4′,5′-dehydratase (SSM synthase; see above candidate proteins), yielding the SSM produced as a major end product in some strains of *M. inoyensis* (see Table 2.15). An SSM 4′,5′-reductase (GEN-C_{1a} synthase; see above candidate proteins) would catalyze the conversion to the GEN-C_{1a} product found in the cultures of *M. echinospora*. Finally, a GEN-C_{1a} 6′-*N*-methyltransferase (SGM synthase; GenN or GenJ) is postulated to synthesize the main product in the SGM producer *M. sagamiensis*.

2.2.3. Fortimicin-Related Pseudodisaccharides

2.2.3.1. The Genetics and Biochemistry of the Biosynthetic Pathway for FOR and IST. The FORs (or astromicins; see Figure 2.19) have been described in cultures of *M. olivasterospora* ATCC 21819 first in 1977.[175,176,185] Later, very

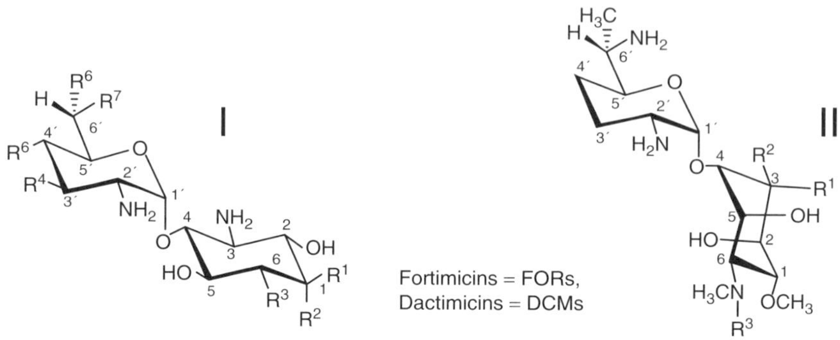

FMs-I	R^1	R^2	R^3	R^4	R^5	R^6	R^7
FOR-FU-10	OH	H	OH	OH	OH	H	OH
FOR-AO	OH	H	$NHCH_3$	OH	OH	H	OH
FOR-KL_1	OH	H	$NHCH_3$	OH	OH	H	NH_2
FOR-KK_1	OH	H	$NHCH_3$	OH	OH	CH_3	NH_2
FOR-AP	OH	H	$NHCH_3$	H	H	CH_3	NH_2
FOR-KH	OCH_3	H	$NHCH_3$	H	H	CH_3	NH_2
FOR-KR	H	OCH_3	$NHCH_3$	H	H	CH_3	NH_2

FORs-II	R^1	R^2	R^3
FOR-B	NH_2	H	H
3-epi-FOR-B	H	NH_2	H
FOR-A	NH_2	H	$COCH_2NH_2$
DCM	NH_2	H	$COCH_2NHCH{=}NH$
EDCM	H	NH_2	$COCH_2NHCH{=}NH$

Figure 2.19. Structures of fortimicin (FOR) family AGAs.

similar compounds were found to be produced in other actinomycete genera and species—for example, the formimidoyl-FOR-A, called dactimicin (DCM; SF-2052) from *Dactylosporangium matsuzakiense* SF-2052.[186–188] Most interesting among these are the ISTs (see Figure 2.20) that have been isolated from cultures of *S. tenjimariensis* SS-939, which use an alternative cyclitol pathway (see below; cf. Figure 2.23a).[189] Others, like the sporaricins (SPOs, KA 6606 complex), are produced by *Saccharopolyspora hirsuta* ssp. *kobensis* ATCC 27875, or sannamycin (SAN) is produced by *S. sannanensis* KC-7083 (see Figure 2.21).[190–194] The similarities between the biosynthetic pathways for the FOR/IST group of AGAs had been pointed out in bioconversion studies in strains of *M. olivasterospora*, *S. tenjimariensis*, and *S. sannanensis* using various FOR precursors (see below) and components of the FOR and IST complexes.[195–197]

In the recently reported sequence of the chromosomal genome of the actinobacterium *Frankia* sp. CcI3 (accession code CP000249; locus-tag nos. Francci3_3357 through Francci3_3375), for the first time an AGA gene cluster appeared by mere genomic sequencing. This cluster strongly relates to the *for*-cluster of

I

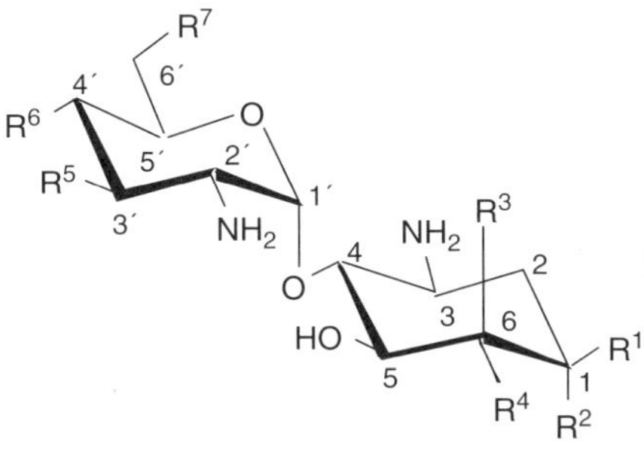

Istamycins = ISTs

II

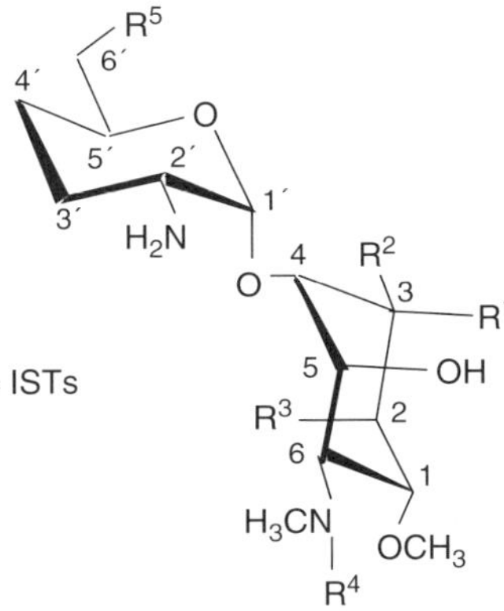

ISTs-I	R^1	R^2	R^3	R^4	R^5	R^6	R^7
IST-FU-10	OH	H	$NHCH_3$	OH	OH	H	H
IST-AO	OH	H	$NHCH_3$	OH	OH	H	OH
IST-KL_1	OH	H	$NHCH_3$	OH	OH	H	NH_2
IST-KK_1	OH	H	$NHCH_3$	OH	OH	H	NH_2
IST-AP	OH	H	$NHCH_3$	H	H	H	NH_2
IST-Y_0	OCH_3	H	$NHCH_3$	H	H	H	NH_2
IST-X_0	H	OCH_3	$NHCH_3$	H	H	H	NH_2

ISTs-II	R^1	R^2	R^3	R^4
IST-B_0	N	NH_2	H	CH_3
IST-B	H	NH_2	$COCH_2NH_2$	CH_3
IST-B_1	H	NH_2	$COCH_2NHCHO$	CH_3
IST-B_3	H	NH_2	$COCH_2NHCH{=}NH$	CH_3
IST-A_0	NH_2	H	H	CH_3
IST-A	NH_2	H	$COCH_2NH_2$	CH_3
IST-A_1	NH_2	H	$COCH_2NHCHO$	CH_3
IST-A_2	NH_2	H	$COCH_2NHCONH_2$	CH_3
IST-A_3	NH_2	H	$COCH_2NHCH{=}NH$	CH_3
IST-C_0	NH_2	H	H	CH_2CH_3
IST-C	NH_2	H	$COCH_2NH_2$	CH_2CH_3
IST-C_1	NH_2	H	$COCH_2NHCHO$	CH_2CH_3

Figure 2.20. Structures of istamycin (IST) family AGAs.

M. olivasterospora in both gene content and gene order; a comparison is outlined in Table 2.18. Therefore, we here postulate that *Frankia* sp. CcI3 is able to produce FORs or a closely related AGA complex.

Odakura et al.[198] used a high producer derivative of *M. olivasterospora* MK-70 (DSM 43868), strain KY11250, to isolate 25 mutants blocked in at least 10 different steps of biosynthesis, as juged from complementation patterns in cosynthesis or biotransformation studies using *scyllo*-inosose, *scyllo*-inosamine

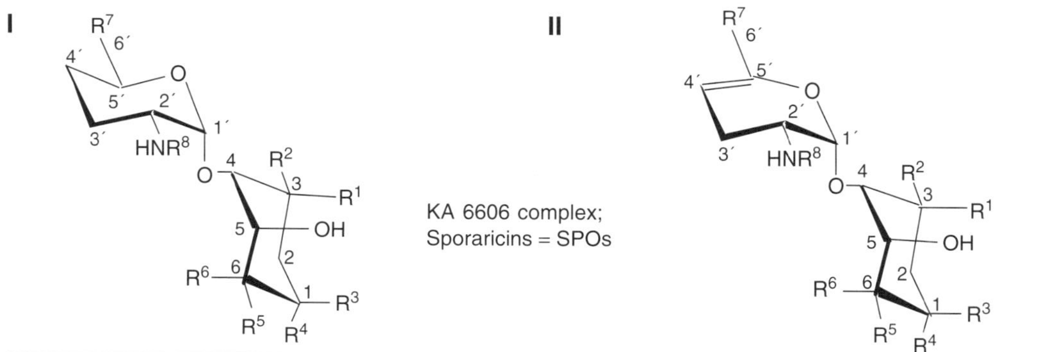

KA 6606 no.	R^1	R^2	R^3	R^4	R^5	R^6	R^7	R^8	A/B
V	NH_2	H	H	OCH_3	H	$NHCH_3$	$C(CH_3)NH_2$	H	A
VII	NH_2	H	OCH_3	H	H	$NHCH_3$	$C(CH_3)NH_2$	H	A
IX	NH_2	H	OCH_3	H	H	$NHCH_3$	$C(CH_3)NH_2$	H	A
XII	NH_2	H	OCH_3	H	H	$NHCH_3$	CH_2OH	H	A
XIII	NH_2	H	OCH_3	H	H	$NHCH_3$	$C(CH_3)NH_2$	H	B
XIV	NH_2	H	OCH_3	H	$NHCH_3$	H	CH_2NH_2	H	A
XV	NH_2	H	OCH_3	H	H	$NHCH_3$	CH_2NH_2	H	B
XVI	H	NH_2	OCH_3	H	$NHCH_3$	H	$C(CH_3)NH_2$	$CONH_2$	A
XVII	H	NH_2	OCH_3	H	$NHCH_3$	H	$C(CH_3)NH_2$	H	A
XVIII	H	NH_2	OCH_3	H	$NHCH_3$	H	$C(CH_3)NH_2$	H	A
XIX	H	NH_2	OCH_3	H	$NHCH_3$	H	CH_2NH_2	H	A
I (= SPO-A)	H	NH_2	OCH_3	H	$N(CH_3)COCH_2NH_2$	H	$C(CH_3)NH_2$	H	A
II (= SPO-B)	H	NH_2	OCH_3	H	$NHCH_3$	H	$C(CH_3)NH_2$	H	A
III (= SPO-C)	H	NH_2	OCH_3	H	$N(CH_3)COCH_2NHCONH_2$	H	$C(CH_3)NH_2$	H	A
IV (= SPO-D)	H	NH_2	OCH_3	H	$N(CH_3)COCH_2NHCHO$	H	$C(CH_3)NH_2$	H	A
(FI-SPO-A)	H	NH_2	OCH_3	H	$N(CH_3)COCH_2NHCH{=}NH$	H	$C(CH_3)NH_2$	H	A
VI (= SPO-E)	NH_2	H	OCH_3	H	$NHCH_3$	H	$C(CH_3)NH_2$	H	A

Figure 2.21. Structures of sporamycin (SPO) family AGAs.

TABLE 2.18. Proteins Encoded in the *Frankia* sp. CcI3 *fra*-Cluster (Accession Code CP000249; Locus Tag Numbers 3357–3383) and Their FOR/IST Protein Equivalents

Locus Tag Francci3	aa	FTM/Ist Equivalent	% aa Identified	CDD Number[a]	Putative Function
3357	384	IstL	64	26181	Oxidoreductase, rad. SAM
3358	668	H/I	78	11977	FTM exporter component
3359	1112	J	62	12168	FTM exporter component
3360	545	Q	71	16494	oxidoreductase, 6′-DH
3361	305	W	64	—	Oxidoreductase (?)
3362	433	M	73	7651	GT, FTM-FU-10 synthase
3363	423	S	73	15279	3-/6-AT
3364	367	E	64	25397	Oxidoreductase, 6-DH
3365	246	O	81	23935	MT, FTM-AP, 1-*O*-MT
3366	374	L2	74	26181	6′-Epimerase or 3′,4′-reductase, rad. SAM
3367	248	D	48	11828	*N*-Acetyl hexosaminyl deacetylase, amidase
3368	268	P	56	25797	3′-PT
3369	444	B	77	25439	AT
3370	618	K	76	22645	6′-*C*-MT, rad. SAM
3371	488	Z	69	7951	Oxidoreductase, FTM-A synthase
3372	284	FmrO	57	27198	16S rRNA MT, FTM-resistance
3373	266	N	70	25711	6-*N*-MT
3374	339	G	63	6445	*myo*-Inositol 2-DH
3375	640	IstL3	50	10406	1-*O*-Methyl-epimerase, rad. SAM
3376	154	X	69	32311	Alkylperoxidase
3377	327	T	66	30661	(Methyltransferase?)
3378	386	FosA	72	41735	TGT, tRNA queosine ribosyl-transferase
3379	236	FosE	71	30182	Queuosine biosynthesis pr. QueC
3380	212	FosF	72	30947	Rad. SAM oxidoreduction QueE
3381	124	FosG	79	29761	Queuosine biosynthesis pr. QueD
3382	288	FmrO	56	27198	16S rRNA MT
3383	125	FosC	83	31123	NADPH-dep. nitrile reductase QueF

[a]Refers to the CDD database.[255]

and isolated intermediates obtained from the fermentations with this same set of mutant strains. The results of this study allowed to propose a FOR pathway, which still forms the basis of what we postulate in the following section (cf. Figure 2.23).

In a quite extensive study using this collection of mutants of *M. olivasterospora* KY11250 for complementation with shotgun-cloned DNA from the wild-type strain, Hasegawa and co-workers identified a number of gene loci, called "*fms*," and mapped those on the DNA segment cloned in cosmid pGLM990, which was isolated on the basis of hybridization with the shotgun-cloned DNA fragments mentioned above.[199–203] In Table 2.18 we have introduced the alternative *fms*-nomenclature for the *for* genes which are known or probaly congruent with *fms* loci identified by Hasegawa and co-workers.[201] Some of these genes have been subcloned and used for enzyme characterization; however, only the *fms14* gene and a segment around the *fmrO* resistance gene including the *fms11* gene had been sequenced in this study (see below).[204] The same approach was used to complement the *S. sannaensis* mutant SN13 blocked in the glycyltransfer reaction between SAN-B and SAN-A and thereby to isolate the *sms13* gene.[205,206]

2.2.3.2. *The Biosynthetic Pathways for FOR and IST.*

2.2.3.2.1. Formation of the Initial Cyclitol. The first steps of FOR production are the same as in the biosynthesis of STR: in short, the first step in FOR biosynthesis is postulated to be the formation of a *myo*-inositol monophosphate (D-*myo*-inositol-3-phosphate or L-*myo*-inositol-1-phosphate) via the cellular L-*myo*-inositol-1-phosphate synthase as in the STR pathway (Ca pathway; see Section 2.2.1.2). As in the *str-/sts*-clusters, no gene for this enzyme has been found in the *for*-cluster. As a second step in FOR biosynthesis the dephosphorylation of D-*myo*-inositol-3-phosphate via an inositolmonophosphate phosphatase has to follow. A putative gene product with this activity is that of the ForA protein (cf. Tables 2.17 and 2.18). The cyclitol is postulated to be first converted via two enzymes, a cluster-encoded *myo*-inositol 3-dehydrogenase (ForG; member of the GFO/IDH/MocA oxidoreductase family) and the L-glutamine:*scyllo*-3-inosose 3-aminotransferase (ketocyclitol aminotransferase I; ForS), to *scyllo*-inosamine (3-deoxy-3-amino-*scyllo*-inositol).

In contrast, in the IST pathway the cyclitol moiety is formed via the first steps of the general 2DOS pathway; that is, it needs only two steps to form a monoaminocyclitol unit (cf. Figure 2.10): in short, by first cyclizing the glucose-6-phosphate precursor by the same type cyclase (2-deoxy-*scyllo*-inosose synthase; IstC) as in the 2DOS pathways to 2-deoxy-*scyllo*-inosose. This is then directly used again by an L-glutamine-dependent aminotransferase (ketocyclitol aminotransferase I; IstS), belonging to the same subfamily of class V aminotransferases, as do all the cyclitol ATs, to yield 2-deoxy-*scyllo*-inosamine.

We have confirmed the ForG activity being encoded by an essential gene by knocking out the *forG* gene, which resulted in loss of FOR production, and also by heterologous overproduction of the ForG protein in both *E. coli* K-12 and *S. lividans* 66, which exhibited the expected activity of forming *scyllo*-inosose from *myo*-inositol in the presence of NAD_+ (B. Schuermann, U. F. Wehmeier, and W. Piepersberg, unpublished). Also, the ForE/IstE enzymes from the biosynthetic toolkit used by the FOR/IST family of pseudodisaccharidic AGAs belongs to this enzyme family (cf. Tables 2.17 and 2.19 and Figure 2.23). However, from the work of Odakura and colleagues it seems likely that ForE does not act on a monomeric substrate, rather on a pseudodisaccharidic one (see below).[198]

TABLE 2.19. Proteins Encoded in the Genomic Area Covering the *ist*-Cluster of *S. tenjimariensis* ATCC 31603 (Accession Code AJ845083)[a]

ORF /Locus Tag[a]	Gene Product[b] Symbol	aa	CDD Number[c]	Putative Function
StenF24.16c		87	—	cHP; unknown
StenF24.17c		88	—	Unknown
StenF24.18c		71	—	Isocitrate lyase (fragmentary)
StenF24.19	IssC	251	26181	Transcriptional regulator
StenF24.20	IssB	229	30191	ATP/GTP-binding protein
StenF24.21c	IssA	388	26554	Predicted hydrolase or acyltransferase
StenF24.22	ImrA	220	27198	16S rRNA MT, IST resistance
StenF24.23c	IstA	397	—	Transposase
StenF24.24c	IstD	238	11828	*N*-Acetyl hexosaminyl deacetylase or amidase
StenF24.25c	IstP	276	25797	APH(3′), IST resistance
StenF24.26c	IstB	448	25439	6′-AT
StenF24.27c	IstJ	621	11977	Transmembrane efflux protein
StenF24.28c	IstF	135	—	Unknown
StenF24.29c	IstG	277	10371	6′-*N*-MT
StenF24.30c	IstZ	484	7951	*N*-Formimidoyl IST A synthase
StenF24.31	IstH	617	11977	Transmembrane efflux protein
StenF24.32	IstW	320	—	Glucose-6-phosphate 1-DH
StenF24.33c	IstQ	517	16494	6′-DH; IM-AO 6′-DH, flavoprotein
StenF24.34c	IstC	396	17039	2-Deoxy-*scyllo*-inosose synthase (cyclase)
StenF24.35c	IstL3	348	10406	Metallo cofactor/biotin biosynthesis
StenF24.36c	IstL	327	26181	IST KH epimerase
StenO22.6	IstI	483	11977	Transmembrane efflux protein
StenO22.7c	IstS	422	15279	L-Glutamine:ketocyclitol 3- and 6-AT
StenO22.8c	IstN	257	25711	Cyclitol 6-*N*-MT
StenO22.9c	IstX	150	25931	Decarboxylase/oxidoreductase
StenO22.10	IstM	331	7651	4-GT, IM-FU-10 synthase
StenO22.11	IstL2	385	26181	(Fe-S) 3′, 4′-reductase
StenO22.12	IstO	247	—	IM-AP 1-*O*-MT
StenO22.13	IstE	331	25397	Aminocyclitol 6-DH

[a]The coding sequences (ORFs) are given with their locus tag number as found on the respective cosmids analyzed.
[b]The list of gene products translated from the nucleotide sequence corresponds to the protein ID numbers CAH60141–CAH60162.
[c]Refers to the CDD database.[255]

2.2.3.2.2. Formation and Modification of the Pseudodisaccharidic Core Unit. The characteristic design of the biosynthetic pathway for the FOR/IST- or 1,4-aminocyclitol-type AGAs is that it is mainly devoted to the formation and modification of a paromamine-related pseudodisaccharide. Hence, when the monoaminocyclitol starter units are provided as sketched above, a glycosyltransfer reaction (fifth step) catalyzes by use of an "M"-type glycosyltransferase (ForM/IstM) its condensation in the 4-position with a D-glucosamine moiety. The likely donor substrate for this reaction is again UDP-(*N*-acetyl)-D-glucosamine and probably comes from the general pool of sugar nucleotides of the cell. Therefore, the first pseudodisaccharide could be a 2′-*N*-acetylated derivative of the "FU-10" intermediate and must be deacetylated to be converted into FOR-FU-10 (or "IST-FU-10"); for this step a putative deacetylase of the "D"-family of enzymes, ForD, is only encoded in the *for*-cluster without having a counterpart from the *ist*-cluster. However, this function could likely be also provided from a cellular housekeeping enzyme, such as the deacetylase MshB used in the biosynthesis of mycothiol (see Section 2.2.2.1.4). The next known intermediate of "AO"-type for the first time contains the 1,4- (3,6-)diaminocyclitol called fortamine. Therefore, the introduction of the 6-aminomethyl group is postulated to occur at this stage; it probably involves the ForE/IstE 6-dehydrogenases (see above), and again the "S"-type aminotransferases, which we here postulate to be bifunctional as they are in the synthesis of 2DOS (see above paragraph on 2DOS).

The 6-*N*-methyltransferase to yield the completed fortamine unit are likely to be the ForN/IstN proteins. The next stage in the pathway is the introduction of the fourth amino group in the 6′-position of the "AO" precursors. This modification step is already well known from the pathways of the major families of 2DOS-AGAs and involves the "Q"/"B"-pair of 6′-dehydrogenases/6′-aminotransferases which are also conserved in the FOR/IST producers—that is, are encoded in the *for-*/*ist*-clusters, respectively (cf. Figure 2.22, Tables 2.17–2.19) and yields an intermediate of type "KL_1." The following conversion into a "KK_1" intermediate is specific for the FOR pathway only and involves a C-methylation catalyzed by the coenzyme methylcobalamine-dependent 6′-C-methyltransferase ForK; this enzyme was characterized *in vitro*.[183]

The next common step in the FOR/IST pathways is postulated to be the 3′, 4′-didehydroxylation, a biosynthetic feature already known from the GEN pathway (see paragraph on GENs), but discussed in detail here: We hypothesize that this conversion is achieved by five successive reactions and yields an "AP"-type metabolite (cf. Figure 2.23): (i) In their study using mutants of *M. olivasterospora* being blocked at various stages in FOR biosynthesis for complementation with cloned DNA from the wild-type strain, Hasegawa and co-workers found that surprisingly the vector plasmid alone could complement strain AN73-1, which was blocked in the 3′, 4′-dehydroxylation steps.[200,201] It turned out that the vector contained a NEO-resistance gene encoding an APH(3′) enzyme, which was shown to be responsible for the complementation of mutant AN73-1. The complementing gene (called *fms8*, equivalent of *forP*) clearly encodes a member of the AphA/APH(3′) subfamily of AGA kinases, ForP. The involvement of 3′-phosphorylation in the dehydroxylation mechanism remains to be verified. (ii) We propose that 3′-phosphorylation is the first reaction in this process and prepares the hexosamine unit for a phospholyase step by which the 3′-OH is removed and probably a 3′–4′ double bond is introduced. A candidate gene encoding a

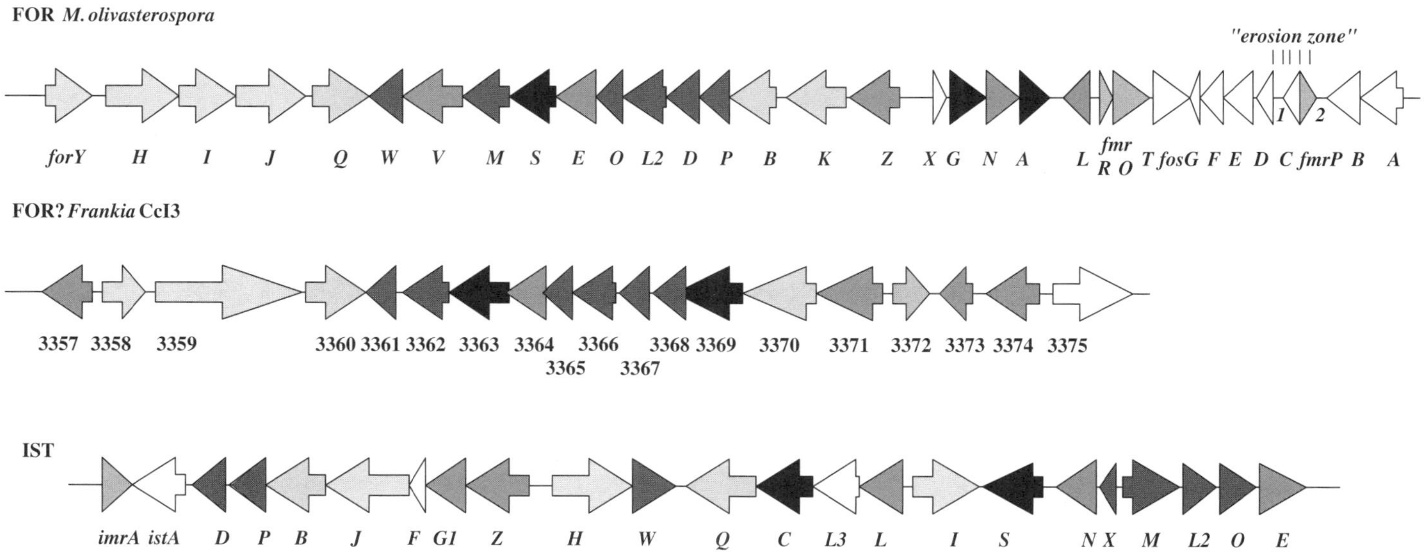

Figure 2.22. Biosynthetic gene clusters for fortimicin (FOR), the putative FOR cluster from *Frankia* sp. CcI3 (*fra* cluster, cf. Table 18) and istamycin (IST) AGAs. Late cyclitol modification, for all other colors see Figure 2.9. See color plates.

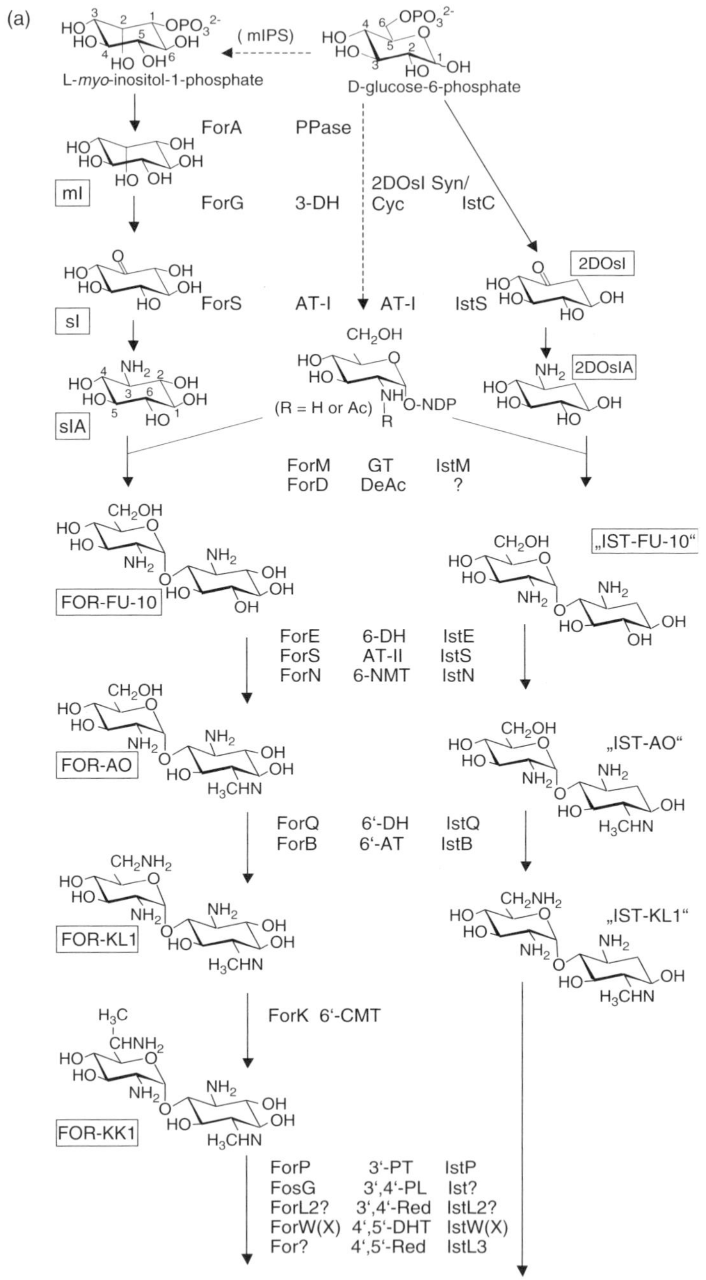

Figure 2.23. Biosynthetic pathway for IST/FOR-related AGAs part B. Other explanations and abbreviations are as given in the legend to Figure 2.3.

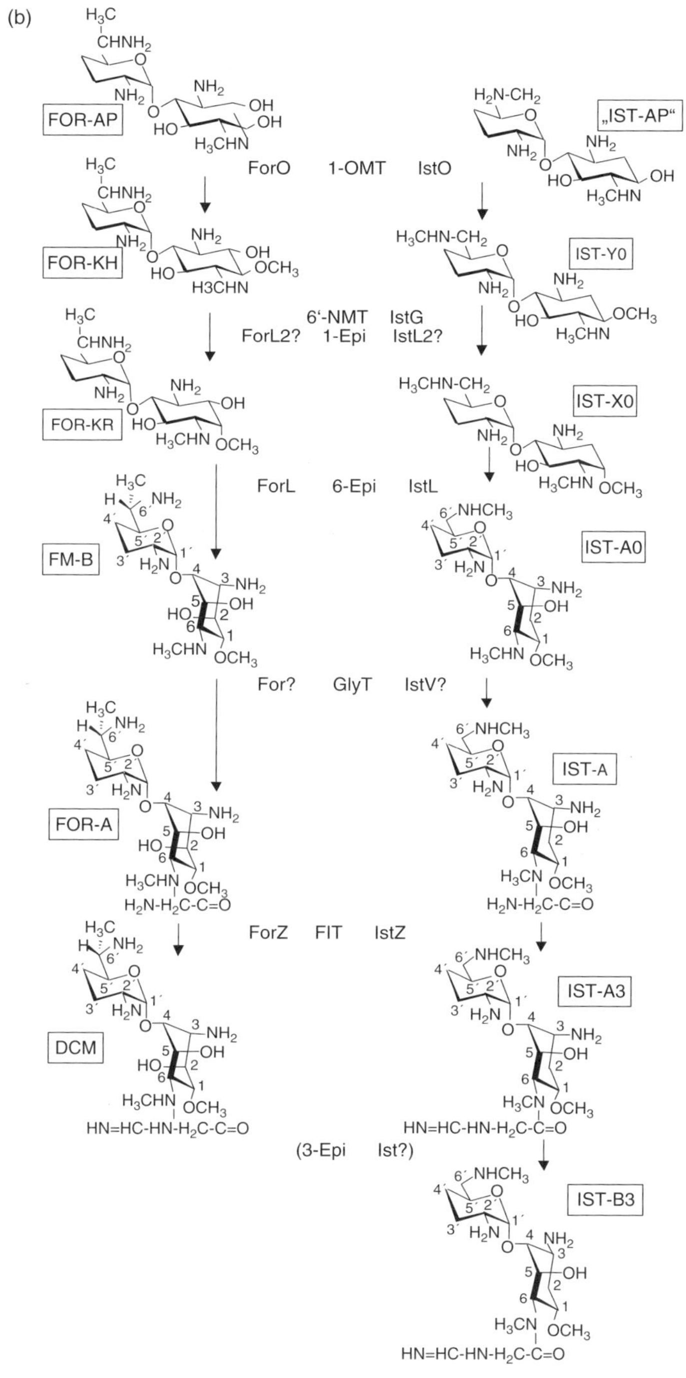

Figure 2.23. *(continued)*

phospholyase is *fosG*, which encodes a protein significantly related to the 6-pyruvoyl tetrahydropterin synthases (2-amino-4-oxo-6-[(1S, 2R)-1,2-dihydroxy-3-triphosphooxypropyl]-7,8-dihydroxypteridine triphosphate lyases) that might involve a similar mechanism. It should be noted that no equivalent to *fosG* is found in the *ist*-cluster. (iii) The following hypothetical reduction of the 3′, 4′ double bond on the hexosamine moiety by 3′-deoxy-3′, 4′-dehydro-FOR-KK_1 3′, 4′-reductase could possibly be attributed to the ForL2/IstL2 proteins belonging to the radical SAM superfamily of proteins; alternatively the ForW/IstW proteins or others could be possible oxidoreductase candidates. (iv) Removing the 4′-hydroxyl group is assumed to be achieved by a 4′, 5′-dehydratation/reduction reaction sequence for which again the ForW/IstW proteins and/or others could be responsible. There are several so-called "radical SAM" Fe–S-type oxidoreductases encoded in both the *for*- and *ist*-gene clusters; whether this fact has an implication for the dehydroxylation steps is unknown; for example for the putative oxidoreductase IstL3 no counterpart is encoded by the *for*-cluster (cf. Tables 2.17 and 2.19). The "AP"-type intermediates are then postulated to be further shaped in a ninth step by 1-*O*-methylation, which is proposedly catalyzed by the putative *O*-methyltransferases ForO/IstO. This yields intermediates designated "FOR-KH" and "IST-Y_0," respectively. As judged from the record of intermediates analyzed from mutant strains, the two pathways differ here sligtly again: The IST precursors so far postulated did not carry a C-methyl group at C-6′; therefore, this type of modification seems obviously to be replaced in this stage by a 6′-*N*-methyltransfer to the "IST-AP" intermediate by the putative methyltransferase IstG. The enzyme pair catalyzing the respective epimerization of the 1-*O*-methyl group to synthezize the "FOR-KR"/"IST-X_0" pair of late precursors is likely to be ForL/IstL; this was already suggested from complementation (by "*fms11*" the equivalent to *forL*; cf. Table 2.17) of a mutant blocked in this step.[199–203] The next step is postulated to be 6-epimerization of the aminomethyl group;[198] the mapped "*fms12*" locus complementing a mutant blocked in this step would suggest that this is identical with the *forL2* gene, which is also conserved in the *ist*-cluster. Both ForL/IstL and ForL2/IstL2 protein pairs are members of the "radical SAM" superfamily, and an elucidation of their possible involvement in the two epimerization steps suggested here will have to be clarified in the future. The products of the last step, "FOR-B" and "IST-A_0," are observed as regular components of the excreted AGA complexes in fermentations with their producers. The last two modification steps are transfer reactions for introduction of a glycyl residue to the 6-aminomethyl group and further formimidoylation of the glycyl-amino group and yield the main components of in the product mixtures "FOR-A"/"DCM" and "IST-A"/"IST-A_3" (cf. Figure 2.23). Though the glycyltransferase genes were mapped by mutant complementations (see above; "*fms13*" and "*sms13*" loci),[206] it is not possible to assign this function unequivocally to a given protein pair so far. Therefore, at present we cannot clearly suggest any candidate gene/protein for this step in the FOR producer and propose that the *istV* gene is the best candidate for this locus in the IST producer. The FOR-A glycylaminoformimidoyltransferase ForZ (Fms14 or FIT; FOR-A or dactimicin synthase) was enzymologically studied *in vitro* and turned out to be an unusual oxidase;[207] its counterpart is IstZ (IST-A3 synthase) in *S. tenjimariensis*.

2.2.3.3. Possible Involvement of tRNA Modification in the Biosynthesis or Resistance of FOR and GEN. A highly conserved set of genes, which recently have been shown to be involved in the modification of guanine residues in the wobble position of tRNAs with GU(N) anticodons specific for Asp, Asn, His, and Tyr, appear in the *gen*- (*genO, genW, genAFG*), *for*- (*fosGFE, fosC, fosA*), and *fra*- (locus tags 3378–3381, 3383) gene clusters (cf. Tables 2.16–2.18). In these GU(N) anticodons the guanines are converted into queuine (Q) (7-(((4,5-*cis*-dihydroxy-2-cyclopenten-1-yl)amino)methyl)-7-deazaguanosine) in most groups of eubacteria (except for actinomycetes) and eucarya, or into archeosine in archaea.[258,259] In the *gen*-cluster these genes are scattered over the whole gene cluster, whereas in the *for*- and *fra*-clusters they are clustered at the distal end. The equivalent Q biosynthetic proteins are QueC (unknown function), QueD (unknown function), QueE (unknown function), QueF (NADPH-dependent nitrile reductase),[260] and a TGT (queuine tRNA-ribosyltransferase).[261] A QueA (*S*-adenosylmethionine:tRNA ribosyltransferase-isomerase or Q-tRNA synthase[262]) enzyme also described in other organisms does not seem to be encoded by the *AGA* producers for GEN or FOR or the FOR-related AGA encoded by the *fra*-cluster, since neither the *gen*- and *for*-gene clusters nor the full-length genome of *Francia* sp. CcI3 do contain a homologous gene. Intriguing also is the fact that only in the *gen*-, *for*-, and *fra*-gene clusters a second 16S rRNA methyltransferase resistance gene is located and physically linked to the major part of the *que*-related genes (cf. Tables 2.16–2.18). These puzzling facts could mean that either (i) in the producers of GEN and FOR-type AGAs the Que-like gene products are involved in biosynthesis of the antibiotics (as we suggest above for some of them), or (ii) a Q pathway is necessary for other, as yet unknown reasons, or (iii) wobble base modification of some tRNAs is necessary for undisturbed translation in cases where bases in the 16S rRNA controlling codon–anticodon interaction are methylated by AGA resistance enzymes of the GmrA/FmrO type. At present we favor the last hypothesis, which of course has to be either proven or disproven in the future—for example, by knock-out mutations in the *que*-related genes.

2.2.4. Monosubstituted 2DOS-AGAs

2.2.4.1. Producers, Product Profiles, Genetics, and Biosynthesis of Apramycin (APR). Apramycin (APR, see Figure 2.24) is a translational inhibitor, as are the other well-investigated 2DOS-AGAs.[208] It is produced together with tobramycin (TOB) and various nebramycin factors (NEBs), such as KAN-B- and TOB-carbamates (NEB-4 and NEB-5; see paragraph on KANs), as part of the nebramycin complex in various actinomycetes, e.g. *Streptomyces* sp. (former "*tenebrarius*") DSM 40477, *Streptoall. hindustanus* DSM 44523, *Saccharopolyspora hirsuta* DSM 43463, and others (cf. Figure 2.24, Table 22). Later, another component of the APR family, called saccharocin (KA-5685,

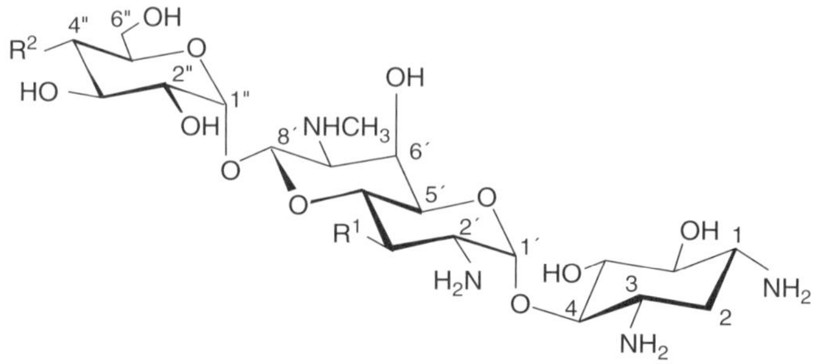

	R^1	R^2
Apramycin	H	NH_2
Oxyapramycin	OH	NH_2
Saccharocin (KA-5685)	H	OH

Figure 2.24. Structures of apramycin (APR) family AGAs.

4″-deamino-4″-hydroxyapramycin; cf. Figure 2.24), was isolated from cultures of *Saccharopolyspora hirsuta* or *Saccharopolyspora* sp. AC-3440.[209,210] Also, co-resistance with the KAN/TOB-group of AGAs is observed in the producers, which is based on 16S rRNA methyltransferase-encoding gene(s).[211]

Therefore, it was apparent that these quite different actinomycete strains are exceptional in that they are in different genera. Each seems to contain either one mixed functional *tob-/apr*-cluster or two 2DOS-AGA gene clusters, one for the production of TOB (and other nebramycin factors) and the second for the production of APR. The second type of genomic situation turned out to be the case: We cloned and sequenced the *apr*- and *tob*-clusters from *S*. sp. ("*tenebrarius*") DSM 40477 (cf. Tables 2.14 and 2.20) and could not find any close physical linkage between the two genomic regions; for comparison the *apr*-cluster from *Streptoall. hindustanus* DSM 44523 (cf. Table 2.21) was analyzed in the same way and found to be almost identical in gene content and order relative to the *apr*-cluster from *S*. sp. ("*tenebrarius*") DSM 40477 (see Figure 2.26; cf. Tables 20 and 21). This raised several questions that we will discuss in more detail below: (i) Is there a physical link between both gene clusters? (ii) Is there a functional interaction between the two gene sets (e.g., in biosynthesis, gene regulation, etc.)? (iii) Or is there a cooperation between both antibiotics in their putative ecological effect(s), making it an advantage to co-produce both compounds as a mixture? Both the *tob*- and *apr*-clusters, though containing an independent set of genes for 2DOS production, seem to share part of the functions encoded for both pathways (see below).

Recently, in another attempt to isolate an AGA biosynthetic gene cluster from *S*. sp. ("*tenebrarius*") H6 an incomplete set of five putative sugar conversion

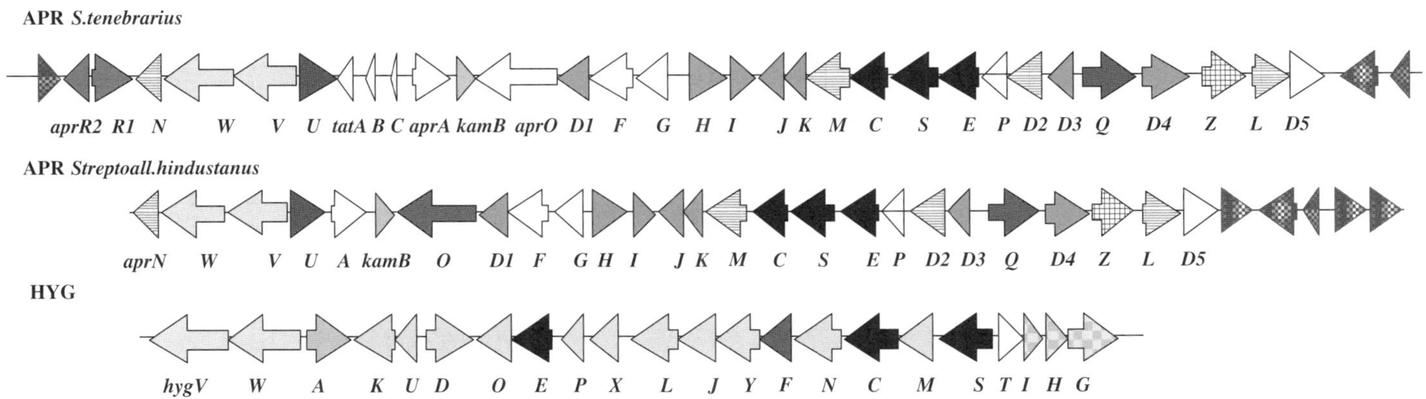

Figure 2.25. Biosynthetic gene clusters for APR and HYG-B. octose synthase, transfer, modification (APR), 8′Glycosyltransfer/modification (APR), HYG-specific genes; all other colors see Figure 2.9.

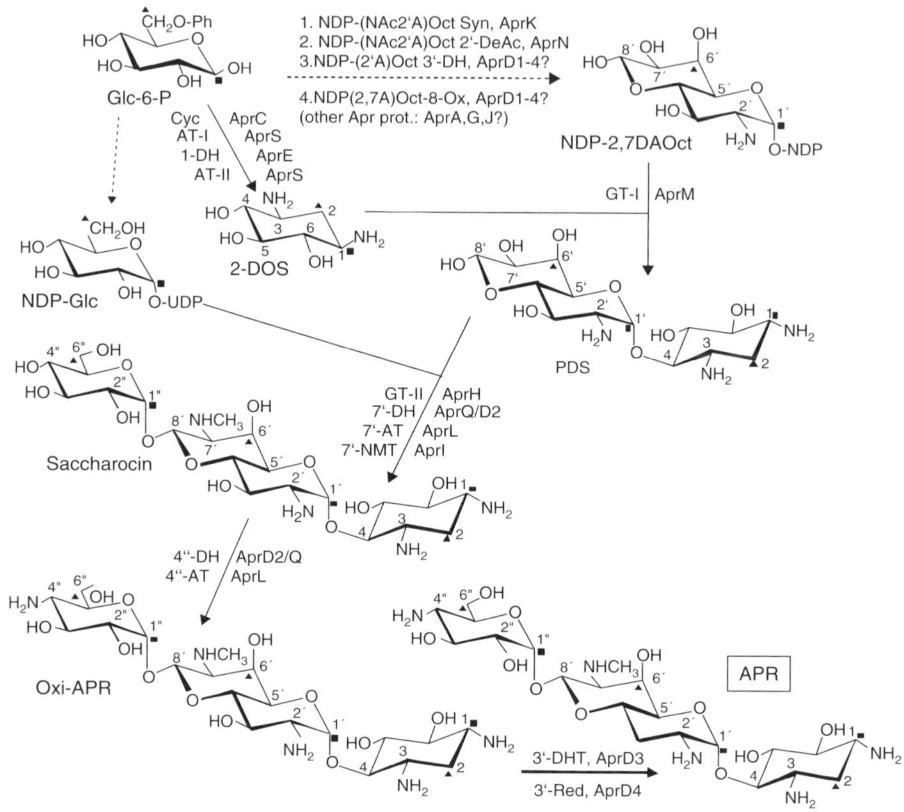

Figure 2.26. Biosynthetic pathway for APR. Other explanations and abbreviations are as given in the legend to Figure 2.3.

genes around an "*orfE*" gene was cloned (accession code AY131228.1)[212] by using the strategy of designing heterologous primers for the isolation of genes equivalent to the highly conserved *strE*-related genes, encoding dTDP-glucose dehydratase (also involved in the STR, SPC, and KAS pathways; see Section 2.2.1.1).[77,133] The function of the *orfE* gene was studied by targeted gene disruption. The resulting mutant failed to produce TOB and KAN-B, but still produced APR, suggesting that the *orfE* gene and adjacent genes are essential for the biosynthesis of TOB and KAN-B in *S.* sp. ("*tenebrarius*") H6. However, this assembly of genes shows neither similarity to the *tob*- and *apr*-clusters, nor does it contain genes functionally absolutely required for either the production of TOB or APR. Therefore, the suggestion of its being involved in either the TOB or APR pathways seems highly questionable at present.

2.2.4.2. The Biosynthetic Pathway for APR. Earlier attempts to propose an APR biosynthesis pathway on the basis of biogenetic studies are scarce. Labeling

TABLE 2.20. Proteins Encoded in the Genomic Area Covering the *apr*-Cluster of *Streptomyces* sp. DSM 40477 (Accession Code AJ629123)[a]

ORF /Locus Tag[a]	Gene Product[b] Symbol	aa	CDD Number[c]	Putative Function
SteO08.1	AprX	205	12871	Lipoprotein
SteO08.2c	AprR2	244	14108	Transcriptional regulator
SteO08.3	AprR1	370	14108	Transcriptional regulator
SteO08.4c	AprN	244	11828	*N*-Acetyl-hexosaminyl deacetylase or amidase
SteO08.5c	AprW	657	10852	ABC transporter
SteO08.6c	AprV	621	10852	ABC transporter
SteO08.7	AprU	335	12010	Kinase
SteO08.8c	TatA	181	25792	Transposase
SteO08.9c	TatB	93	—	Transposase (fragmentary)
SteO08.10c	TatC	84	—	Transposase (fragmentary)
SteO08.11	AprA	373	9586	Unknown
SteO08.12	KamB	155	27198	16S rRNA MT; APR resistance
SteO08.13c	AprO	785	11268	Hydrolase or phosphorylase
SteO08.14c	AprD1	312	23188	NDP-hexose DH or epimerase
SteO08.15	AprF	418	9229	Unknown
SteO08.16c	AprG	296	27577	Unknown, hydrolase?
SteO08.17	AprH	359	13285	Glycosyltransferase; GT I or II
SteO08.18	AprI	262	25197	Oxidase or *N*-MT
SteO08.19c	AprJ	238	25595	Phosphosugar mutase
SteO08.20c	AprK	189	12222	NDP-heptose or -octose synthase
SteO08.21c	AprM	438	7651	Glycosyltransferase; GT II or I
SteO08.22c	AprC	384	17039	2-Deoxy-*scyllo*-inosose synthase,cyclase
SteO08.23c	AprS	454	15279	L-Glutamine:ketocyclitol AT I + II
SteO08.24c	AprE	373	25397	Aminocyclitol 1-DH
SteO08.26c	AprP	249	3140	Amidohydrolase
SteO08.27c	AprD2	329	10808	UDP-*N*-acetylglucosamine 4,6 dehydratase
SteO08.28c	AprD3	260	25396	Short-chain oxidoreductase
SteO08.29c	AprY	285	—	Unknown
SteO08.31	AprQ	503	16494	6-DH
SteO08.32	AprD4	457	10760	Fe–S oxidoreductase
SteO08.33	AprZ	462	28900	Aminoglycoside-phosphate phosphatase
SteO08.34	AprL	373	15279	AT
SteO08.35	AprD5	348	23188	Epimerase/dehydratase

[a]The coding sequences (ORFs) are given with their locus tag numbers as found on the respective cosmids analyzed.
[b]The list of gene products translated from the nucleotide sequence corresponds to the protein ID numbers CAF33026–CAF33058.
[c]Refers to the CDD database.[255]

studies have revealed that all the C atoms of D-glucose appear in the C-1 to C-6 atoms of the unusual octose moiety and that the C-7 and C-8 atoms could be derived from C-2 and C-3 of pyruvate (Pearce and Rinehart, unpublished).[122] This suggests a formation of the octose as a transketolase/transaldolase by-product of the activities of an extended pentose phosphate cycle. Likely precursors for APR are 2DOS, an NDP-activated D-glucose, and a possibly NDP-activated octose. The two *apr*-gene clusters reflect this suggestion to some extent and can be interpreted as described below.

At present, all availabale data and evidences taken together allow us to postulate the following biosynthetic pathway for the APRs (Figure 2.26): In a first branch 2DOS is formed by the already described pathway elucidated earlier (see Figure 2.10) by use of the conserved gene products of "C"-, "S"-, and "E"-type

TABLE 2.21. Proteins Encoded in the Genomic Area Covering the *apr*-Cluster of *Streptoalloteichus hindustanus* DSM 44523 (Accession Code AJ875019)[a]

ORF /Locus Tag[a]	Gene Product[b] Symbol	aa	CDD Number[c]	Putative Function
ShiN01.1c	AprN	170	11828	*N*-Acetyl hexosaminyl deacetylase or amidase
ShiN01.2c	AprW	626	10852	ABC transporter
ShiN01.3c	AprV	600	10852	ABC transporter
ShiN01.4	AprU	336	12010	Kinase
ShiN01.5	AprA	373	9586	Unknown
ShiN01.6	KamB	155	27198	16S rRNA MT; APR resistance
ShiN01.7c	AprO	780	11268	Hydrolase or phosphorylase
ShiN01.8c	AprD1	312	23188	NDP-hexose DH or epimerase
ShiN01.9c	AprF	425	9229	Unknown
ShiN01.10c	AprG	298	27577	Unknown, hydrolase?
ShiN01.11	AprH	358	13285	Glycosyltransferase; GT I or II
ShiN01.12	AprI	262	25197	Oxidase or N-MT
ShiN01.13c	AprJ	237	25595	Phosphosugar mutase
ShiN01.14c	AprK	163	12222	NDP-heptose or -octose synthase
ShiN01.15c	AprM	469	7651	Glycosyltransferase; GT II or I
ShiN01.16c	AprC	386	17039	2-Deoxy-*scyllo*-inosose synthase, cyclase
ShiN01.17c	AprS	424	15279	L-Glutamine:ketocyclitol AT I + II
ShiN01.18c	AprE	338	25397	Aminocyclitol 1-DH
ShiN01.19c	AprP	272	3140	Amidohydrolase
ShiN01.20c	AprD2	319	10808	UDP-*N*-acetylglucosamine 4,6-dehydratase
ShiN01.21c	AprD3	248	25396	Short-chain oxidoreductase
ShiN01.22	AprQ	503	16494	6-DH
ShiN01.23	AprD4	456	10760	Fe–S oxidoreductase
ShiN01.24	AprZ	466	28900	Aminoglycoside-phosphate phosphatase
ShiN01.25	AprL	373	15279	AT
ShiN01.26	AprD5	345	23188	Epimerase/dehydratase
ShiN01.27		449	25528	Phenylacetyl-CoA ligase

[a]The coding sequences (ORFs) are given with their locus tag numbers as found on the respective cosmids analyzed.
[b]The list of gene products translated from the nucleotide sequence corresponds to the protein ID numbers CAI47637–CAI47662.
[c]Refers to the CDD database.[255]

enzymes. In a second branch we postulate that an activated octose, possibly a two-ring ADP-2-aminooctose which had undergone a second intramolecular pyranosidic ring formation via oxidation and ring closure in the side chain of the preexisting pyranosidic 2-aminooctose-8-phosphate, is formed from a hexose or hexosamine via reactions analogous to the pentosephosphate shunt. Evidence for this hypothesis comes from the fact that both *apr*-clusters investigated conservatively encode the AprK protein, which is a member of the ADP-heptose synthase protein family (RfaE-related; conserved domain family CDD28834); we suggest that enzymes with this functional category can also catalyze nucleotidylyltransfer to octoses. However, the exact nature of the octosamine precursor as well as of the enzymology involved in its biosynthesis remain enigmatic and highly speculative. In order to explain the origin of the 2-amino group of the octosyl residue, we speculate that the octose or 2-aminooctose precursor could be started to be made from *N*-acetyl-D-glucosamine and intermediates of the pentosephosphate cycle via transketolase (or transaldolase) reactions. This would explain why no specific genes/enzymes for this conversion or later introduction of another amino group are provided by the *apr*-clusters themselves and why a gene, *aprN*, encoding a putative *N*-acetylaminosugar deacetylase is conserved in both *apr*-clusters. Also, a putative phosphosugar mutase, AprJ, a possible UDP-*N*-acetylglucosamine 4,6-dehydratase, AprD2, and other proteins with as yet unknown functions—for example, AprA, AprF, and AprG (cf. Tables 2.19 and 2.20)—could be involved in the formation of the NDP-2-aminooctose building block. In the third phase the two precursors could be condensed in a first glycosyltransfer reaction to a pseudodisaccharide intermediate of the structure given in Figure 2.26; the likely glycosyltransferase involved could be AprM. This enzyme is closely related to the conserved "M"-type paromamine synthases (see above) and, therefore, gives another hint to the suggestion that the co-substrate in this reaction is a preformed 2-aminosugar unit. The fourth phase in this pathway is the recruitment and incorporation of a neutral hexose, probably UDP-D-glucose, into the PDS precursor via a second glycosyltransfer, a putatively AprH-catalyzed step. The resulting pseudotrisaccharide could then become modified by transamination of the 7′-position—for example, by use of the dehydrogenases AprQ or AprD2 and the aminotransferase AprL—and 7′-*N*-methylation via the putative methyltransferase AprI. The product of this conversion is identical with saccharocin, is a first antibiotically active one, and could become excreted with low efficiency by the postulated APR exporter system (AprV,W). In the subsequent fifth biosynthetic phase, saccharocin would be further modified to oxi-APR (3′-hydroxy-APR) by 4″-transamination (same candidate enzymes as for 7′-amino transfer; AprQ/D2 and AprL) and to APR by 3′-dehydroxylation. The latter step was already discussed in detail in the context of the biosynthesis of LIV (see Section 2.2.2.1.6) and TOB (see Section 2.2.2.2.3). The postulated dehydratase/oxidoreductase proteins involved are likely to be the AprD3/AprD4 enzymes, which are conserved only in the complement of gene products encoded by the *liv*-cluster (LivW/LivY; see above). However, there is no satisfactory explanation to this hypothetical pathway since both the 7′- and 4″-transamination are unlikely to be catalyzed by the same enzyme (AprL) and an aminotransferase coding gene is lacking in the *apr*-clusters. Alternatively, either the 7′ or 4″ amino group originate from an already aminated sugar molecule as a precursor.

Finally, we propose that APR becomes phosphorylated by the putative APR-kinase AprU and thereby inactivated, as a second resistance mechanism in addition to the 16S rRNA methylation by KamB, during biosynthesis or thereafter. Because AprZ is significantly similar to the StrK protein, a member of the protein family of extracellular alkaline phosphates and a STR-phosphate-specific dephosphorylase (see Section 2.2.1.2), this modification is urgently suggested by presence of the conserved *aprZ* gene in the biosynthetic cluster.[213] AprU is a member of the large kinase family comprising all the antibiotic and protein kinases.[16,43] As in the STR producers, the postulated APR-phosphate would exported via the ABC transport system AprV/AprW and set free by dephosphorylation outside the cells via the phospatase AprZ.

2.2.4.3. Producers, Product Profiles, Genetics, and Biosynthesis of Hygromycin B (HYG-B). Hygromycin B (HYG-B; Figure 2.27) was first described in 1953 as a product of *Streptomyces hygroscopicus* ssp. *hygroscopicus* DSM 40578 and its structure elucidated later.[214–216] Also, a second heterogenous cyclitol-, sugar-(6-deoxyhexose), polyketide-, and aromatic amino acid-(chorismate-)based and a (mono-)aminocyclitol-containing antibiotic, HYG-A (Fig. 27B), was detected in culture fluids of *S. hygroscopicus* ssp. *hygroscopicus* DSM 40578.[217,218,256] In addition, HYG-B-related AGAs isolated from *S. rimofaciens* and from *Saccharopolyspora hirsuta* were designated as destomycins (A, B, C),[219–221] those from *S. eurocidius* were designated the SS-56 complex of compounds (SS-56-A through -D),[222–224] and those from *Streptoverticillium eurocidius* were designated the A-365 complex of compounds (-I and -II; the latter is identical with HYG-B).[221]

Both HYG-B (and its relatives) and HYG-A act on distinct locations and steps during translation on the bacterial ribosome;[225,226,257] that is, the binding site of HYG-B on the ribosomal 30S subunit (on the 16S rRNA molecule) has been identified, characterized, and shown to be involved in the ribosomal decoding site, but has been shown to be distinct from that for other 2DOS-AGAs.[227,228] In contrast, HYG-A is a peptidyltransferase inhibitor and binds to a site overlapping with that for chloramphenicol- and macrolide-binding near the polypeptide exit channel on the 50S subunit.[257]

Since we regard HYG-A not to be an AGA within the scope of our chapter, we mainly focus on HYG-B herein and only briefly mention some biosynthetic aspects of this compound for reasons of comparison below. The ribosomes of the HYG-B producer are sensitive to its own product, and resistance to HYG-B is acquired by the expression of the antibiotic modifying enzyme HYG-B 7″-phosphotransferase.[229,230] The resistance gene encoding APH(7″), *hygA* or *hph* gene, seems to be widespread in natural bacterial populations and located on mobile "environmental cassettes," as detected in DNA samples directly isolated from materials (e.g., soils) in the biosphere.[231] No *in vivo* feeding studies with labeled precursors, in order to predict the biogenesis of the HYG-B molecule,

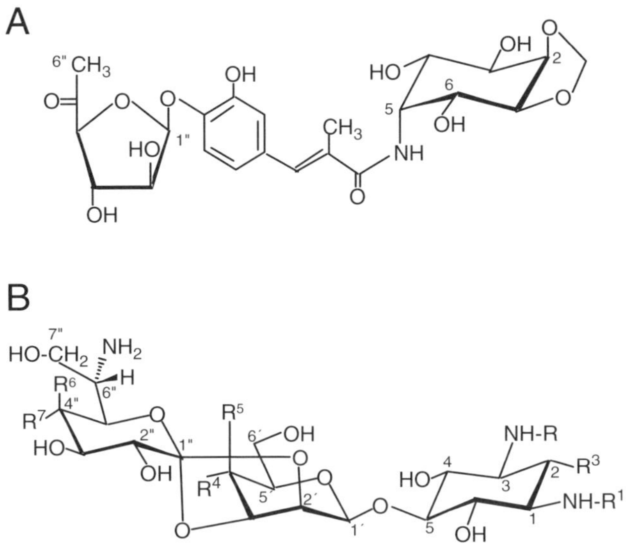

Hyg-B and related 2DOS-AGAs.

	R^1	R^2	R^3	R^4	R^5	R^6	R^7
Hygromycin B	H	CH_3	H	OH	H	OH	H
Destomycin A	CH_3	H	H	OH	H	OH	H
Destomycin B	CH_3	CH_3	H	H	OH	H	OH
Destomycin C	CH_3	CH_3	H	OH	H	OH	H
A-16316-C	CH_3	CH_3	H	H	OH	OH	H
SS-56-A	H	H	H	H	OH	OH	H
SS-56-B	H	H	H	OH	H	OH	H
SS-56-C	H	H	OH	OH	H	OH	H
A-396-I (SS-56-D)	H	H	H	OH	H	OH	H
1-*N*-Amidino-1-*N*-demethyl-2-hydroxy-destomycin A	$C(NH)NH_2$	H	OH	OH	H	OH	H

Figure 2.27. Structures of (A) hygromycin A (HYG-A) family AGAs and (B) hygromycin B (HYG-B).

seem to have been reported in the literature; therefore, at present the sole basis for a postulate on the biosynthetic pathway are the functions encoded in the HYG-B production gene cluster of *S. hygroscopicus* ssp. *hygroscopicus* DSM 40578 (*hyg*; cf. Table 2.3, and Figure 2.20).

2.2.4.3.1. The Biosynthetic Pathway for HygB. The *hyg*-gene cluster encodes the full enzymological capacity for an active 2DOS pathway (cf. Figure 2.10),—that is, "C"-, "S"-, and "E"-type gene products, activation, and/or modification of hexosyl and heptosyl sugar precursors and two glycosyltransferases, HygD and HygF, for the two condensation steps (cf. Table 2.22, Figure 2.28). We have to assume and postulate that the HYG-B pathway follows the following route and thereby passes the following phases:

TABLE 2.22. Proteins Encoded in the Genomic Area Covering the *hyg*-Cluster of *Streptomyces hygroscopicus* subsp. *hygroscopicus* DSM 40578 (Accession Code AJ628642)[a]

ORF /Locus Tag[a]	Gene Product[b] Symbol	aa	CDD Number[c]	Putative Function
ShyG17.8		474	12627	Transposase
ShyG17.9c	HygV	616	10852	ABC transporter
ShyG17.10c	HygW	604	10852	ABC transporter
ShyG17.11	HygA	332	25797	APH(7"), HYG-B-resistance
ShyG17.12c	HygK	319	25396	NDP-heptose-(or hexose) DH
ShyG17.13c	HygU	162	10116	Dehydratase or phosphatase
ShyG17.14	HygD	401	7651	Glycosyltransferase, GT II
ShyG17.15c	HygO	287	29587	NDP-heptose synthase
ShyG17.16c	HygE	339	25397	Aminocyclitol 1-DH
ShyG17.17c	HygP	198	10153	Phosphoheptose isomerase
ShyG17.18c	HygX	258	24783	Dioxygenase or hydroxylase (orthoester formation?)
ShyG17.19c	HygL	377	15279	(NDP-) heptose AT
ShyG17.20c	HygJ	308	23188	(NDP-) heptose (or -hexose) DH or epimerase
ShyG17.21c	HygY	343	12246	Fe–S oxidoreductase; radical SAM; 2′-(3′-)->1"oxidation/condensation
ShyG17.22c	HygF	270	25539	UDP-Gal:2DOS galactosyltransferase, GT I
ShyG17.23c	HygN	351	12122	Heptose-(7-P) 1-kinase
ShyG17.24c	HygC	410	17039	2-Deoxy-*scyllo*-inosose synthase, cyclase
ShyG17.25c	HygM	272	2985	*N*-MT
ShyG17.26c	HygS	434	15279	L-Glutamine:ketocyclitol AT I + II
ShyG17.27	HygT	182	—	Unknown
ShyG17.28	HygI	163	28977	Component of sensor/response regulator
ShyG17.29	HygH	168	—	Component of sensor/response regulator
ShyG17.30	HygG	392	—	Component of sensor/response regulator
ShyG17.31c	HygZ	662	12179	Transposase

[a]The coding sequences (ORFs) are given with their locus tag numbers as found on the respective cosmids analyzed.
[b]The list of gene products translated from the nucleotide sequence corresponds to the protein ID numbers CAF31836–CAF31859.
[c]Refers to the CDD database.[255]

Phase 1. Biosynthesis of 3-*N*-methyl-2DOS by the basic enzyme complement for 2DOS biosynthesis as given above (cf. Figure 2.10) and extended by the additional requirement and presence of a 3-*N*-methyltransferase (3-NMT), putatively represented by the HygM protein (see Table 2.22). In his investigation and setup of test conditions for aminocyclitol NMTs from the HYG-B and SPC producers, Walker has described an activity in extracts of *S. hygroscopicus* ssp. *hygroscopicus* ATCC 27438 that acts on the aminocyclitol monomer and should originate from the HygM enzyme.[104] Surprisingly, however, the extracts of this AGA producer methylated both

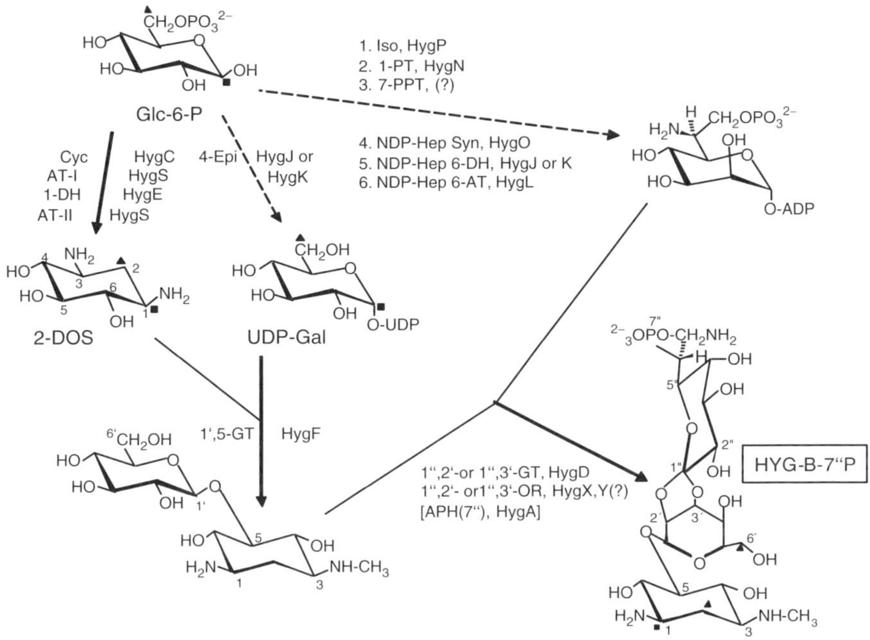

Figure 2.28. Biosynthetic pathway for HYG-B. Other explanations and abbreviations are as given in the legend to Figure 2.3.

the N^1- and N^3-amino groups of 2-deoxystreptamine, streptamine, and 2-*epi*-streptamine. Thus, the identity of the measured *N*-methyltransferase remains questionable.

Phase 2. Recruitment and conversion of an NDP-activated and neutral hexose from primary metabolism; this is most likely UDP-D-glucose, which by the action of an UDP-glucose 4-epimerase, either the HygJ or HygK protein, respectively, becomes converted into the suggested precursor UDP-D-galactose.

Phase 3. Formation activation and modification of a putative NDP-heptose. We suggest that the heptose moiety is derived from *sedo*-heptulose-7-phosphate, which is the universal source for heptoses in all organisms having a functional pentosephosphate cycle. Also, for C-7 cyclitols (e.g., valienamine) or for secondary metabolic sugar units with longer C chains, such as octoses (e.g., lincosamine, or the octose unit in APR, see Section 2.2.4.2), this seems to be the preferred precursor in actinomycete pathways.[2,134,232] In recent years the biosynthetic route for the ADP-heptose precursor of the lipopolysaccharide inner core component in the gram-negative proteobacterium *E. coli* has been elucidated.[233] This involves isomerizion of D-*sedo*-heptulose-7-phosphate to D-alpha,beta-D-heptose-7-phosphate by GmhA, phosphorylation to D-alpha-D-heptose-1,7-diphosphate by HddA, dephosphorylation by GmhB to D-alpha-D-heptose-1-phosphate, and nucleotidylation to ADP-D-alpha-D-heptose by HldE. The use of at least part of this pathway in *S. hygroscopicus* ssp. *hygroscopicus* DSM 40578 for the biosynthesis of the heptose moiety of HYG is evident from protein similarities. The initial isomerization step could be catalyzed by HygP, the 1-phosphotransfer by HygN, and the nucleotide activation (putative adenylyltransfer) by HygO. However, a protein related to a heptose-1,7-diphosphate 7-phosphatase is lacking among the *hyg*-cluster-encoded gene products. Alternatively, another type of phosphatase (e.g., the possible phosphatase HygU) could remove the 7-phospho group from the precursor. Another and presently preferred possibility to solve this puzzle could be that in case of the HYG-B pathway the 7″-phospho group stays in the precursor and is incorporated into the final pseudotrisaccharide to avoid its becoming toxic to the producing cell. A strong hint for this could come from the fact that 7″-phosphorylation is also the natural resistance mechanism of the producer, guaranteed by the *hygA*-dependent expression of the APH(7″) enzyme (see above; cf. Figure 2.28); a similar dependence of protection from self-toxicity of the end product by a phosphate group preformed during biosynthesis is also known from the STR pathway (see above). The introduction of the 6-amino group into the heptose unit could occur on the precursor level or after condensation; here we postulate that this group is introduced into the NDP-activated heptose by use of one of the putative *hyg*-encoded dehydrogenases, such as HygJ or HygK, and an aminotransferase, possibly HygL.

Phase 4. Condensation of the three units to a pseudotrisaccharide (PTS) and its further transglycosylation and oxidation to form an ortho-ester type of binding between the C-7 and C-6 sugar units: First, the putative glycosyltransferase HygF is postulated to link the neutral hexose to the 5-hydroxyl of the 3-*N*-methyl-2DOS moiety to form the PDS intermediate. Second, it seems likely that another *O*-glycosyltransferase reaction, postulated to be catalyzed by HygD, transfers the 7-aminoheptose to the PDS unit, either to the 2′ or 3′ positions of the hexose moiety. Obviously, ortho ester synthesis is a very rare type of reaction in biochemistry, and a structurally analogous case which had been analyzed enzymologically cannot be presented. Therefore, also the participation of *hyg*-encoded proteins, probably an oxidoreductase(s), in this step has to remain merely speculative (cf. Figure 2.28).

Since HYG-A and HYG-B are produced in the same strain and both contain an aminocyclitol unit, *neo*-inosamine-2 and 2DOS, respectively, it is interesting to note that both pathways do not seem to share any biosynthetic steps. This becomes evident when analyzing both the results of a recent biogenetic study using stable isotope incorporation[256] and reporting the HYG-A production gene cluster from *S. hygroscopicus* ssp. *hygroscopicus* NRRL 2388 (*hyg1* to *hyg29* cluster; accession code DQ314862). The *neo*-inosamine-2 moiety of HYG-A is obviously synthesised via the Ca route (cf. Section 2.2.1.2.1; Figure 2.3). However, in contrast to the gene clusters for STRs, SPCs, KAS, and FORs, the HYG-A production gene cluster seems to encode also the enzyme, Hyg18, for the first biosynthetic step in cyclitol formation, the L-*myo*-inositol-1-phosphate synthase (D-*myo*-inositol-3-phosphate or L-*myo*-inositol-1-phosphate-forming). The putative enzymes for the next steps in aminocycltitol biosynthesis are also encoded in the *hyg1-hyg29* cluster, with Hyg25 (L-*myo*-inositol-1-phosphate phosphatase), Hyg17 (*myo*-inositol 2-dehydrogenase), and Hyg8 (2-ketocyclitol 2-aminotransferase). The protein Hyg25 is closely related (44-45% identity in the protein sequences) to the StsD and KasE proteins used in the pathways for STRs and KAS, respectively (see Section 2.2.1.1); the Hyg17 protein has significant similarity (40% identity in the protein sequences) to the ForG enzyme encoded in the *for*-clusters (see Section 2.2.3.2). In contrast, the putative cyclitol aminotransferase Hyg8 (a class III aminotransferase) is not related at all to the typical and highly conserved AGA aminotransferases of class V ("S"-enzymes; see Section 2.3.2, Figure 2.29a). Thus, we have evidence for the HYG-A pathway and respective gene cluster (*hyg1-hyg29*) to have arisen from a more recent and independent evolution by collecting genes from various other secondary metabolic gene clusters, such as STR-related AGAs, A201A-type nucleoside antibiotics, and polyketides.[226] Noteworthy in our context is the fact of having two distinct aminocyclitol pathways in one and the same strain since it means that both enzyme complements (especially the cyclitol dehydrogenases and aminotransferases) obviously do not cross-react with the alternative and closely related substrates in the 2DOS and *neo*-inosamine-2 pathways.

2.3. SPECIAL ASPECTS OF THE GENETICS AND PHYSIOLOGY OF AGA PRODUCTION

2.3.1. Genomics of AGA Gene Clusters

Recently, Piepersberg proposed a scenario for the large-scale genome rearrangement in higher (i.e., filamentous and differentiating) actinomycetes, mainly based on the genome structure of *S. coelicolor* A3(2).[14,49] In these organisms, also in the avermectin producer *S. avermitilis*,[15] an interesting chromosomal organization leads to largely split gene pools, falling into two categories (though there seems to be considerable fluctuation between these two): first, those present mostly in an equal or inversed order in the much more stable central core regions (mostly housekeeping genes) as in other related bacterial genomes; second, those which are more or less randomly distributed to the two flanking variable arms of the linear streptomycete chromosomes. These latter regions also contain most of the secondary metabolic genes and gene clusters. Therefore, if a collection of genes isolated from another higher actinomycete strain belongs to either of these two gene pools, the stable and the variable one, this fact can be used as an indicator for their as yet unknown genomic location in either the core or the variable segments of the respective chromosome. If we apply this method to the prediction of the most likely genomic location for the actinomycete AGA clusters discussed here, we can conclude that (i) these clusters are not located in a commonly conserved chromosomal region, (ii) the *kan*- and *neo*-clusters can lie in either the central core segments or at the variable arms borders, (iii) most of the other clusters are instead located in a variable arm of the host strains (the close presence of the rRNA operon (*rrnE*) to the *gen*-cluster in *M. echinospora* does not contradict this suggestion since up to three of the six rRNA operons lie outside the core regions of the two completely analyzed streptomycete genomes).[14,15] We may also conclude that (iv) the *hyg*-cluster can lie at one of the immediate ends of the bearing chromosome in *S. hygroscopicus* ssp. *hygroscopicus* because two of the directly adjacent genes (locus tags ShyG17.32c and ShyG17.33c) are strongly related to two chromsonal end genes in both *S. coelicolor* A3(2) and *S. lividans* 66 (SCO0004 and SCO0003, respectively). Moreover, (v) the *str-/sts*-gene cluster in *S. griseus* DSM 40236 (accession code AJ862849 lies in a chromosomal region stably conserved in both *S. coelicolor* A3(2) and *S. avermitilis*, which explains its observed genetic stability. Conversely, the instability of the *str-/sts*-gene cluster in *S. glaucescens* GLA.0 is due to its location in the extremely unstable region at one end of the chromosome.[234] These facts are in good agreement with the suggestion that an intrachromosomal migration between the variable and stable gene pools is possible and frequently used in higher actinomycetes.[49]

2.3.2. Evolution of Gene Clusters and Biosynthetic Pathways for AGA Production

Since this is not a major aspect within the scope of this chapter, we only briefly summarize some important facts derived from the informational content of the AGA gene clusters described and hypotheses-based thereon, which shed light on the evolution of AGAs and their biosynthetic pathways. Evolutionary forces become visible on various different levels, such as:

A. *Genetic Drift by the Mere Accumulation of Point Mutations.* This is a trivial aspect in all related gene sets, such as the equivalents of essential housekeeping genes/enzymes in the genomes of all living organisms. These relationships lead to the so-called clusters of orthologous groups of genes.[235] This is especially evident when the conserved "core genes/enzymes," which form the basis of the general paromamine (neamine) pathway (see paragraph on NEOs)—that is, the gene set encoding the "C"-, "S"-, "E"-, "M"-, "Q"-, and "B"-type biosynthetic proteins—are regarded with respect to their interrelationships. The phylograms in Figure 2.29 should illustrate this and other evolutionary forces discussed below. A nice example of the change in genetic drift speed (the number of nucleotide changes per time unit and reflected in the number of aa replacements) after a gene or gene set has moved from a high to a low G+C organism from an actinomycete donor (e.g., related to *S. fradiae* NEO-producer) to a firmicute receptor (*B. circulans*, BTR-producer) is seen when comparing these protein phylograms for the producers of NEO-type AGAs: In each case the respective Btr proteins appear to be more deeply branching relative to their relatives in actinomycete NEO producers than those from several other actinomycete, non-NEO AGA producers—for example, GEN or KAN producers (cf. Figure 2.29; indicated by black arrows). We interpret this, however, not as true evolutionary distance, but rather the result of the enhanced evolutionary speed forced by the necessity of adapting a gene to a quite different DNA composition and codon usage.

B. *Genetic Shift.* This might preferably be induced by gene duplication and further aquisition of structural changes reshaping the substrate recognition properties of the gene product. This process might come strongly under selective pressure after a gene product (enzyme) has acquired a second function—that is, catalyzing the same type of reaction on a pair of closely related substrates. This is, for instance, the case whenever in an evolving pathway a very similar reaction turns out to be useful in a later stage again and, therefore, becomes being repeated in the conversion of a later stage intermediate. An example is the probably bifunctionality of the NeoB-type aminotransferases which, unlike their KanB/GenB-like counterparts, have probably acquired two functions; namely catalysts for both the 6′ and 6‴-aminotransfer, after the possible evolutionary precursor of RIB-type became further glycosylated with a second 2-aminohexose unit, requiring further modification (see Section 2.2.1.4). A further evolutionary step would be to duplicate a gene for a bifunctional enzyme (genetic shift) to allow for further independent evolution by genetic drift and thereby foster a more efficient adaptation of the two existing enzymes to their individual functions.

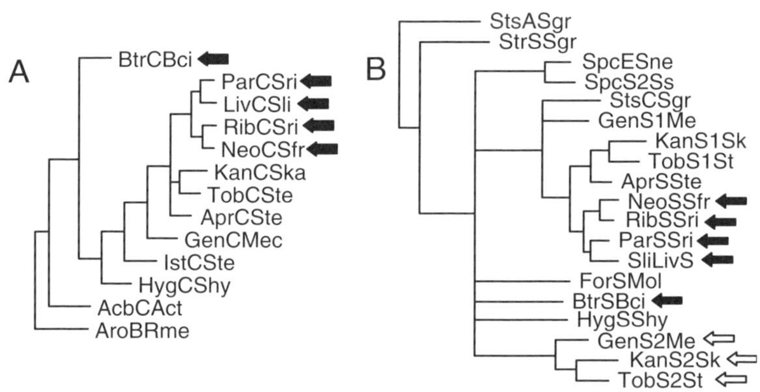

Figure 2.29A. Phylograms of the conserved "C and S1"-type enzymes involved in the 2DOS/paromamine/neamine subpathway of 2DOS-containing and related AGAs. All phylogenetic trees (phylograms) were calculated by the PAUP3.1.1 program by use of 500-times bootstrapping and giving nodes with higher than 50% matches within the bootstrap results; all trees were rooted by including an outgroup as given below. **A**. Phylogram of the sugar-phosphate cyclase ("C"-enzymes) involved in the 2DOS-formation. As an outgroup the DHQ-synthase from *Sinorhizobium meliloti* (Rme; CAC47255.1) and the *2-epi-5-epi*-valionone synthase AcbC from *Actinoplanes* sp. (CAA 77208) were used. **B**. Phylogram of the L-glutamine-dependent ketocyclitol ("S or S1") aminotransferases of class IV. As an outgroup the class IV aminotransferases (SMATs) StrS and StsA encoded by the *str*-cluster of *S. griseus* were used.

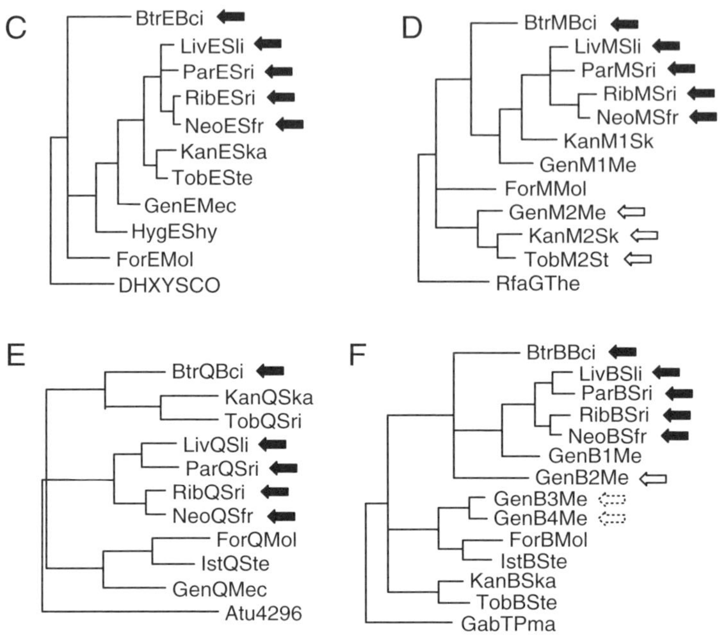

Figure 2.29B. Phylograms of the conserved "E, M, Q, B"-type enzymes involved in the 2DOS/paromamine/neamine subpathway of 2DOS-containing and related AGAs. For details see Figure 2.29a. (**C**). Phylogram of the "E"-type dehydrogenases related to zinc-dependent alcohol dehydrogenases (COG0604, COG1062, COG1063, COG1064).[235] As an outgroup the putative zinc-dependent alcohol dehydrogenase SCO2402 of *S. coelicolor* A3(2) (DHXYSCO; accession. code CAB62734.1) was used. (**D**). Phylogram of the "M"-type glycosyltransferases. As an outgroup the putative glycosyltransferase RfaG of *Thermoanaerobacter tengcongensis* MB4T (accession code AAG44711.1) was used. (**E**). The "Q"-type aminoglycoside 6′ (and/or 6‴)-dehydrogenases are choline dehydrogenase-related flavine-dependent oxidoreductases (COG2303); as an outgroup the putative dehydrogenase Atu4296 of *Agrobacterium tumefaciens* C58 (accession code AAL45090.1) was used. (**F**). Phylogram of the class III ("B"-type) 6'-aminotransferases. As an outgroup the class III aminotransferase of the GabT family (GabTPma) of *Prochlorococcus marinus* MIT 9313 was used (PMT0103; accession code NP893936.1). The filled arrows indicate protein species coming from the producers of NEO-family AGAs; the entries marked by open arrows are probable descendants that evolved after a duplication of the ancestral "1"-type gene in the same gene cluster and acquisition of a different function ("genetic shift"); the dotted open arrows mark proteins that have been evolved from coupled events of horizontal recruitment of an "outside" gene and its subsequent duplication and acquisition of two additional and different functions.

Again the "B"-type aminotransferase family gives an example in the GEN producer, where the *genB*-type genes have been even multiplied to four descendants. The phylogenetic relationships of the four *genB* genes could have arisen as follows (cf. Figure 2.29b): The *genB1* gene could have been the originally resident precursor and could have duplicated at some stage to form the *genB2* gene; these are more related to the *neoB* subfamily. The ancestor of the *genB3* and *genB4* genes in turn should have come later via horizontal gene exchange and a subsequent recombination event from a producer of FOR/IST-like AGAs and should have duplicated thereafter in the *gen*-cluster; this seems reasonable since the GenB3/4 proteins are very close relatives and in a common subfamily with the ForB/IstB proteins. The differing functions of the four GenB representatives are not clear at the moment, but it seems reasonable to assume that they have had an essential role in the development of the uniquely branching late pathway leading to the large variety of components in the GEN complex (see Section 2.2.2.2.7).

C. *Mixing of Whole Gene Clusters or Part Thereof.* After horizontal transfer of larger genomic segments between two different AGA producers and subsequent homologous recombination, other types of reorganization between the two related gene sets can result in mixed-type pathways. Intriguingly, the occurence of transposase genes or remnants thereof in or adjacent to AGA production gene clusters is observed in several cases (see, e.g., the *spc*-, *rib*-, *tob*-, *ist*-, and *hyg*-clusters). Later, such a process could be repeated several times and lead to very complicated interrelations between basically related gene clusters. This could, for instance, have been the case in the hypothetical evolution of the *gen*-gene cluster from an early fusion and more recently recurring recombination events between each a *kan*- and a *for*-type ancestral gene cluster. The evidence for this hypothesis, in addition to the above-mentioned facts, comes from the following observations: (i) The "S"-type aminotransferases and the "M"-type glycosyltransferases (Figure 2.29b) are likely to be used mainly as cyclitol aminotransferases (AT-I and-II) and pseudodisaccharide-forming hexosaminyltransferases (GT-I) in all the 2DOS- and FOR-related pathways, where they slightly differ in substrate profiles. They have obviously further evolved in the kanamycin/gentamicin producers after gene duplication to form a new subfamily of neutral sugar ring aminotransferases and aminocyclitol 6-glycosyltransferases (GT-II), respectively (cf. Figures. 2.10 and 2.16). (ii) A particularly drastic example of this type of evolution can be seen in the subfamily "B"-type class III aminotransferases (see above; Figure 2.29b). (iii) In the *for*-cluster even on the nucleotide level a segment can be found which looks like an evolutionary "erosion zone" containing several fragmentary reading frames (i.e., degraded by deletion events) with high degrees of sequence identity (up to higher than 90%) to several segments of the *gen*-cluster (see Figure 2.22). This indicates that recent recombination probably between chromosomal different segments of strains of *Micromonospora* sp. harboring ancestral *gen*- and *for*-clusters took place.

2.3.3. Export of AGAs Out of the Producing Cells

AGA transporters found to be encoded in the production gene clusters of their producers are practically exclusively exporter-type systems. Otherwise they may belong to the large ABC drug exporter family or an ion-driven major facilitator superfamily (MFS; with efflux-type transporter subfamilies).[236,237] For instance, in the *str*-, *neo*- and *apr*-gene clusters, ABC-exporter genes are found (see Tables 2.3, 2.8, and 2.19). In other cases or even in the same gene clusters, (e.g., in the *kan*-cluster), genes encoding members of the pH-gradient-dependent efflux protein family are found (cf. Table 2.13). Only in a minority of cases experimental data are available for the phenotypic properties of AGA transporter systems, when tested outside of their producers. Three ABC transporter genes, *kasKLM*, have been found to be probably responsible for self-resistance in a kasugamycin producer in addition to a KAS 2′-acetyltransferase (encoded by the *kac* gene; cf. Table 2.7).[20]

Most of the putatively AGA-transporting ABC transporters are composed of two proteins that are members of the CDD10852,[255] for example, StrW/V, AprW/V, or NeoU/T (cf. Tables 2.3, 2.4, 2.8, and 2.20). These proteins are related ABC-type multidrug transport systems and both subunits contain permease components with six transmembrane-spanning segments (TMSs) in the first half and an ATPase domain at their C-termini. The ATPase domains contain a signature motif, Q-loop, and H-loop/switch region in addition to the Walker A motif/P-loop and Walker B motif commonly found in a number of ATP- and GTP-binding and -hydrolyzing proteins.

The "H", "I", and "J"genes/proteins, conserved in the *gen*-, *for*-, *ist*-, and *fra*-clusters, represent an efflux-type exporter system (MF) belonging to the drug resistance transporter EmrB/QacA-related superfamily in the MFS1 subfamily. However, the transporters encoded by the AGA clusters are structurally new in that most of the representatives found encoded in the AGA producers were created from three protein components with similar primary structures. To our knowledge, this type of transporter has not yet been described, neither phenotypically nor biochemically. In case of the *fra*- and *gen*-cluster the "H" and "J"genes are fused to form a single ORF (see Tables 2.16 and 2.18). Transmembrane transport complexes with similar components are encoded, for example, in *S. avermitilis* MA 4680 (SAV2013, Q82LJ7). In the *fra*-cluster we have identified in the genome of the actinomycete bacterium *Frankia* sp. CcI3 a gene assembly emerged (accession code CP000249; see Section 2.2.2.2) encoding also two putative MFS transporter proteins (locus tag numbers 3358 and 3359; see Table 2.18) most closely related to GenH/I, ForH/I/J, and IstI/J, respectively. We can assume, from the predicted topology models of these transporters, that they are members of a new transporter family. All computer models predict 13 or 14 TMSs for ForI and IstI, respectively, 17 TMSs for GenH and the *Frankia* Protein 3358, and 2 TMSs for ForJ/IstJ, GenI and *Frankia* 3359.

Interestingly, the last 150–200 aa at the C-termini of the proteins IstJ and ForJ are obviously located outside of the cells and may form an extracellular loop or functional domain. The GenH protein also contains a long extracellular domain. All models for the GenH proteins predict that the protein contains a 300-aa-long domain (from aa 538 to aa 836) which is obviously located outside. The function of this domain, or its real topological location, remains mysterious. One can speculate that these extracellular domains might have a catalytical function or act as a binding domain.

2.3.4. Regulation of Genes and Biosynthetic Enzymes for AGA Production

This topic was last reviewed recently by us and others.[7,238,239] Therefore, we here give only a short summary of what is known: In *S. griseus* DSM 40236 at least some of the operons of the *str-/sts*-cluster are regulated by the DNA-binding protein StrR, a pathway-specific regulator encoded in the same cluster: The *strR* gene itself is regulated by the complex A-factor-dependent autoregulator cascade, which seems to be typical for streptomycetes, frequently synthesizing and using butyrolactone autoregulators.[240] Putative StrR-type regulators are also encoded in the *spc*-cluster of *S. netropsis* NRRL 2820 (SpcR), and in the *kas*-cluster of *S. kasugaensis* DSM 40819 (KasT). Other putative transcriptional regulators, belonging to various regulator families, are also located in several of the AGA gene clusters reported; but no experimental evidence proves their cluster-specific activities. In a series of clusters, mainly for the production of 2DOS-AGAs, genes are conserved for a new type of secondary metabolic sensor-response (phospho-relay) regulator system. The *neoI/H/G*-related genes, also present in the *rib-, par-, kan-, hyg-*, and *liv*-clusters, could encode a set of three protein components each that could form membrane-spanning complexes of a novel class of sensor kinase/response regulator systems, "three-component regulator systems," as compared to the well-known two-component response regulators. Evidence for this speculation comes from the following facts: In the *neo-*, *rib-*, *par-*, and *liv*-clusters (cf. Tables 2.8–2.11), the clusters for the aminoglycosides kanamycin (*kan;* cf. Table 2.13) and hygromycin B (*hyg*; cf. Table 2.22), and the clusters for the cinnamycin gene *S. cinnamoneus* ssp. *cinnamoneus* DSM 40005,[242] three strongly conserved reading frames have been found, which regularly mostly occur in a common operon. The exception to this regular occurrence was the position of *parI* lying separately from *parHG* (cf. Figure 2.9 and Tables 2.8–2.11). Equivalent genes are absent from other groups of AGA clusters in actinomycetes—for example, such as those for the production of GEN-, FOR-, TOB-, APR-, and STR/SPC-related AGAs. Genes more distantly related to *neoHG*, also in tandem, occur in other actinobacteria, such as *S. coelicolor* A3(2) and *S. avermitilis* MA-4680, and on the *Gordonia westfalica* DSM44215 plasmid pKB1 (ORF5 and ORF4; NC005307.1), and they encode the pairs of hypothetical sensor-domain and His-kinase proteins SCO2435/2436, SAV1743/SAV1744, and pKB1Gw5/pKB1Gw4, respectively. A single gene encoding the NeoH-related protein Nfa41130 also is present in the genome of *Nocardia (N.) farcinica* IFM 10152, whereas distantly NeoI-related proteins were also encoded by single genes in *Mycobacterium leprae* TN (ML0069), *M. tuberculosis* H37Rv (Rv3849), *N. farcinica* IFM 10152 (Nfa5160), and *Arthrobacter nicotinovorans* plasmid pAO1 (ORF155, TrEMBL:Q8GAD8). In the cinnamycin producer these three genes had been preliminarily called *cinORF12, cinORF13*, and *cinORF14* (sequence incomplete) and evidently did not contribute to the heterologous production of cinnamycin.[242] The structural analysis of these proteins revealed that (i) the NeoI (CinORF12) family proteins are in the range of 132–175 aa long, obviously cytoplasmically localized proteins, and do contain a possible helix-turn-helix (HTH) DNA-binding motif in the area of positions 51–90 in NeoI (they have a distant relationship to the helix-turn-helix XRE-family-like proteins, which are prokaryotic DNA-binding proteins belonging to the xenobiotic response element family of transcriptional regulators; (conserved domain motif cd00093.2); (ii) the NeoH (CinORF13)-related proteins contain a strongly conserved, hydrophilic and histidine-rich central domain with the characteristic motif H-Q-e-h-i-v-l-H-E-i-g-H (position 89–100 in NeoH; bold letters indicate uniformly conserved residues) preceded by a single predicted trans-membrane helix (position 42–62 in NeoH); (iii) the NeoG (CinORF14) family proteins consist of an N-terminal transmembrane domain with 7(6) predicted trans-membrane helices (first 230 to 250 aa) and a C-terminal hydrophilic domain without uniformly conserved residues and no similarity to any known domain structures.

Though other membrane-associated functions, such as transport, could be attributed to these three proteins, the most likely and attractive speculation is that they form an interesting new regulatory complex (see above). This would further extend our currently known collection of such systems which is already now highly diversified and reflects the different needs of signalling mechanisms for bacteria with largely differing ecology.[241] This attractive postulate is based on the following facts and observations: (1) The occurrence of these genes/protein products does not correlate with the structures of the metabolic products formed by the respective gene clusters and any other functional needs, such as biosynthetic-, resistance-, or transport-associated; this is underlined by their absence from the related *gen*-, *for*-, *tob*-, and *apr*-gene clusters; (2) the encoded proteins are highly conserved and have putative domain structures fitting best with the above hypothesis; (3) the model would predict that: (i) the NeoG-type proteins, which are built similarly to the UhpB-related sensor proteins (N-terminal trans-membrane domain plus C-terminal sensor H-kinase domain, but lacking the H-kinase motif), would form a complex H-kinase together with the membrane-associated NeoH-type proteins, which presumably would complement the H-phosphorylation site, and (ii) these two proteins interact, on the cytoplasmic side of the membrane, with the postulated transcriptional regulators of NeoI-type to control pathway-specific gene expression within the particular antibiotic clusters in response to some unknown signals. Future physiological analysis and transcriptional studies in wild-type and knock-out mutant strains will help to clarify this working hypothesis.

2.4. CONCLUSIONS AND OUTLOOK

2.4.1. Conclusions

The main recent progress in AGA research reported in this chapter was in the molecular genetics and biochemistry of AGA biosynthsis in their producers. From this gain of knowledge we conclude that it can be used in the future for renewed research efforts in the molecular biology, biotechnology and medical use of AGAs. In detail, it will help us to better understand (1) the biochemical tool/kit and biosynthetic capacity for AGA production in their producers, (2) the origin of the complex product patterns seen in AGA fermentations, (3) the regulatory mechanisms used to control AGA biosynthesis and couple it to other metabolic changes (such as cell differentiation) during the life cycle of the producing bacteria; especially, to define the internal and environmental signals steering gene regulation in production gene clusters, (4) the mode of AGA export and extracellular release of active end products from inactive (modified) precursors, (5) the AGA/target site interactions, (6) and the ecological meaning of AGAs—that is, the mechanisms of dissemination of AGA gene clusters and the ecological value these natural products have for their producers. (7) Further, we conclude, that the evolutionary origin of AGAs is quite old and might reach into the phases of the hypothetical RNA world, where they could have been together with other small molecules, such as peptides, ligands of RNA complexes in a primordial translational machinery.[243] This "molecular fossil" hypothesis would be in favor with an earlier suggestion that AGAs might have preexisted the evolutionary invention of their recent biosynthetic pathways.[49] The modern pathways, therefore, could be regarded as a reinvention in form of a more stable and inheritable "blueprint" for making these molecules after a DNA/protein world had been started. The deep evolutionary branching in the AGA relationship is reflected by the fact that there are almost no common gene/enzyme families used for the production of the major subclasses—for example, the STR-related AGAs as compared to 2DOS-containing and related AGAs. The only gene/enzyme family joining all of them are the "S"-type genes encoding L-glutamine dependent cyclitol aminotransferases (see phylogram B in Figure 2.29a). The search via "genetic screening" for further occurences of AGA-related gene clusters in larger numbers of other microorganisms and the investigation of their compositions and functions could bring us additional information on this issue and even producers of new and so far unknown AGAs. The chance detection of an AGA clusters for the production of a FOR/IST-related product in *Frankia* sp. CcI3 (accession code CP000249; locus-tag numbers 3357–3375) by genome sequencing encourage us to follow this strategy in future.

2.4.2. Outlook: Pathway Engineering of Gene Clusters and Biosynthetic Pathways for AGA Production

As other typical antibiotic families (e.g., macrolides, tetracyclins, betalactams) which have been developed in contact with a distinct target site, especially those acting on the translational machinery (because of its unique ubiquity, stability, and conservativeness), the AGAs have undergone a natural pathway engineering probably lasting hundreds of million years.[49,244] In view of the present state of AGA evolution, gained from observations of different structure–target site interactions and efficiencies in related families, comparisons of the producing gene clusters/enzymology of AGAs (bearing a record of their successful attempts for improvement) will enable us to hypothesize on the existing potential for future improvements.

Like many other typical bacterial secondary metabolites, AGAs tend to be produced as complex product mixtures released from their producing cells. We interpret this as a strong sign for ongoing evolution of the respective pathways and ligand–target site interactions. In other words the biochemical interactions leading to the biological activity of the AGAs have not reached an optimum as yet in the ecological interplay their producers produce them for. For most of the AGAs we have treated here, this is the interaction with part of the A-site or decoding site of the small ribosomal subunit of bacterial (70S) ribosomes.[244–246] Besides NEOs, KANs, GENs, and APR, pseudodisaccharides such as neamine and FORs built from paromamime-like structural elements, the binding site for the paromamine-containing 2DOS-AGAs or "A-site AGAs" are formed by helix 44 of 16S rRNA (H44; nucleotides 1400–1410 and 1490–1500 near the 3′ end of 16S rRNA).[247] This interaction can be demonstrated now *in vitro* by complex formation between these AGAs, (e.g., paromomycin) and a short synthetic (27-nt–long RNA) analog of this part of H44 (see respective chapter in this book).[248,249] This is also supported by the two resistance mechanisms involving 16S rRNA methylation, one of which results in co-resistance of GENs/KANs by methylating G-1405 to 7-methylguanosine and the other one in co-resistance of APR/TOB/KANs by methylating A-1408 to 1-methyladenosine.[253,254] Other AGAs, like STRs, HYG-B, SPCs, and KAS, have distinctly different binding sites in 30S ribosomal subunits—that is, in 16S rRNA molecules.

In the late phases of that pathways leading to all subfamilies of 2DOS-AGAs, a tendency is obvious to remove either the 3′-OH or 4′-OH or both. These hydroxyls make contact and confront the 16S rRNA nucleotides A-1492 and A-1493 in H44, whereas the C-6′-bound groups confront A-1408 and C-1409. Also, a strong conservation of the modification at the 6′-C atom by transamination and in part by *C*- or *N*-methylation is seen in the paromamine unit. As specifically considered from the complex situation regarding modifications in the GEN pathway (see Section 2.2.2.2.6), this part of the 2DOS-AGA surface is still a subject of ongoing biochemical evolution. Therefore, it will make sense in semisynthetic or biocombinatorial approaches to preferentially reshape the chemical groups in the 3′, 4′, and 6′ positions of the hexosamine moiety in the paromamine or paromamine-like units in the future. One well-known advantage of the removal of the 3′-hydroxyl, known for some time, is that removing the 3-hydroxyl negates the effectiveness of the APH(3′) resistance mechanism which is widespread among clinically relevant bacteria. Also, another and maybe more important aspect of AGA pharmacology is avoidance of the nephro- and/or ototoxic side effects of some of the most important members of the 2DOS-AGAs.[250–252] Other targets of reshaping the structure and the changing the overall chemistry of the 2DOS-related AGA molecules by biocombinatorial strategies could be (i) methylations in other positions or (ii) the replacement of the amino groups by other groups or (iii) amino group insertions into other positions. A natural model system for suggestions like these could be the 2′-desamnination/hydroxylation process in the biosynthesis of the KAN-A component, whenever its enzymology is clarified. In other words, nature will reveal more about the possibility of creating such a natural model system and will provide us with many more similar ideas after we meticulously investigate the biosynthetic biochemistry of AGAs.

ACKNOWLEDGMENTS

We thank K. Welzel (Combinature Biopharm, AG, Berlin, Germany) for the construction and screening of the genomic cosmid libraries. We are grateful to all former and present members of our group who contributed to this work.

REFERENCES

1. Piepersberg, W. Molecular biology, biochemistry, and fermentation of aminoglycoside antibiotics. In *Biotechnology of Industrial Antibiotics*, 2nd edition; Strohl, W. R.; New York: Marcel-Dekker, 1997; pp. 81–163.
2. Piepersberg, W.; Distler, J. Aminoglycosides and sugar components in other secondary metabolites. In *Biotechnology*, 2nd ed., *Vol. 7, Products of Secondary Metabolism;* Rehm, H.-J.; Reed, G.; Pühler, A.; Stadler, P.; Series Eds.; Kleinkauf., H.; v. Döhren, H.; Volume Eds.; Weinheim: VCH-Verlagsgesellschaft, 1997; pp. 397–488.
3. Walker, J. B. *Methods Enzymol.* **1975**, *43*, 429–470.
4. Rinehart, K. L.; Jr.; Stroshane, R. M. *J. Antibiot.* **1976**, *29*, 319–353.
5. Umezawa, H.; Hooper, I. R. *Aminoglycoside Antibiotics*; Berlin: Springer-Verlag, 1982.
6. Piepersberg, W. *Crit. Rev. Biotechnol.* **1994**, *14*, 251–285.
7. Piepersberg, W.; Diaz-Guardamino, P. M.; Zhang, C. S.; Stratmann, A.; Wehmeier, U. F. Recent advances in biosynthesis and regulation of aminoglycosides. In *Microbial Secondary Metabolites*; Fierro, F.; Martin, J. F.; Eds.; Kerala, India: Research Signpost, 2002; pp. 1–26.
8. Wehmeier, U. F. *Biocat. Biotrans.* **2003**, *21*, 279–284.
9. Wehmeier, U. F.; Piepersberg, W. *Appl. Microbiol. Biotechnol.* **2004**, *63*, 613–625.
10. Newton, G. L.; Fahey, R. C. *Arch. Microbiol.* **2002**, *178*, 388–394.
11. Newton, G. L.; Ta, P.; Fahey, R. C. *J. Bacteriol.* **2005**, *187*, 7309–7316.
12. Piepersberg, W. Metabolism and cell individualization. In *Secondary Metabolites: Their Function and Evolution*, Ciba Foundation Symposium. 171; Chadwick, D.; Whelan, J., Eds.; Chichester: John Wiley and Sons, 1992; pp. 294–304.
13. Kieser, T.; Bibb, M. J.; Buttner, M. J.; Chater, K. F.; Hopwood, D. A. *Practical Streptomyces Genetics. 2000*, The John Innes Foundation.
14. Bentley, S. D.; Chater, K. F.; Cerdeno-Tarraga, A. M.; et al. *Nature* **2002**, *417*, 141–147.
15. Ikeda, H.; Ishikawa, J.; Hanamoto, A.; Shinose, M.; Kikuchi, H.; Shiba, T.; Sakaki, Y.; Hattori, M.; Omura, S. *Nat. Biotechnol.* **2003**, *21*, 526–531.
16. Distler, J.; Braun, C.; Ebert, A.; Piepersberg, W. *Mol. Gen. Genet.* **1987**, *208*, 204–210.
17. Jung, Y. G.; Kang, S. H.; Hyun, C. G.; Yang, Y. Y.; Kang, C. M.; Suh, J. W. *FEMS Microbiol. Lett.* **2003**, *219*, 285–289.
18. Lyutzkanova, D.; Distler, J.; Altenbuchner, J. *Microbiology* **1997**, *143*, 2135–2143.
19. Ikeno, S.; Tsuji, T.; Higashide, K.; Kinoshita.; N.; Hamada, M.; Hori, M. A. *J. Antibiot.* **1998**, *51*, 341–352.

20. Ikeno, S.; Aoki, D.; Sato, K.; Hamada, M.; Hori, M.; Tsuchiya, K. S. *J. Antibiot.* **2002**, *55*, 1053–1062.
21. Ikeno, S.; Yamane, Y.; Ohishi, Y.; Kinoshita, N.; Hamada, M.; Tsuchiya, K. S.; Hori, M. *J. Antibiot.* **2000**, *53*, 373–384.
22. Huang, F.; Haydock, S. F.; Mironenko, T.; Spiteller, D.; Li, Y.; Spencer, J. B. *Org. Biomol. Chem.* **2005**, *3*, 1410–1418.
23. Kudo, F.; Yamamoto, Y.; Yokoyama, K.; Eguchi, T.; Kakinuma, K. *J. Antibiot.* **2005**, *58*, 766–774.
24. Ota, Y.; Tamegai, H.; Kudo, F.; Kuriki, H.; Koike-Takeshita, A.; Eguchi, T.; Kakinuma, K. *J. Antibiot.* **2000**, *53*, 1158–1167.
25. Kudo, F.; Numakura, M.; Tamegai, H.; Yamamoto, H.; Eguchi, T.; Kakinuma, K. *J. Antibiot.* **2005**, *58*, 373–379.
26. Herbert, C. J.; Giles, I. G.; Akhtar, M. *FEBS Lett.* **1983**, *160*, 67–71.
27. Huang, F.; Li, Y.; Yu, J.; Spencer, J. B. *Chem. Commun.* **2002**, *23*, 2860–2861.
28. Kharel, M. K.; Subba, B.; Basnet, D. B.; Woo, J. S.; Lee, H. C.; Liou, K.; Sohng, J. K. *Arch. Biochem. Biophys.* **2004**, *429*, 204–214.
29. Yanai, K.; Murakami, T. *J. Antibiot.* **2004**, *57*, 351–354.
30. Kharel, M. K.; Basnet, D. B.; Lee, H. C.; Liou, K.;Woo, J. S.; Kim, B. G.; Sohng, J. K. *FEMS Microbiol. Lett.* **2004**, *230*, 185–190.
31. Unwin, J.; Standage, S.; Wellington, E. *Gentamycin Antibiotics Patent: GB (0303767.8)-A*; 18-FEB-2003; Coventry, England: The University of Warwick.
32. Kharel, M. K.; Basnet, D. B.; Lee, H. C.; Liou, K.; Moon, Y. H.; Kim, J. J.; Woo, J. S.; Sohng, J. K. *Mol. Cells.* **2004**, *31*, 71–78.
33. Schatz, A.; Bugie, E.; Waksman, S. A. *Clin. Orthop. Relat. Res.* **2005**, 3–6.
34. Waksman, S. A.; Schatz, A. *Proc. Natl. Acad. Sci. USA* **1943**, *29*, 74–79.
35. Waksman, S. A.; Reilly, H. C.; Johnstone, D. B. *J. Bacteriol.* **1946**, *52*, 393–397.
36. Waksman, S. A. *Science* **1953**, *118*, 259–266.
37. Waksman S. A. *Antimicrobial. Agents Chemother.* **1965**, *5*, 9–19.
38. Weisblum, B.; Davies, J. *Bacteriol Rev.* **1968**, *32*, 493–528.
39. Gorini, L. Streptomycin and misreading of the genetic code. In *Ribosomes*; Nomura, M.; Tissieres, A.; Lengyel, P., Eds.; New York: Cold Spring Harbor Laboratory, 1974; pp. 791–803.
40. Benveniste, R.; Davies, J. *Proc. Natl. Acad. Sci. USA* **1973**, *70*, 2276–2280.
41. Walker, J. B. *Methods Enzymol.* **1975**, *43*, 628–632.
42. Davis, B. D. *Microbiol. Rev.* **1987**, *51*, 341–350.
43. Piepersberg, W.; Distler, J.; Heinzel, P.; Perez-Gonzalez, J. A. *Actinomycetol.* **1988**, *2*, 83–98.
44. Walker J. B.; Walker M. S. *Methods Enzymol.* **1975**, *43*, 632–634.
45. Walker, J. B.; Walker, M. S. *Methods Enzymol.* **1975**, *43*, 634–637.
46. Walker, J. B. *Dev. Ind. Microbiol.* **1980**, *21*, 105–113.
47. Piepersberg, W. Streptomycin and related aminoglycosides. In *Biochemistry and Genetics of Antibiotic Biosynthesis*; Vining, L.; Stuttard, C., Eds.; Boston: Butterworth-Heinemann, 1995; pp. 531–570.
48. Egan, S.; Wiener, P.; Kallifidas, D.; Wellington, E. M. *Appl. Environ. Microbiol.* **1998**, *64*, 5061–5063.

49. Piepersberg, W. Endogenous antimicrobial molecules: An ecological perspective. In *Molecular Medical Microbiology*; Sussman, M., Ed; London: Academic Press, 2001; pp. 561–583.
50. Beyer, S.; Distler, J.; Piepersberg, W. *Mol. Gen. Genet.* **1996**, *250*, 775–784.
51. Beyer, S.; Mayer, G.; Piepersberg, W. *Eur. J. Biochem.* **1998**, *258*, 1059–1067.
52. Distler, J.; Ebert, A.; Mansouri, K.; Pissowotzki, K.; Stockmann, M.; Piepersberg, W. *Nucleic Acids Res.* **1987**, *15*, 8041–8056.
53. Heinzel, P.; Werbitzky, O.; Distler, J.; Piepersberg, W. *Arch. Microbiol.* **1988**, *150*, 184–192.
54. Stumpp, T.; Jennen, D.; Lyutzkanova, D.; Starodubtseva, L.; Altenbuchner, J. Cloning of the spectinomycin biosynthesis genes. In *Microbial Secondary Metabolites*; Fierro, F.; Martin, J. F., Eds.; Kerala, India: Research Signpost: 2002, pp. 27–41.
55. Schatz, A.; Waksman, S. A. *Proc. Natl. Acad. Sci. USA* **1945**, *31*, 129–137.
56. Waksman, S. A.; Reilly, H. C.; Harris, D. A. *J. Bacteriol.* **1948**, *56*, 259–269.
57. Chen, Y. M.; Walker, J. B. *Biochem. Biophys. Res. Commun.* **1977**, *77*, 688–692.
58. Lucher, L. A.; Chen, Y. M.; Walker, J. B. *Antimicrob. Agents Chemother.* **1989**, *33*, 452–459.
59. Horner, W. H. *J. Biol. Chem.* **1964**, *239*, 578–581.
60. Horner, W. H. *J. Biol. Chem.* **1964**, *239*, 2256–2258.
61. Horner, W. H.; Thaker, I. H. *Biochim. Biophys. Acta* **1968**, *165*, 306–308.
62. Horner, W. H.; Russ, G. A. *Biochim. Biophys. Acta* **1971**, *237*, 123–127.
63. Okuda, T.; Ito, Y. Biosynthesis and mutasynthesis of aminoglycoside antibiotics. In *Aminoglycoside Antibiotics*; Umezawa, H.; Hooper IR., Eds.; Berlin: Springer-Verlag, 1982; pp. 111–203.
64. Jackson, M.; Crick, D. C.; Brennan, P. J. *J. Biol. Chem.* **2000**, *275*, 30092–30099.
65. Bachhawat, N.; Mande, S. C. *J. Mol. Biol.* **1999**, *291*, 531–536.
66. Norman, R. A.; McAlister, M. S.; Murray-Rust, J.; Movahedzadeh, F.; Stoker, N. G.; McDonald, N. Q. *Structure* **2002**, *10*, 393–402.
67. Movahedzadeh, F.; Smith, D. A.; Norman, R. A.; Dinadayala, P.; Murray-Rust, J.; Russell, D. G.; Kendall, S. L.; Rison, S. C.; McAlister, M. S.; Bancroft, G. J.; McDonald, N. Q.; Daffe, M.; Av-Gay, Y.; Stoker, N. G. *Mol. Microbiol.* **2004**, *51*, 1003–1014.
68. Nigou, J.; Dover, L. G.; Besra, G. S. *Biochemistry* **2002**, *41*, 4392–4398.
69. Ahlert, J.; Distler, J.; Mansouri, K.; Piepersberg, W. *Arch. Microbiol.* **1997,** *168*, 102–113.
70. Distler, J.; Mansouri, K.; Mayer, G.; Stockmann, M.; Piepersberg, W. *Gene* **1992**, *115*, 105–111.
71. Pissowotzki, K.; Mansouri, K.; Piepersberg, W. *Mol. Gen. Genet.* **1991**, *231*, 113–123.
72. Candy, D. J.; Blumsom, N. L.; Baddiley, J. *Biochem. J.* **1964**, *91*, 31–35.
73. Candy, D. J.; Baddiley, J. *Biochem. J.* **1965**, *96*, 526–529.
74. Bruton, J.; Horner, W. H. *J. Biol. Chem.* **1966**, *241*, 3142–3146.
75. Bruton, J.; Horner, W. H. *Biochim. Biophys. Acta* **1969**, *184*, 641–642.
76. Grisebach, H. *Adv. Carbohydr. Chem. Biochem.* **1978**, *35*, 81–126.

77. Stockmann, M.; Piepersberg, W. *FEMS Microbiol. Lett.* **1992**, *90*, 185–190.
78. Doumith, M.; Weingarten, P.; Wehmeier, U. F.; Salah-Bey, K.; Benhamou, B.; Capdevila, C.; Michel, J. M.; Piepersberg, W.; Raynal, M. C. *Mol. Gen. Genet.* **2000**, *264*, 4774–4785.
79. Elling, L.; Rupprath, C.; Gunther, N.; Romer, U.; Verseck, S.; Weingarten, P.; Drager, G.; Kirschning, A.; Piepersberg, W. *Chembiochem.* **2005**, *6*, 1423–1430.
80. Kniep, B.; Grisebach, H. *FEBS Lett.* **1976**, *65*, 44–46.
81. Kniep, B.; Grisebach, H. *J. Antibiot.* **1980**, *33*, 416–419.
82. Kniep, B.; Grisebach, H. *Eur. J. Biochem.* **1980**, *105*, 139–144.
83. Maier, S.; Grisebach, H. *Biochim. Biophys. Acta.* **1979**, *586*, 231–241.
84. Hunter, G. D.; Hockenhull, J. D. *Biochem. J.* **1955**, *59*, 268–272.
85. Silverman, M.; Rieder, S. V. *J. Biol. Chem.* **1960**, *235*, 1251–1254.
86. Bruton, J.; Horner, W. H.; Russ, G. A. *J. Biol. Chem.* **1967**, *242*, 813–818.
87. Munro, M. H.; Taniguchi, M.; Rinehart, K. L., Jr.; Gottlieb, D.; Stoudt, D.; Rogers, T. O. *J. Am. Chem. Soc.* **1975**, *97*, 4782–4783.
88. Fernandes, P. B.; Vojtko, C. M.; Bower, R. R.; Weisz, J. *J Antibiot.* **1984**, *37*, 1525–1527.
89. Kumada, Y.; Horinouchi, S.; Uozumi, T.; Beppu, T. *Gene* **1986**, *42*, 221–224.
90. Meier-Dieter, U.; Starman, R.; Barr, K.; Mayer, H.; Rick, P. D. *J. Biol. Chem.* **1990**, *265*, 13490–13497.
91. Ohnuki, T.; Imanaka, T.; Aiba, S. *J. Bacteriol.* **1985**, *164*, 85–94.
92. Ohnuki, T.; Imanaka, T.; Aiba, S. *Antimicrob. Agents Chemother.* **1985**, *27*, 367–374.
93. Mason, D. J.; Dietz, A.; Smith, R. M. *Antibiot Chemother.* **1961**, *11*, 118–122.
94. Wiley, P. F.; Argoudelis, A. D.; Hoeksema, H. *J. Am. Chem. Soc.* **1963**, *85*, 2652.
95. Hoeksema, H.; Knight, J. C. *J. Antibiot.* **1975**, *28*, 240–241.
96. Wiley, P. F.; Argoudelis, A. D.; Hoeksema, H. *J. Am. Chem. Soc.* **1963**, *85*, 2652–2659.
97. Rosenbrook, W., Jr. *J. Antibiot.* **1979**, *32*, 211–227.
98. Karwowski, J. P.; Jackson, M.; Bobik, T. A.; Prokop, J. F.; Theriault, R. J. *J. Antibiot.* **1984**, *37*, 1513–1518.
99. McAlpine, J. B.; Brill, G. M.; Spanton, S. G.; Mueller, S. L.; Stanaszek, R. S. *J. Antibiot.* **1984**, *37*, 1519–1524.
100. Peeters, M.; Van Dyck, E.; Piot, P. *Antimicrob. Agents Chemother.* **1984**, *26*, 608–609.
101. Mitscher, L. A.; Martin, L. L.; Feller, D. R. *Chem. Commun.* **1971**, 1541–1542.
102. Stroshane, R. M.; Taniguchi, M.; Rinehart, K. L., Jr.; Rolls, J. P.; Haak, W. J.; Ruff, B. A. *J. Am. Chem. Soc.* **1976**, *98*, 3025–3027.
103. Floss, H. G.; Chang, C. J.; Mascaretti, O.; Shimada, K. *Planta Med.* **1978**, *34*, 345–380.
104. Walker, J. B. *Appl. Environ Microbiol.* **2002**, *68*, 2404–2410.
105. Liu, H. W.; Thorson, J. S. *Annu. Rev. Microbiol.* **1994**, *48*, 223–256.
106. Umezawa, H.; Okami, Y.; Hashimozo, T.; Suhara, Y.; Hamada, M.; Takeuchi, T. *J. Antibiot.* **1965**, *18*, 101–103.

107. Hamada, M.; Kondo, S.; Yokoyama, T.; Miura, K.; Iinuma, K. *J. Antibiot.* **1974**, *27*, 81–83.

108. Iinuma, K.; Kondo, S.; Maeda, K.; Umezawa, H. *J. Antibiot.* **1975**, *28*, 613–615.

109. Olesker, A.; Mercier, D.; Gero, S. D.; Pearce, C. J.; Barnett, J. E. *J. Antibiot.* **1975**, *28*, 490–491.

110. Suzukake, K.; Hori, M. *J Antibiot.* **1977**, *30*, 132–140.

111. Fukagawa, Y.; Sawa, T.; Takeuchi, T.; Umezawa, H. *J. Antibiot.* **1968**, *21*, 50–54.

112. Fukagawa, Y.; Sawa, T.; Takeuchi, T.; Umezawa, H. *J. Antibiot.* **1968**, *21*, 182–184.

113. Fukagawa, Y.; Sawa, T.; Takeuchi, T.; Umezawa, H. *J. Antibiot.* **1968**, *21*, 185–188.

114. Fukagawa, Y.; Sawa, T.; Homma, I.; Takeuchi, T.; Umezawa, H. *J. Antibiot.* **1968**, *21*, 358–360.

115. Fukagawa, Y.; Sawa, T.; Homma, I.; Takeuchi, T.; Umezawa, H. *J. Antibiot.* **1968**, *21*, 410–412.

116. Waksman, S. A.; Lechevalier, H. A.; Harris, D. A. *J. Clin. Invest.* **1949**, *28*, 934–939.

117. Lechevalier, H. A. *CRC Crit. Rev. Microbiol.* **1975**, *3*, 359–397.

118. Koster, H.; Liebermann, B.; Reuter, G. *Z. Allg. Mikrobiol.* **1975**, *15*, 437–445.

119. Hessler, E. J.; Jahnke, H. K.; Robertson, J. H.; Tsuji, K.; Rinehart, K. J., Jr.; Shier, W. T. *J. Antibiot.* **1970**, *23*, 464–466.

120. Tsunakawa, M.; Hanada, M.; Tsukiura, H.; Tomita, K.; Tomatsu, K.; Hoshiya, T.; Miyaki, T.; Konishi, M.; Kawaguchi, H. *J. Antibiot.* **1985**, *38*, 1302–1312.

121. Tsunakawa, M.; Hanada, M.; Tsukiura, H.; Konishi, M.; Kawaguchi, H. *J. Antibiot.* **1985**, *38*, 1313–1321.

122. Rinehart, K. L., Jr. *J. Antibiot.* **1979**, *32*, 32–46.

123. Yagisawa, M.; Huang, T. S.; Davies, J. E. *J. Antibiot.* **1978**, *31*, 809–813.

124. Aboshanab, K. M. Ph.D thesis, Department of Natural Science, Bergische University Wuppertal, 2005.

125. Suzukake, K.; Tokunaga, K.; Hayashi, H.; Hori, M.; Uehara, Y.; Ikeda, D.; Umezawa, H. **1985**, *38*, 1211–1218.

126. Kakinuma, K.; Ogawa, Y.; Sasaki, T.; Seto, H.; Otake, N. *J Antibiot.* **1989**, *42*, 926–933.

127. Daum, S. J.; Rosi, D.; Goss, W. *J. Am. Chem. Soc.* **1977**, *99*, 283–284.

128. Furumai, T.; Takeda, K.; Kinumaki, A.; Ito, Y.; Okuda, T. *J. Antibiot.* **1979**, *32*, 891–899.

129. Walker, J. B. *J. Bacteriol.* **1995**, *177*, 818–822.

130. Kudo, F.; Tamegai, H.; Fujiwara, T.; Tagami, U.; Hirayama, K.; Kakinuma, K. *J. Antibiot.* **1999**, *52*, 559–571.

131. Nield, B. S.; Willows, R. D.; Torda, A. E.; Gillings, M. R.; Holmes, A. J.; Nevalainen, K. M.; Stokes, H. W.; Mabbutt, B. C. *Protein Sci.* **2004**, *13*, 1651–1659.

132. Tamegai, H.; Nango, E.; Koike-Takeshita, A.; Kudo, F.; Kakinuma, K. *Biosci. Biotechnol. Biochem.* **2002**, *66*, 1538–1545.

133. Crueger, A.; Piepersberg, W.; Distler, J.; Stratmann, A. German and international patent no. Le A 30515; Bayer AG; **1995**, EP A 0730 029, DE 195 07 214.

134. Stratmann, A.; Mahmud, T.; Lee, S.; Distler, J.; Floss, HG.; Piepersberg, W. *J. Biol. Chem.* **1999**, *274*, 10889–10896.

135. Shier, W. T.; Schaefer, P. C.; Gottlieb, D.; Rinehart, K. L., Jr. *Biochemistry* **1974**, *13*, 5073–5078.
136. Pearce, C. J.; Barnett, J. E.; Anthony, C.; Akhtar, M.; Gero, S. D. *Biochem. J.* **1976**, *159*, 601–606.
137. Pearce, C. J.; Akhtar, M.; Barnett, J. E.; Mercier, D.; Sepulchre, A. M.; Gero, S. D. *J. Antibiot.* **1978**, 31, 74–81.
138. Takeda, K.; Kinumaki, A.; Furumai, T.; Yamaguchi, T.; Ohshima, S.; Ito, Y. *J. Antibiot.* **1978**, *31*, 247–249.
139. Takeda, K.; Aihara, K.; Furumai, T.; Ito Y. *J. Antibiot.* **1978**, *31*, 250–253.
140. Takeda, D.; Okuno, S.; Ohashi, Y.; Furumai T. *J. Antibiot.* **1978**, *31*, 1023–1030.
141. Takeda, K.; Kinumaki, A.; Okuno, S.; Matsushita, T.; Ito Y. *J. Antibiot.* **1978**, *31*, 1039–1045.
142. Takeda, K.; Kinumaki, A.; Hayasaka, H.; Yamaguchi, T.; Ito Y. *J. Antibiot.* **1978**, *31*, 1031–1038.
143. Takeda, K.; Aihara, K.; Furumai, T.; Ito, Y. *J. Antibiot.* **1979**, *32*, 18–28.
144. Truman, A. W.; Huang, F.; Llewellyn, N. M.; Spencer, J. B. *Angrew. Che. Int. Ed. Engl.* **2007**, *46*, 1462–1464.
145. Kudo, F.; Kawabe, K.; Kuriki, H.; Eguchi, T.; Kakinuma, K. *J. Am. Chem. Soc.* **2005**, *127*, 1711–1718.
146. Baud, H.; Betencourt, A.; Peyre, M.; Penasse, L. *J. Antibiot.* **1977**, *30*, 720–723.
147. Waksman, S. A.; Harris, D. A. *J. Clin. Invest.* **1949**, *26*, 934–939.
148. Autissier, D.; Barthelemy, P.; Mazieres, N.; Peyre, M.; Penasse, L. *J. Antibiot.* **1981**, *34*, 536–543.
149. Yukita, T.; Nishida, H.; Eguchi, T.; Kakinuma, K. *J. Antibiot.* **2003**, *56*, 497–500.
150. Li, Y.; Llewellyn, N. M.; Giri, R.; Huang, F.; Spencer J. B. *Chem. Biol.* **2005**, *12*, 665–675.
151. Umezawa, H.; Ueda, M.; Maeda, K.; Yagiashita, K.; Kondo, S.; Okami, Y.; Utahara, R.; Osato, Y.; Nitta, K.; Takeuchi, T. *J. Antibiot.* **1957**, *10*, 181–188.
152. Takeuchi, T.; Hikiji, T.; Nitta, K.; Yamazaki, S.; Abe, S.; Takayama, H.; Umezawa, H. *J. Antibiot.* **1957**, *10*, 107–114.
153. Cron, M. J.; Fardig, O. B.; Johnsohn, D. L.; Palermiti, F. M.; Schmitz, H.; Hooper, I. R. *Ann. N. Y. Acad. Sci.* **1958**, *76*, 27–30.
154. Murase, M.; Ito, S.; Fukatsu, T.; Umezawa, H. *Progr. Antimicrob. Anticancer Chemother.* **1970**, *2*, 1098–1110.
155. Claes, P. J.; Vanderhaeghe, H.; Compernolle, F. *Agents Chemother.* **1973**, *4*, 560–563.
156. Kojima, M.; Inouye, S.; Nida, T. *J. Antibiot.* **1975**, *28*, 48–55.
157. Umezawa, H. *J. Japan Med. Assoc.* **1967**, *58*, 1328–1334.
158. Murakami, T.; Nojiri, C.; Toyama, H.; Hayashi, E.; Katumata, K.; Anzai, H.; Matsuhashi, Y.; Yamada, Y.; Nagaoka, K. *J. Antibiot.* **1983**, *36*; 1305–1311.
159. Demydchuk, J.; Oliynyk, Z.; Fedorenko, V. *J. Basic Microbiol.* **1998**, *38*, 231–239.
160. Weinstein, M. J.; Lüedemann, G. M.; Oden, E. M.; Wagman, G. H. *Antimicrobial Agents Chemother.* **1964**, *10*, 24–32.
161. Cooper, D. J. *Pure Appl. Chem.* **1971**, *28*, 455–467.

162. Cooper, D. J.; Daniels, P. J., Yudis, M. D.; Marigliano, H. M.; Traubel, T. *J. Chem. Soc.* **1971**, *17*, 2876–2879.

163. Cooper, D. J.; Yudis, M. D.; Marigliano, H. M.; Guthrie, R. D.; Bukhari, S. T. *J. Chem. Soc.* **1971**, *18*, 3126–3129.

164. Cooper, D. J.; Weinstein, J.; Waitz, J. A. *J. Med. Chem.* **1971**, *14*, 1118–1120.

165. Daniels, P. J.; Weinstein, J.; Nagabhushan, T. L. *J. Antibiot.* **1974**, *27*, 889–893.

166. Daniels, P. J.; Luce, C.; Nagabhushan, T. L. *J. Antibiot.* **1975**, *28*, 35–41.

167. Weinstein, M. J.; Marquez, J. A.; Testa, R. T.; Wagman, G. H.; Oden, E. M.; Waitz, J. *J. Antibiot.* **1970**, *23*, 551–554.

168. Cooper, D. J.; Yudis, M. D.; Marigliano, H. M.; Traubel, T. *J. Chem. Soc.* **1971**, *17*, 2876–2879.

169. Wagman, G. H.; Weinstein, M. J. *Annu. Rev. Microbiol.* **1980**, *34*, 537–557.

170. Waitz, J. A.; Moss, E. L., Jr.; Oden, E. M.; Weinstein, M. J. *J. Antibiot.* **1970**, *23*, 559–565.

171. Berdy, J.; Pauncz, J. K.; Vajna, Z. M.; Horvath, G.; Gyimesi J. *J. Antibiot.* **1977**, *30*, 945–954.

172. Okachi, R.; Kawamoto, I.; Takasawa, S.; Yamamoto, M.; Sato, S. *J. Antibiot.* **1974**, 27, 793–800.

173. Nara, T.; Kawamoto, I.; Okachi, R.; Takasawa, S.; Yamamoto, M. *J Antibiot.* **1975**; *28*, 21–28.

174. Egan, R. S.; DeVault, R. L.; Mueller, S. L.; Levenberg, M. I.; Sinclair, A. C.; Stanaszek, R. S. *J. Antibiot.* **1975**, *28*, 29–34

175. Egan, R. S.; Stanaszek, R. S.; Cirovic, M.; Mueller, S. L.; Tadanier, J.; Martin, J. R.; Collum, P.; Goldstein, A. W.; De Vault, R. L.; Sinclair, A. C.; Fager, E. E.; Mitscher, L. A. *J. Antibiot.* **1977**, *30*, 552–563.

176. Nara, T.; Yamamoto, M.; Kawamoto, I.; Takayama, K.; Okachi, R.; Takasawa, S.; Sato, T.; Sato, S. *J. Antibiot.* **1977**, *30*, 533–540. Testa, R. T.; Tilley, B. C. *J. Antibiot.* **1976,** *29*, 140–146.

177. Sato, S.; Iida, T.; Okachi, R.; Shirahata, K.; Nara, E. *J. Antibiot.* **1977**, *30*, 1025–1027.

178. Weinstein, M. J.; Luedemann, G. M.; Oden, E. M.; Wagman, G. H. *Antimicrobial. Agents Chemother.* **1963**, *161*, 1–7.

179. Marquez, J. A.; Wagman, G. H.; Testa, R. T.; Waitz, J. A.; Weinstein, M. J. *J. Antibiot.* **1976**, *29*, 483–487.

180. Kelemen, G. H., Cundliffe, E.; Financsek, I. *Gene* **1991**, *98*, 53–60.

181. Taylor, H. D.; Schmitz, H. *J. Antibiot.* **1976**, *29*, 532–535.

182. Kase, H.; Odakura, Y.; Takazawa, Y.; Kitamura, S.; Nakayama, K. Biosynthesis of sagamicin and related aminoglycosides. In *Trends in Antibiotic Research. Genetics; Biosyntheses, Actions, and New Substances*; Umezawa, H.; Demain, A. L.; Hata, T.; Hutchinson, C. R., Eds.; Tokyo: Japan Antibiotcs Research Association, 1982; pp. 195–212.

183. Kuzuyama, T.; Seki, T.; Dairi, T.; Hidaka, T.; Seto, H. *J. Antibiot.* **1995**, *48*, 1191–1193.

184. Testa, R. T.; Tilley, B. C. *J. Antibiot.* **1979**, *32*, 47–59.

185. Okachi, R.; Takasawa, S.; Sato, T.; Sato, S.; Yamamoto, M.; Kawamoto, I.; Nara, T. *J. Antibiot.* **1977**, *30*, 541–551.

186. Inouye, S.; Ohba, K.; Shomura, T.; Kojima, M.; Tsuruoka, T.; Yoshida, J.; Kato, N.; Ito, M.; Amano, S.; Omoto, S.; Ezaki, N.; Ito, T.; Niida, T.; Watanabe, K. *J. Antibiot.* **1979**, *32*, 1355–1356.

187. Shomura, T.; Kojima, M.; Yoshida, J.; Ito, M.; Amano, S.; Totsugawa, K.; Niwa, T.; Inouye, S.; Ito, T.; Niida, T. *J. Antibiot.* **1980**, *33*, 924–930.

188. Ohba, K.; Tsuruoka, T.; Mizutani, K.; Kato, N.; Omoto, S.; Ezaki, N.; Inouye, S.; Niida, T.; Watanabe, K. *J. Antibiot.* **1981**, *34*, 1090–1100.

189. Okami, Y.; Hotta, K.; Yoshida, M.; Ikeda, D.; Kondo, S.; Umezawa H. **1979**, *32*, 964–966.

190. Iwasaki, A.; Itoh, H.; Mori, T. *J. Antibiot.* **1979**, *32*, 180–186

191. Deushi, T.; Iwasaki, A.; Kamiya, K.; Kunieda, T.; Mizoguchi, T.; Nakayama, M.; Itoh, H.; Mori, T.; Oda, T. *J. Antibiot.* **1979**, *32*, 173–179.

192. Deushi, T.; Nakayama, M.; Watanabe, I.; Mori, T. *J. Antibiot.* **1979**, *32*, 187–192.

193. Deushi, T.; Yamaguchi, T.; Kamiya, K.; Iwasaki, A.; Mizoguchi, T.; Nakayama, M.; Watanabe, I.; Itoh, H.; Mori, T. *J. Antibiot.* **1980**, *33*, 1274–1280

194. Deushi, T.; Iwasaki, A.; Kamiya, K.; Mizoguchi T.; Nakayama, M.; Itoh, H.; Mori, T. *J. Antibiot.* **1979**, *32*, 1061–1065.

195. Dairi, T.; Hasegawa, M. *J. Antibiot.* **1989**, *42*, 934–943.

196. Hotta, K.; Morioka, M.; Okami, Y. *J. Antibiot.* **1989**, *42*, 745–751.

197. Hotta, K.; Morioka, M.; Tohyama, H.; Okami, Y. Biosynthesis of istamycins by *Streptomyces tenjimariensis*. In *Trends in Actinomycetology in Japan*; Koyama Y., Ed; Tokyo: Society for Actinomycetes, 1989, pp. 61–64.

198. Odakura, Y.; Kase, H.; Itoh, S.; Satoh, S.; Takasawa, S.; Takahashi, K.; Shirahata, K.; Nakayama, K. *J. Antibiot.* **1984**, *37*, 1670–1680.

199. Hasagawa, M.; Dairi, T.; Ohta, T.; Hashimoto, E. *J. Bacteriol.* **1991**, *173*, 7004–7011.

200. Dairi, T.; Ohta, T.; Hashimoto, E.; Hasegawa, M. *Mol. Gen. Genet.* **1992**, *232*, 262–270.

201. Dairi, T.; Ohta, T.; Hashimoto, E.; Hasegawa, M. *Mol. Gen. Genet.* **1992**, *236*, 39–48.

202. Hasegawa, M. *Actinomycetol.* **1991**, *5*, 126–131.

203. Hasegawa, M. *Gene* **1992**, *115*, 85–91.

204. Ohta, T.; Hasegawa, M. *J. Antibiot.* **1993**, *46*, 511–517.

205. Ohta, T.; Hashimoto, E.; Hasegawa, M. *J. Antibiot.* **1992**, *45*, 289–291.

206. Ohta, T.; Hashimoto, E.; Hasegawa, M. *J. Antibiot.* **1992**, *45*, 1167–1175.

207. Dairi, T.; Yamaguchi, K.; Hasegawa, M. *Mol. Gen. Genet.* **1992**, *236*, 49–59.

208. Perzynski, S.; Cannon, M.; Cundliffe, E.; Chahwala, S. B.; Davies, J. *Eur. J. Biochem.* **1979**, *99*, 623–628.

209. Awata, M.; Satoi, S.; Muto, N.; Hayashi, M.; Sagai, H.; Sakakibara, H. *J. Antibiot.* **1983**, *36*, 651–655.

210. Kamiya, K.; Deushi, T.; Iwasaki, A.; Watanabe, I.; Itoh, H.; Mori, T. *J. Antibiot.* **1983**, *36*, 738–741.

211. Holmes, D. J.; Drocourt, D.; Tiraby, G.; Cundliffe, E. *Gene* **1991**, *102*, 19–26.

212. Du, Y.; Li, T.; Wang, Y. G.; Xia, H. *Curr. Microbiol.* **2004**, *49*, 99–107.

213. Mansouri, K.; Piepersberg, W. *Mol. Gen. Genet.* **1991**, *228*, 459–469.

214. Pittenger, R. C.; Wolfe, R. N.; Hoehn, P. N.; Daily, W. A.; McGuire, J. M. *Antibiot. Chemother.* **1953**, *3*, 1268–1278.

215. Mann, R. L.; Gale, R. M.; Van Abeele, R. F. *Antibiot. Chemother.* **1953**, *3*, 1279–1282.

216. Neuss, N.; Koch, K. F.; Molloy, B. B.; Day, W.; Huckstep, L. L.; Dorman, D. E.; Roberts, J. D. *I. Helv. Chim. Acta.* **1970**, *53*, 2314–2319.

217. Mann, R. L.; Bromer, W. W. *J. Am. Chem. Soc.* **1958**, *80*, 2714–2716.

218. Habib, S. E.; Scarsdale, J. N.; Reynolds K. A. *Antimicrob. Agents Chemother.* **2003**, *47*, 2065–2071.

219. Kondo, S.; Iinuma, K.; Naganawa, H.; Shimura, M.; Sekizawa, Y. *J. Antibiot.* **1975**, *28*, 79–82.

220. Shimura, M.; Sekizawa, Y.; Iinuma, K.; Naganawa, H.; Kondo, S. *J. Antibiot.* **1975**, *28*, 83–84.

221. Ikeda, Y.; Kondo, S.; Kanai, F.; Sawa, T.; Hamada, M.; Takeuchi, T.; Umezawa, H. *J. Antibiot.* **1985**, *38*, 436–438.

222. Shoji J.; Kozuki S.; Mayama M.; Kawamura Y.; Matsumoto K. *J. Antibiot.* **1970**, *23*, 291–294.

223. Tamura, A.; Furuta, R.; Kotani, H. *J. Antibiot.* **1975**, *28*, 260–265.

224. Inouye, S.; Shomura, T.; Watanabe, H.; Totsugawa, K.; Niida, T. *J Antibiot.* **1973**, *26*, 374–385.

225. Guerrero, M. D.; Modolell, J. *Eur. J. Biochem.* **1980**, *107*, 409–414.

226. Ganoza, M. C.; Kiel, M. C. *Antimicrob. Agents Chemother.* **2001**, *45*, 2813–2819.

227. Brown, C. M.; McCaughan, K. K.; Tate, W. P. *Nucleic Acids Res.* **1993**, *21*, 2109–2115.

228. Pfister, P.; Risch, M.; Brodersen, D. E.; Bottger, E. C. *Antimicrob. Agents Chemother.* **2003**, *47*, 1496–1502.

229. Zalacain, M.; Malpartida, F.; Pulido, D.; Jimenez, A. *Eur. J. Biochem.* **1987**, *162*, 413–418.

230. Pardo, J. M.; Malpartida, F.; Rico, M.; Jimenez, A. *J. Gen. Microbiol.* **1985**, *131*, 1289–1298.

231. Nield, B. S.; Willows, R. D.; Torda, A. E.; Gillings, M. R.; Holmes, A. J.; Nevalainen, K. M.; Stokes, H. W.; Mabbutt, B. C. *Protein Sci.* **2004**, *13*, 1651–1659.

232. Brahme, N. M.; Gonzalez, J. E.; Rolls, J. P.; Hessler, E. J.; Mizsak, S.; Hurley L. H. *J. Am. Chem. Soc.* **1984**, *106*, 7878–7883.

233. Kneidinger, B.; Marolda, C.; Graninger, M.; Zamyatina, A.; McArthur, F.; Kosma, P.; Valvano, M. A.; Messner, P. *J. Bacteriol.* **2002**, *184*, 363–369.

234. Birch, A.; Hausler, A.; Hutter, R. *J. Bacteriol.* **1990** *172*, 4138–4142.

235. Tatusov, R. L.; Fedorova, N. D.; Jackson, J. D.; Jacobs, A. R.; Kiryutin, B.; Koonin, E. V.; Krylov, D. M.; Mazumder, R.; Mekhedov, S. L.; Nikolskaya, A. N.; Rao, BS.; Smirnov, S.; Sverdlov, A. V.; Vasudevan, S.; Wolf, Y. I.; Yin, J. J.; Natale, D. A. *BMC Bioinform.* **2003**, *4*, 41.

236. Fath, M. J.; Kolter, R. *Microbiol. Rev.* **1993**, *57*, 995–1017.

237. Saier, M. H., Jr. *Microbiol. Mol. Biol. Rev.* **2000**, *64*, 354–411.

238. Yamazaki, H.; Ohnishi, Y.; Horinouchi, S. *J. Bacteriol.* **2003**, *185*, 1273–1283.

239. Horinouchi, S.; Onaka, H.; Yamazaki, H.; Kameyama, S.; Ohnishi, Y. *Actinomycetol.* **2000**, *14*, 37.

240. Choi, S. U.; Lee; C. K.; Hwang, Y. I.; Kinosita, H.; Nihira, T. *Arch. Microbiol.* **2003**, *180*, 303–307.

241. Galperin, M. Y. *BMC Microbiol.* **2005**, *5*, 35–53.

242. Widdick, D. A.; Dodd, H. M.; Barraille, P.; White, J.; Stein, T. H.; Chater, K. F.; Gasson, M. J.; Bibb, M. J. *Proc. Natl. Acad. Sci.* **2003**, *100*, 4316–4321.

243. Noller, H. F. *Nature* **1991**, *353*, 302–303.

244. Chadwick, D.; Whelan, J. *Secondary Metabolites*, Function and Evolution. Ciba Foundation Symposium 171; Chichester: John Wiley and Sons, 1992.

245. Moazed, D.; Noller, H. F. *Nature* **1987**, *327*, 389–394.

246. Carter, A. P.; Clemons, W. M.; Brodersen, D. E.; Morgan-Warren, R. J.; Wimberly, B. T.; Ramakrishnan, V. *Nature* **2000**, *407*, 340–348.

247. Moreau, N.; Jaxel, C.; Le Goffic, F. *Antimicrob. Agents Chemother.* **1984**, *26*, 857–862.

248. Fourmy, D.; Recht, M. I.; Blanchard, S. C.; Puglisi, J. D. *Science* **1996**, *274*, 1367–1371.

249. Woodcock, J.; Moazed, D.; Cannon, M.; Davies, J.; Noller, H. F. *EMBO J.* **1991**, *10*, 3099–3103.

250. Mingeot-Leclercq, M. P.; Tulkens, P. M. *Antimicrob. Agents Chemother.* **1999**, *43*, 1003–1012.

251. Mingeot-Leclercq, M. P.; Glupczynski, Y.; Tulkens, P. M. *Antimicrob. Agents Chemother.* **1999**, *43*; 727–737.

252. Kotra, L. P.; Haddad, J.; Mobashery, S. *Antimicrob. Agents Chemother.* **2000**, *44*, 3249–3256.

253. Beauclerk, A. A.; Cundliffe, E. *J. Mol. Biol.* **1987**, *193*, 661–671.

254. Cundliffe, E. *Annu. Rev. Microbiol.* **1989**, *43*, 207–233.

255. Marchler-Bauer, A.; Anderson, J. B.; Cherukuri, P. F.; DeWeese-Scott, C.; Geer., L. Y.; Gwadz, M., He, S.; Hurwitz, D. I.; Jackson, J. D.; Ke, Z.; Lanczycki, C. J.; Liebert, C. A.; Liu, C.; Lu, F.; Marchler, G. H.; Mullokandov, M.; Shoemaker, B. A.; Simonyan, V.; Song, J. S.; Thiessen, P. A.; Yamashita, R. A.; Yin, J. J., Zhang, D.; Bryant, S. H. *Nucleic Acids Res.* **2005**, *33*, D192-D196.

256. Habib, El-S. E.; Scarsdale, N.; Reynolds, K. A. *Antimicrob. Agents Chemother.* **2003**, *47*, 2065–2071.

257. Poulsen, S. M.; Kofoed, C.; Vester, B. *J. Mol. Biol.*, **2000**, *304*, 471–481.

258. Iwata-Reuyl, D. *Bioorg Chem.*, **2003**, *31*, 24–43.

259. Reader, J. S.; Metzgar, D.; Schimmel P, de Crecy-Lagard V. *J. Biol. Chem.* **2004**, *279*, 6280–61285.

260. Van Lanen, S. G.; Reader, J. S.; Swairjo, M. A.; de Crecy-Lagard, V.; Lee, B.; Iwata-Reuyl, D. *Proc. Natl. Acad. Sci. USA.* **2005**, *102*, 4264–4269.

261. Van Lanen, S. G.; Kinzie, S. D.; Matthieu, S.; Link, T.; Culp, J.; Iwata-Reuyl, D. *J. Biol. Chem.* **2003**, *278*, 10491–10499.

262. Slany, R. K.; Bosl, M.; Crain, P. F.; Kersten, H. *Biochemistry* **1993**, *32*, 7811–7817.

3

MECHANISMS OF AMINOGLYCOSIDE ANTIBIOTIC RESISTANCE

TUSHAR SHAKYA AND GERARD D. WRIGHT

Antimicrobial Research Centre, Department of Biochemistry and Biomedical Sciences, McMaster University, Hamilton, Ontario, Canada, L8N 3Z5

Aminoglycoside Antibiotics: From Chemical Biology to Drug Discovery, Second Edition, by Dev P. Arya

3.1. INTRODUCTION

Resistance to the aminoglycoside antibiotics has had a profound impact on the clinical utility and medicinal chemistry of these antibiotics. The aminoglycoside-aminocyclitols were among the first antibiotics identified during the "Golden Age" of antibiotic discovery (late 1940s–early 1960s). Their excellent characteristics as broad-spectrum antibacterial agents with desirable bactericidal activity against difficult-to-treat gram-negative bacteria and mycobacteria were counterbalanced by inherent oto- and nephrotoxicity. These negative effects were tempered by improved treatment regimens and the discovery of aminoglycoside antibiotics with improved pharmacological properties such as the gentamicins (**1**).[1] Aminoglycosides bind to the tRNA acceptor site of the ribosome where codon–anticodon interactions occurs.[2] Most aminoglycosides disrupt ribosomal discrimination of cognate codon-anticodon pairs resulting in impairment of fidelity of the genetic code and production of missense proteins, which contributes to cell death.[3,4]

The first accounts of enzyme-mediated aminoglycoside resistance were published by Umezawa's group in the 1960s,[5,6] and these reports were the harbinger for the international emergence of high-level aminoglycoside resistance that has been largely spread via R-plasmids, transposons, and integrons. This emergence of transferable aminoglycoside antibiotic resistance spurred the search and development of new natural product and semisynthetic aminoglycoside drugs that retained antimicrobial activity in the face of resistance. Thus the natural product tobramycin (**2**) and the semisynthetic derivative of kanamycin B (**3**), amikacin (**4**), were introduced in the early 1970s. Since the clinical introduction of these compounds, only a limited number of aminoglycosides were successfully introduced into the clinic, mostly in Japan, in response to the emergence of resistance. The most successful of these were isepamicin (**5**) and arbekacin (**6**).

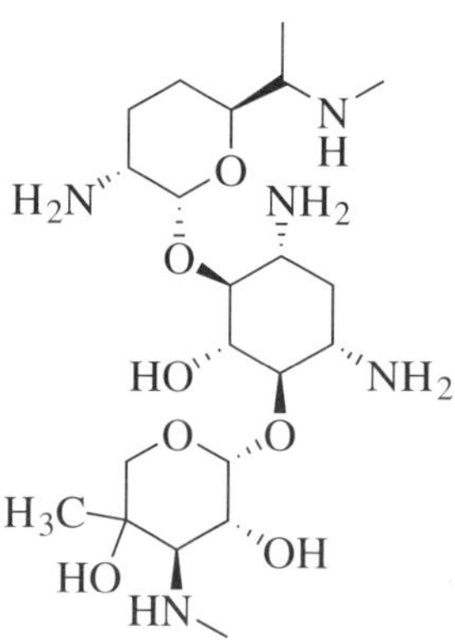

1

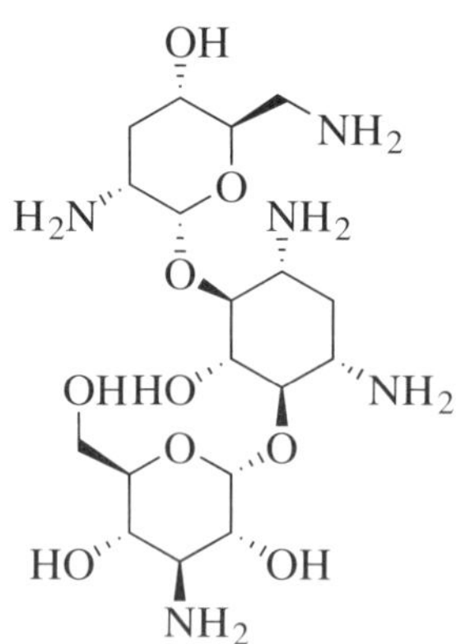

2

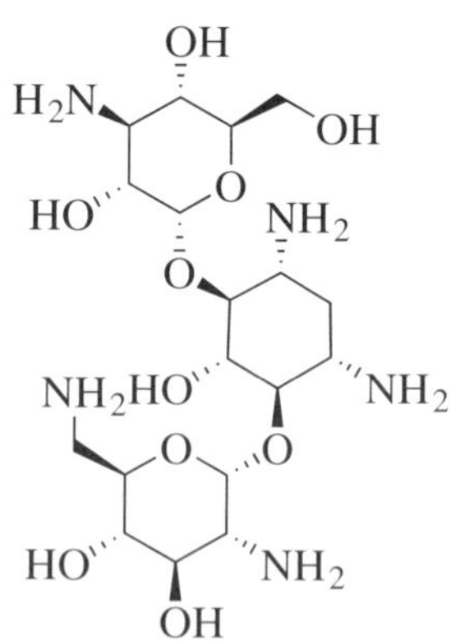

3

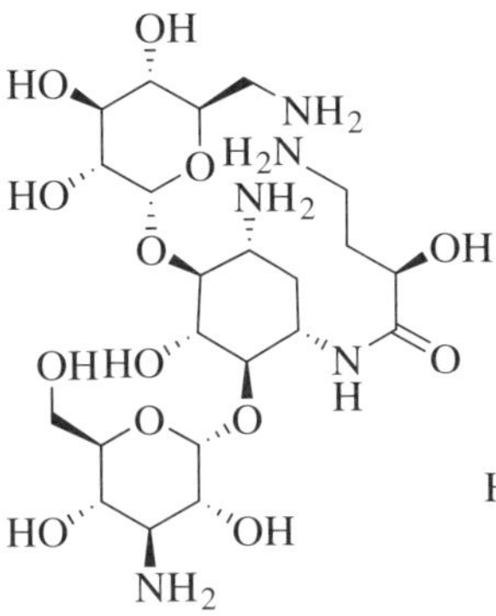

4

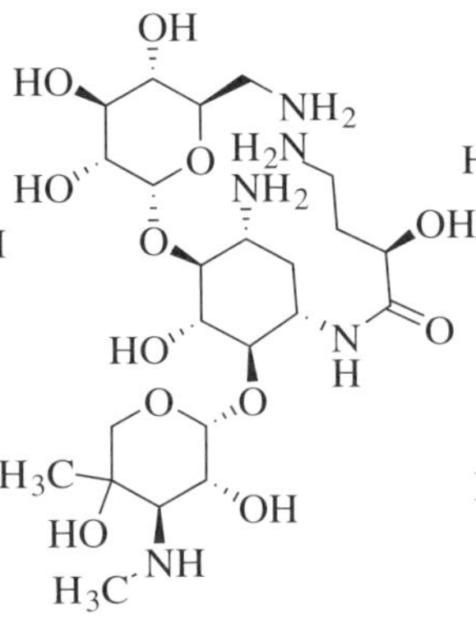

5

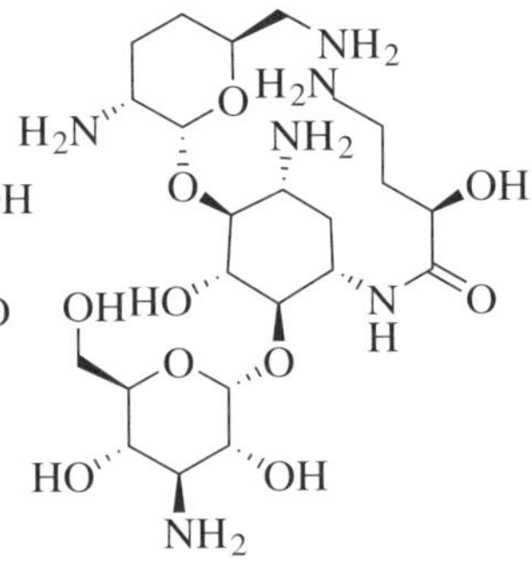

6

There have been no new aminoglycoside antibiotics brought to the clinic in North America for over two decades. One of the most important contributing factors for the decline in interest in these antibiotics has been the emergence and dissemination of resistance. Consequently, the molecular details of antibiotic resistance have been explored intensively; as a result, examples of all major classes of resistance have been well-studied. At the same time, we now know more about the molecular details of aminoglycoside inhibition of translation than ever before. These are understood at atomic resolution thanks to the determination of co-complex structures of aminoglycosides with the ribosome and model RNAs.[2,7–9] By strategic leveraging our knowledge of mode of action and resistance, the next generation of aminoglycoside antibiotics will emerge, and indeed several first efforts in this area have been reported (e.g., references 10–16).

The key to the successful deployment of the next generation of aminoglycoside antibiotics is evasion of existing resistance mechanisms. Aminoglycoside resistance occurs by three methods: (1) modification of the rRNA and ribosomal-protein targets, (2) modification of aminoglycoside transport (import and efflux), and (3) via the synthesis of aminoglycoside-modifying enzymes. The latter have been the most prevalent mechanism in most clinical isolates of resistant bacteria, but the other mechanisms are now emerging as more important especially in niche settings and organisms. In this review we survey relevant aminoglycoside clinical resistance mechanisms and focus on more recent data. Extensive review of the historical literature can be found in references.[17,21]

3.2. AMINOGLYCOSIDE RESISTANCE BY TARGET MODIFICATION

3.2.1. Mutation of Targets

Aminoglycosides interfere with translational fidelity by binding to the 30S ribosomal subunit.[22] Most aminoglycoside antibiotics of the 2-deoxystreptamine aminocyclitol class (e.g., neomycin (**7**), gentamicins (**1**), etc.), bind specifically to the 16S rRNA in the codon-decoding A-site, resulting in impairment of cognate codon–anticodon pairing (Figure 3.1A).[2,7–9] Consequently, point mutations in the 16S rRNA can result in resistance to aminoglycosides.[23] For example, for the 2-deoxystreptamine antibiotics (e.g., **1**–**7**), mutations of A1408 (*E. coli* numbering) confer high-level resistance.[24,25].

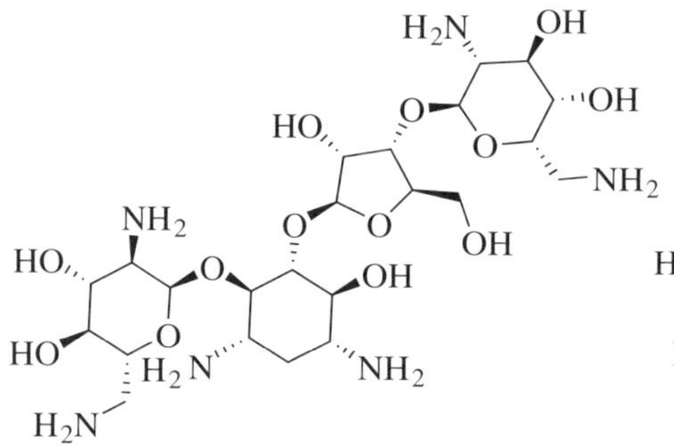

7

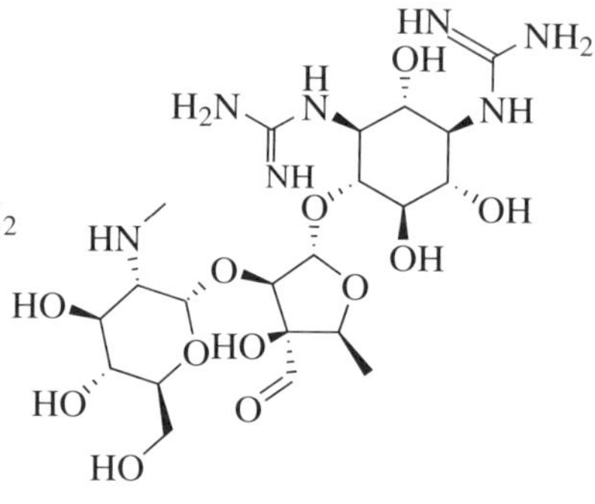

8

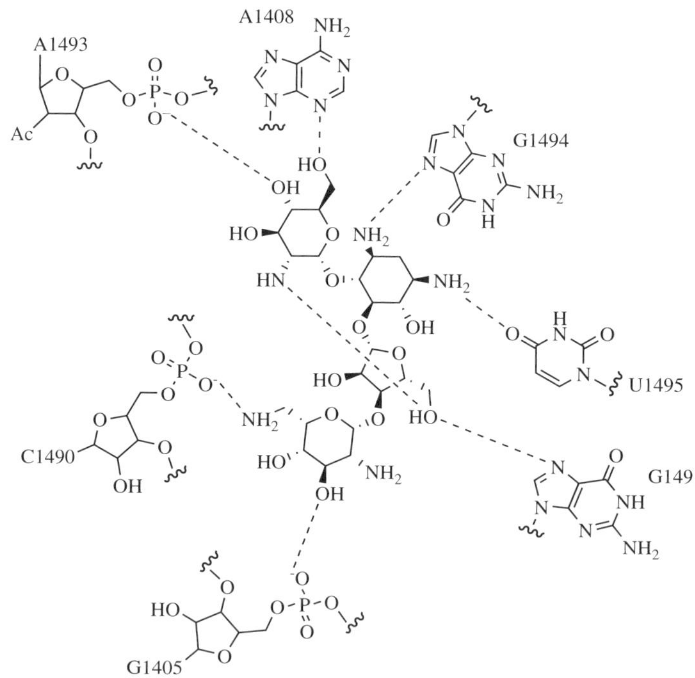

Figure 3.1A. Interaction of aminoglycosides with the acceptor site of the 30S ribosome. Schematic representation of (A) paromomycin and (B) streptomycin interactions with the 16S rRNA and ribosomal protein S12.

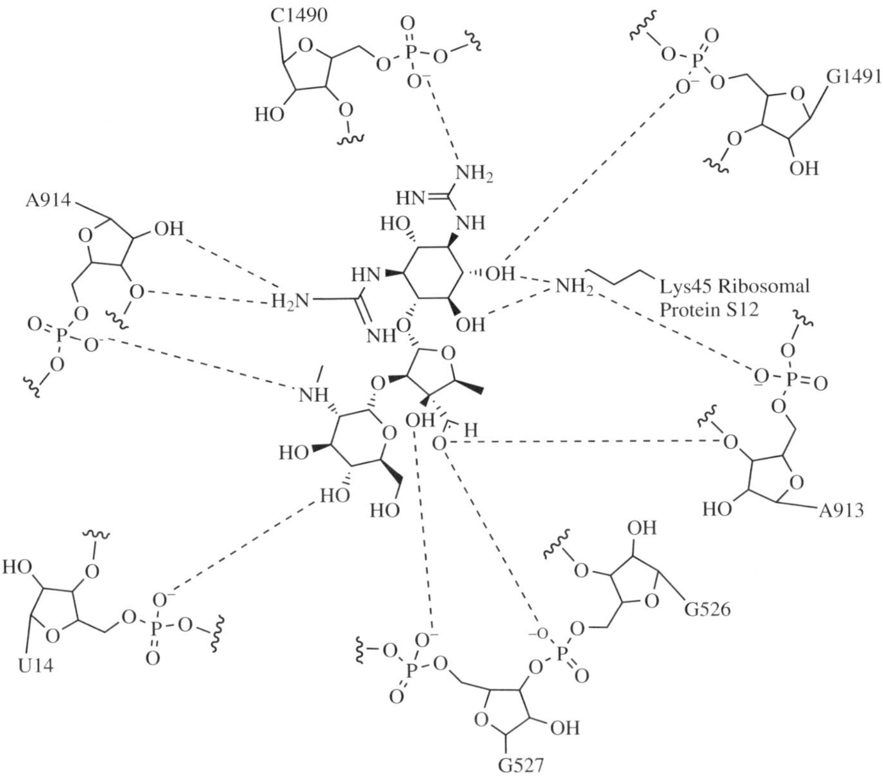

Figure 3.1B. (*continued*)

Antibiotics that incorporate the streptamine aminocyclitol such as streptomycin (**8**) also bind in the codon-decoding site, but make multiple contacts to rRNA and proteins such as S12 (Figure 3.1B). Consequently, mutations in 16S rRNA and the ribosomal protein RpsL (S12) result in high-level resistance that is clinically relevant in *Mycobacterium tuberculosis*.[26,27] Recently, an A1408G mutation, which confers high-level resistance to the 2-deoxysptreptamine antibiotics as discussed above, also was shown to result in streptomycin resistance in the thermophile *Thermus thermophilus*.[28] This is unexpected because A1408 is not proximal to the streptomycin-binding site; however, a parallel observation has been made in a streptomycin resistance mutation in ribosomal protein S4, which also does not contact streptomycin, in *Salmonella typhimurium*.[29] These studies point to the potential emergence of nonobvious target mutations that confer aminoglycoside resistance and the need for improved atomic resolution studies on ribosome-antibiotic structures and associated dynamics to interpret the molecular basis for the emergence of aminoglycoside resistance.

3.2.2. Enzyme-Catalyzed Target Modification

High-level resistance in aminoglycoside producers is frequently the result of ribosomal methylation. The gentamicin producer *Micromonospora purpurea*[30] and the tobramycin producer *Streptomyces tenebrarius*,[31] for example, encode S-adenosylmethionine-dependent methyltransferases that modify G1405 or A1408 of the 16S rRNA to the 7-methyl derivative.[32] The rRNA methylation typically confers very-high-level resistance to aminoglycosides (MICs > 0.5 mg/ml) and until recently was confined to nonpathogenic actinomycetes. In 2003, the first description of a 16S rRNA methyltransferase, RmtA, conferring resistance to 4,6-disubstituted 2-deoxystreptamine aminoglycosides in a clinical isolate of *Pseudomonas aeruginosa* was reported.[33] This was followed by reports of similar genes, termed *arm*, from *Serratia marcescens*,[34] *Klebsiella pneumoniae*,[35] and *Escherichia coli*.[36] The genes in *Enterobacteriaceae* are encoded on transposons[37,38] and in *Pseudomonas* on R-plasmids[39] facilitating dissemination. This mechanism will therefore likely continue to spread to other strains.

3.3. AMINOGLYCOSIDE RESISTANCE BY ALTERED TRANSPORT

3.3.1. Import Mechanisms

Aminoglycosides must traverse the plasma membrane and, in the case of gram-negative bacteria, the outer membrane to gain access to the target ribosomes. Transport across the plasma membrane has been shown to require the proton motive force, and mutants deficient in electron transport chain components fail to transport aminoglycosides and are consequently resistant.[40–43]

Inactivation of the outer-membrane porin protein OprH in *Pseudomonas aeruginosa* has been reported to confer gentamicin resistance.[44] However, this is likely the result of a polar effect on the downstream two-component regulatory system *phoP–phoQ*. These genes are global regulators of a number of genes including some that are involved in lipopolysaccharide modification, a known modulator of aminoglycoside entry.[45,46]

3.3.2. Export Mechanisms

Aminoglycoside efflux is a significant mechanism of aminoglycoside resistance in bacteria of the genera *Pseudomonas, Burkholderia*, and *Stenotrophomonas*. There are five classes of transmembrane efflux systems associated with antibiotic resistance; however, the resistance nodulation division (RND) family is the predominant class (Table 3.1).[47]

The RND family of efflux pumps are restricted to gram-negative organisms[48] and consists of three major components: the RND pump, a periplasmic membrane fusion protein (MFP), and an outer-membrane factor (OMF). Together these three proteins are able to effectively pump a variety of antibiotics, dyes, and ions depending on the type of RND pump associated with the system (Figure 3.2). The

TABLE 3.1. Aminoglycoside Efflux-Mediated Resistance Systems

Organism	Efflux Components	Super Family	Reference
E. coli	AcrAD-TolC	RND	50
	MdfA	MF	51
B. pseudomallei	AmrAB-OprA	RND	49
	BpeAB-OprB	RND	52
P. aeruginosa	MexXY-OprM	RND	53
	MexAB-OprM	RND	54
	EmrE	SMR	55
Lactococcus lactis	LmrA	ABC	56
Acinetobacter baumannii	AdeAB-AdeC	RND	57
S. maltophilia	SmeAB-SmeC	RND	58

first aminoglycoside efflux system, AmrAB-OprA, was reported in *Burkholderia pseudomallei*[49]; since then, several others have been reported in *E. coli* and *Pseudomonas aeruginosa* (Table 3.1).

P. aeruginosa is an opportunistic pathogen and commonly associated with nosocomial (hospital-acquired) infections; it is a major cause of morbidity in burn victims and cystic fibrosis. Because of its multidrug resistance (MDR) profile, effective treatment of *P. aeruginosa* is challenging. The MDR phenotype of this pathogen can be attributed to a series of efflux pumps found in *P. aeruginosa.* The first such pump found in *P. aeruginosa* was the MexAB-OprM system.[54] This system's ability to mobilize aminoglycosides was restricted to low-ionic-strength media[55] and, as a result, is likely not the principal source of aminoglycoside resistance in *P. aeruginosa*.

The *P. aeruginosa* MexXY system is able to effectively pump out a broad range of aminoglycosides resulting in pan-aminoglycoside resistance.[53] The genes *mexXY* are collocated in an operon with *mexZ*, which expresses a negative regulator of *mexXY*. It was originally thought that the OMF associated with MexXY was OprM; however, several other studies have found evidence to show that if the OMFs OprG, OprH, and OprI are not expressed, then the MIC against various aminoglycosides can drop 4- to 16-fold when compared to the wild-type parent strain.[59] Although MexXY is able to function with all the aforementioned OMFs, it is still unclear which one(s) predominates.

The AcrAD–TolC system in *E. coli* is homologous to MexXY–OprM and confers efflux-mediated resistance to aminoglycosides.[50] There are a number of AcrD homologs in other members of the *Enterobacteriaceae*, suggesting that this pan-aminoglycoside resistance system may exist in other gram-negative organisms. The majority of RND pumps like MexAB–OprM and AcrAB–TolC have a multidrug resistance profile and act more effectively against lipophilic and amphiphilic substrates. As a result, the hydrophilic aminoglycosides are poor substrates for these pumps. The substrate specificity in these systems is dictated by two large periplasmic loops, as demonstrated by chimeric AcrB and AcrD

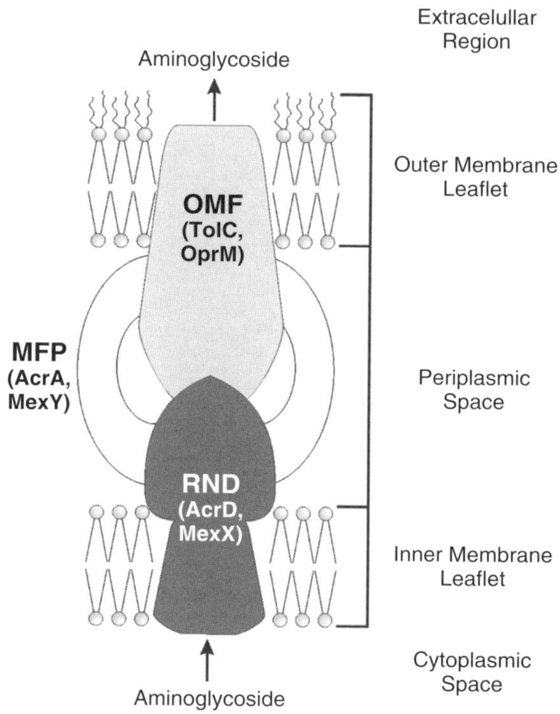

Figure 3.2. A representative RND efflux pump system. This system is comprised of three essential components; The RND pump (e.g., AcrD and MexX) is located on the cytoplasmic membrane and is responsible for the recognition of substrates in the cytosol, including aminoglycosides, and moving them into the periplasmic space. The membrane fusion protein (MFP) (e.g., AcrA and MexY) is responsible for moving the substrate across the periplasm; the final component is the outer membrane factor (OMF) (e.g., TolC and OprM) that provides a conduit for the substrate to the extracellular region of the cell.

proteins,[60] perhaps providing an opportunity to selectively target members of class with small-molecule inhibitors.

3.4. AMINOGLYCOSIDE RESISTANCE BY ENZYMATIC MODIFICATION

The aminoglycoside modifying enzymes are the most prevalent and clinically relevant mechanism of aminoglycoside antibiotic resistance. They are classified into three major families: the aminoglycoside *N*-acetyltransferases (AACs), the aminoglycoside *O*-nucleotidylyltransferases (ANTs), and the aminoglycoside *O*-phosphotransferases (APHs) (Figure 3.3). These enzymes act by acetylating, adenylylating, or phosphorylating their target aminoglycosides on selected amino or hydroxyl groups using either acetylCoA or ATP as co-substrates.

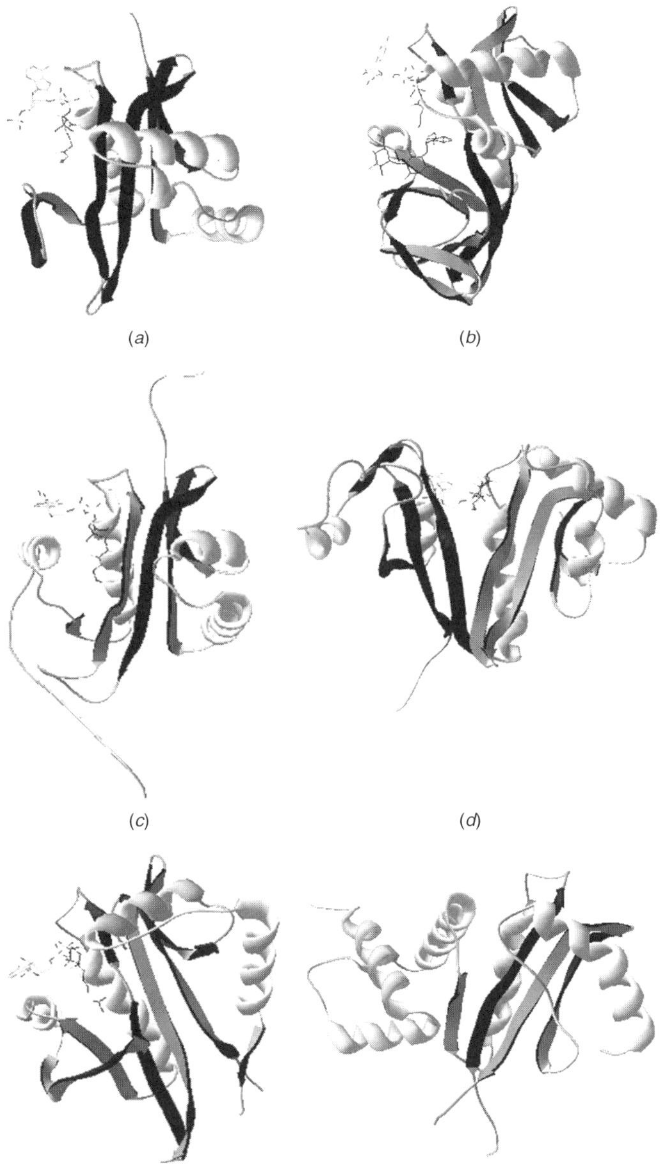

Figure 3.3. AACs are members of the GNAT superfamily. Aminoglycoside acetyltransferase belongs to the GNAT superfamily of acyltransferase enzymes that are characterized by a common acylCoA binding fold. (A) AAC(3)-Ia from *Serratia marcesans* (pdb ID# 1B04), (B) AAC(2′) from *Mycobacterium tuberculosis* (pdb ID# 1M4I), (C) AAC(6′)-Iy from *Salmonella enterica* (pdb ID# 1S60), (D) AAC(6′)-Ii from *Enterococcus faecium* (pdb ID# 2A4N), (E) Gna1 from *Saccharomyces cerevisiae* (pdb ID# 1I12) F) Hat1 (res 142–317) from *Sacchromyces cerevisiae* (pdb ID# 1BOB). The monomeric units of the aforementioned enzymes retain the central motif of 3–4 β-sheets flanked by α-helices.

The susceptible chemical groups are also frequently essential for aminoglycoside binding to the ribosome. For example, the 6′-amino group of the 2-deoxystreptamine aminoglycosides contacts A1408, and modification by acetylation both alters charge and results in a steric block for productive complex formation (Scheme 3.1).

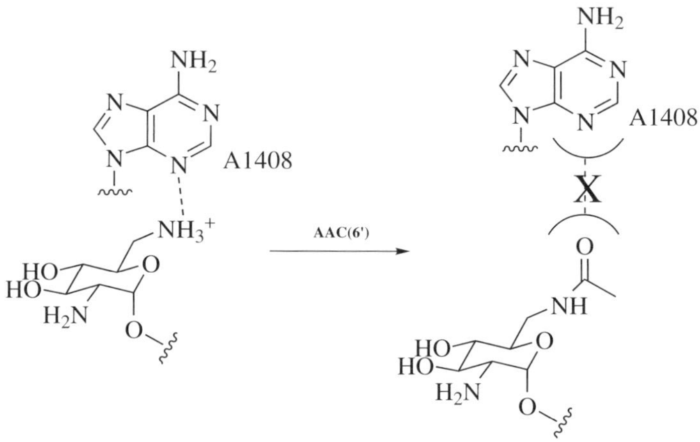

Scheme 3.1.

The large number of these enzymes and associated genes discovered over the past 40 years has necessitated the establishment of a standardized nomenclature,[61] which has been widely adopted. Using this systematic approach, the aminoglycoside resistance enzyme APH(3′)-IIIa is deciphered as "APH," which refers to the enzyme family (aminoglycoside kinase in this case), "(3′)" is the regiospecific site of modification on the antibiotic, "III" refers to the resistance phenotype (i.e., the individual aminoglycosides modified), and "a" is the specific gene.[61] A partial list of known enzymes with their aminoglycoside resistance profiles is provided in Table 3.2.

Aminoglycoside-modifying enzymes have become widespread throughout bacterial communities because many of the encoding genes are found on mobile genetic elements such as integrons, transposons, and plasmids, thus facilitating horizontal gene transfer. Many organisms also harbor chromosomally encoded aminoglycoside resistance genes such as *Enterococcus faecium*, which encodes a ubiquitous gene encoding the aminoglycoside acetyltransferase AAC(6′)-Ii.[62] As a result of this widespread gene dissemination, there are no clinically important bacterial species that have not been associated with aminoglycoside resistance.

The emergence of aminoglycoside-inactivating enzymes has contributed to the diminished use of several aminoglycoside antibiotics. For example, the wide distribution of APH(3′) enzymes have effectively made kanamycin ineffective in clinical settings. With the emergence of genes encoding ANT(2″), AAC(3), and APH(2″), gentamicin, one the few remaining aminoglycosides still in broad

TABLE 3.2. Aminoglycoside-Modifying Enzymes and Their Resistance Profiles

Enzyme		Substrates	
		Aminoglycosides[a]	Donor
***N*-Acetyltransferases**			
AAC(6′)-	I(a-d,e[1],f-z)	T, A, N, D, S, K, I	AcCoA
	II	T, G, N, D, S, K	AcCoA
AAC(3)-	I(a-b)	G, S, F	AcCoA
	II(a-c)	T, G, N, D, S	AcCoA
	III(a-c)	T, G, D, S, K, N, P, L	AcCoA
	IV	T, S, N, D, S, A	AcCoA
	VII	G	AcCoA
AAC(1)-		P, L, R, AP	AcCoA
AAC(2′)-	Ia	T, S, N, D, Ne	AcCoA
***O*-Nucleotidyltransferases**			
ANT(2″)-	I	T, G, D, S, K	ATP
ANT(3′)-	I	St, Sp	ATP
ANT(4′)-	Ia	T, A, D, K, I	ATP
	IIa	T, A, K, I	ATP
ANT(6′)-	I	St	ATP
ANT(9′)-	I	Sp	ATP
***O*-Phosphotransferases**			
APH(3′)-	I	K, Ne, L, P, R	ATP
	II	K, Ne, B, P, R	ATP
	III	K, Ne, L, P, R, B, A, I	ATP
	IV	K, Ne, B, P, R	ATP
	V	Ne, P, R	ATP
	VI	K, Ne, P, R, B, A, I	ATP
	VII	K, Ne	ATP
APH(2″)-	Ia[b]	K, G, T, S, D	ATP
	I(b,d)	K, G, T, N, D	ATP
	Ic	K, G, T	ATP
APH(3″)-	I(a-b)	St	ATP
APH(7″)-	Ia	H	ATP
APH(4)-	I(a-b)	H	ATP
APH(6)-	I(a-d)	St	ATP
APH(9)-	I(a-b)	Sp	ATP

[a]A, amikacin; Ap, apramycin; B, butirosin; D, dibekacin; G, gentamicin; H, hygromycin; I, isepamicin; K, kanamycin; L, lividomycin; N, netilmicin; Ne, neomycin; P, paromomycin; R, ribostamycin; S, sisomicin; Sp, spectinomycin; St, streptomycin; T, tobramycin.
[b]From the bifunctional enzyme AAC(6′)–APH(2″).
Adapted from references 19, 21, and 61.

clinical use is beginning to lose its efficacy as well.[63] Furthermore, these resistance genes are frequently clustered with other aminoglycoside resistance genes in gram-negative bacteria, conferring pan-aminoglycoside resistance in a single species.[64] In gram-positive bacteria, the emergence of AAC(6′)-APH(2″), a bifunctional enzyme that is able to modify almost every 2-deoxystreptamine aminoglycoside known, is also a significant concern. As a result of the emergence of these resistance mechanisms, the clinical utility of aminoglycosides has been challenges over the last few decades.

3.4.1. Aminoglycoside N-Acetyltransferases (AACs)

The AAC family of enzymes is composed of four major subclasses based on regiospecificity of aminoglycoside acetyltransfer: AAC(6′), AAC(2′), AAC(1), and AAC(3) (Figure 3.3). These enzymes are found in both gram-positive and gram-negative bacteria and generally present a very broad aminoglycoside resistance profile (Table 3.1). The AACs form the largest family of aminoglycoside resistance enzymes and are dominated by the large number of AAC(6′) homologs, many of which are chromosomally encoded (Table 3.2).

Four AAC enzymes representing three of the regiospecific subclasses have been crystallized and their 3D structures determined; these are AAC(3)-Ia from *Serratia marcesans*,[65] AAC(6′)-Ii from *Enterococcus faecium*,[66] AAC(6′)-Iy from *Salmonella enterica*,[67] and AAC(2′)-Ic from *Mycobacterium tuberculosis*.[68] The structures of these enzymes revealed not only their similarity to one another, but their affiliation with the GCN5-related *N*-acetyltransferases (GNAT) superfamily.[69] A signature element of the members of this family is a general lack of both amino acid sequence similarity and invariant catalytic residues. Instead, the family is defined by a common 3D-folding pattern formed around an acylCoA-binding pocket. Not surprisingly then, the AAC enzymes share little amino acid sequence homology with other members of the GNAT superfamily such as the histone acetyltransferases. They do, however, share the folding motifs that define this family: an N-terminal α-helix, a central antiparallel β-sheet, and a four-stranded mixed β-sheet flanked by two α-helices at the C-terminus (Figure 3.3).

The GNAT common fold also translates into a tolerance for other acyl-accepting substrates, and AACs have been determined to modify aminoglycosides and other substrates. For example, in addition to aminoglycosides, AAC(6′)-Ii has been shown to be able to acetylate histones, myelin basic protein, and ribonuclease A, all protein substrates of other GNAT enzymes.[66] Similarly, AAC(6′)-Iy from *S. enterica* also has protein acetyltransferase activity,[67] and the proposed biological substrate for AAC(2′)-Ic from *M. tuberculosis* has been implicated in the synthesis of the cellular reducing agent mycothiol.[68] These chromosomally encoded AACs may in fact have had another intended purpose within the cell and evolved the ability to modify aminoglycosides, or perhaps the ability to acylate these antibiotics may be serendipitous. In support of this hypothesis, the binding of aminoglycosides to AACs is nonuniform and can require the intermediacy of water molecules.[67,69] The aminoglycoside-binding pocket is therefore quite plastic and able to accommodate numerous substrates.

The continued evolution of the class is exemplified by the recent report of the ability of a variant of AAC(6′)-Ib, AAC(6′)-Ib-cr, to modify the synthetic fluoroquinolone antibiotic ciprofloxacin (Scheme 3.2.).[70] Substrate plasticity may be a general hallmark of the GNAT class and this could help to explain the plethora of these genes in clinical strains and in bacterial chromosomes alike.

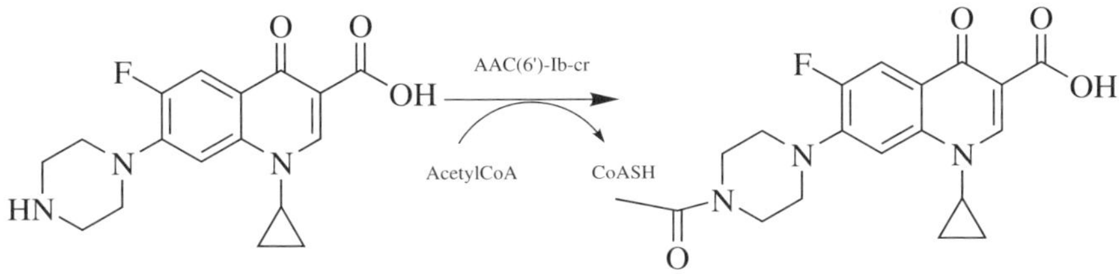

Scheme 3.2.

3.4.2. Aminoglycoside O-Nucleotidyltransferases (ANTs)

There are five members of the ANT family of enzymes, ANT(2″), ANT(4′), ANT(3′), ANT(6′) and ANT(9) (Table 3.2). ANT(2″)-Ia is a highly prevalent cause of gentamicin resistance in North America and, as a result, is one of the most clinically significant members of this family of enzymes,[71] and it is widespread among gram-negative bacteria. The *ant(2″)-Ia* gene has been found on small nonconjugative plasmids, conjugative plasmids, transposons and integrons. The enzyme follows a Theorell–Chance kinetic mechanism in which the Mg–ATP complex first binds to ANT(2″)-Ia followed by the binding of the aminoglycoside, leading to the transfer of the adenyl moiety to the aminoglycoside. The rate-limiting step of this reaction is the release of the adenylylated aminoglycoside.[72–74] There is currently no 3D structure of this enzyme; however, the NMR structure of the aminoglycoside substrate isepamicin (**5**) bound to the enzyme together with a Cr–ATP complex has been reported.[75] The staphylococcal enzyme ANT(4′)-Ia has been crystallized, and the structure has been solved in the presence of substrates.[76] The X-ray structure of this enzyme revealed it to be an obligate dimer, and each subunit interacts with the substrates in an asymmetric fashion. The signature sequence $GSX_{10-12}D/EXD/E$ is located at the N-terminus of ANT enzymes, which are conserved among many NMP-transferases and are mainly responsible for ATP binding. One subunit contributes the residues required for binding ATP and Mg^{2+}, and the other subunit provides Lys149 required for transfer the adenyl moiety to the target aminoglycoside. Both subunits are involved in aminoglycoside binding, and the intersection of the two subunits generates a negatively charged patch on the surface of ANT(4′)-Ia that likely is involved in recruiting the positively charged aminoglycoside to the active-site-binding pocket of the enzyme. The structure of this enzyme also bears significant structural homology to NMP-transferases, specifically rat DNA-polymerase β[77] Figure 3.4), another nucleotidylyltransferase.

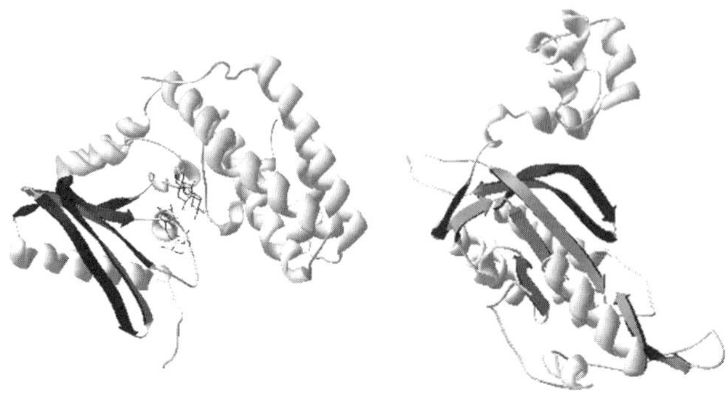

(*a*) (*b*)

Figure 3.4. ANT(4′) from *Staphylococcus aureus* shares structural homology with Rat DNA polymerase β. (A) Monomeric subunit of ANT(4′) from *Staphylococcus aureus* (pdb ID# 1KNY), (B) Rat DNA polymerase β (pdb ID# 1BOB).

3.4.3. Aminoglycoside O-Phosphotransferases (APHs)

The third mechanism of aminoglycoside modification is ATP-dependent phosphorylation of key hydroxyl groups, a reaction catalyzed by the APH family of kinases including APH(2″), APH(3′), APH(3″), APH(7″), APH(4), APH(6), and APH(9) (Table 3.2). The best-studied member of the family is APH(3′)-IIIa, which is generally found in gram-positive pathogens. The enzyme has been crystallized and the 3D structure has been determined bound with nucleotide and aminoglycoside substrates.[78–80] The monomeric subunit of this dimeric protein is structurally very similar to the serine/threonine/tyrosine protein and the inositide kinases that are common to eukaryotic signaling pathways such as murine cAMP-dependent protein kinase and casein kinase (Figure 3.5). Although these prokaryotic and eukaryotic counterparts share less than 5% amino acid sequence homology, they share almost a 40% structural identity as well as many short sequence motifs with protein kinases.[78] The details of the molecular mechanism of phosphate transfer and substrate recognition have been well worked out,[81–87] and this has resulted in efforts to identify small molecule inhibitors of the class. A second homolog, APH(3′)-IIa, has also been crystallized and resolved to 2.1-Å resolution as well, and it shares extensive structural similarities with APH(3′)-IIIa.[88]

Like the protein and inositide kinases, the 3D structures of APHs display two distinct domains: The N-terminal β-sheet region is responsible for ATP binding, and the α-helical C-terminal provides the aminoglycoside recognition site. The active site, where phosphate transfer occurs, lies at the interface of the two domains. APHs also contain the H-G/N-D-XXXX-N sequence motif, which is common among protein kinases and involved in phosphate transfer catalysis. Structural homology to protein kinases is extended to function, as it has been demonstrated that APHs have weak but measurable protein kinase activity.[89]

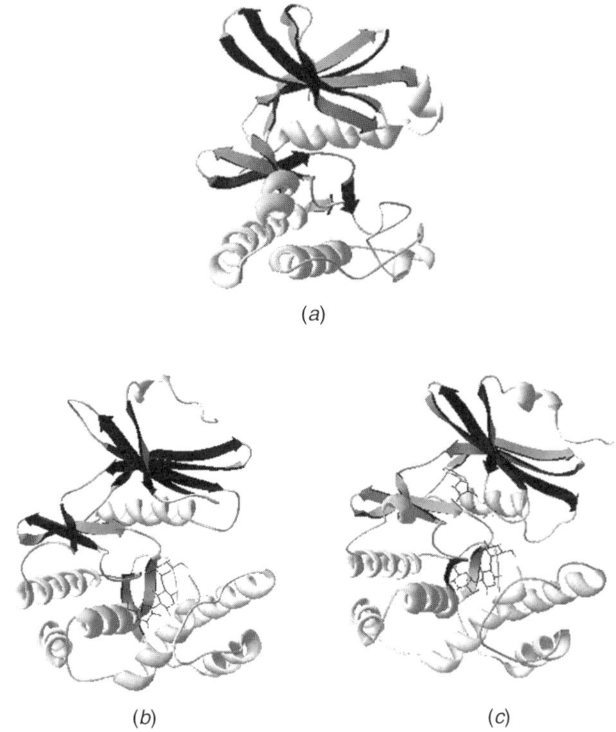

Figure 3.5. Aminoglycoside kinases are members of the protein kinase family. Comparison of c-AMP-dependent protein kinase (cAPK) to APH(3′)-IIa and APH(3′)-IIIa. (A) cAPK (catalytic domain) from *Mus musculus* (pdb ID# 2CPK). (B) APH(3′)-IIa from *Klebsiella pneumoniae* (pdb ID# 1ND4). (C) APH(3′)-IIIa from *Enterococcus faecalis* (pdb ID# 1L8T).

Furthermore, APHs can be inhibited by small molecules well known to inhibit protein and inositide kinases such as wortmanin (**9**).[90,91] It is reasonable to speculate that APHs evolved from protein or inositide kinases, perhaps in an aminoglycoside producer where such signaling proteins are well known.[92,93]

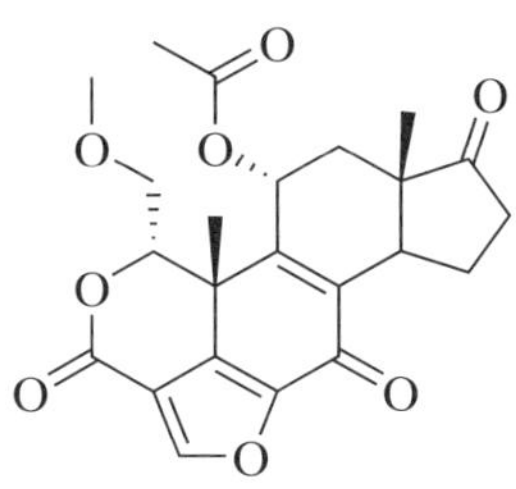

9

3.5. OVERCOMING AMINOGLYCOSIDE RESISTANCE

The impact of aminoglycoside resistance on the use of these bactericidal antibiotics has been significant. Resistance has marginalized these highly successful drugs, and they have been eclipsed by other classes less prone to preexisting resistance and with improved pharmacological properties. As a result, there have been no new additions to the class for a number of years. The emergence of resistance to other antibiotic classes, the value of bactericidal agents in the treatment of immune compromised patients, the availability of atomic resolution detail of the molecular mode of aminoglycoside binding at the ribosomes, and the unparalleled understanding of the molecular mechanism of resistance have coalesced in recent years to offer an opportunity to revisit this class of drug in the design of new chemotherapeutic strategies. Two such approaches are inhibition of resistance and "resistance-proof" aminoglycosides.

3.5.1. Inhibition of Aminoglycoside Resistance Enzymes

A small-molecule inhibitor of an aminoglycoside-inactivating enzyme could "reverse" resistance by blocking the activity of the resistance enzyme and thereby rescue antibiotic activity in the face of resistance. This strategy has been highly successful in the β-lactam antibiotic field, where inhibitors of β-lactamases in combination with β-lactam antibiotics are proven blockbuster drugs, potentiating antibiotic activity in resistant organisms.[94] Two decades ago, the inhibitory activity of 7-hydroxytropolone (**10**) on ANT(2″) was reported and shown to potentiate aminoglycosides in resistant strains.[95] As noted above, the similarity between protein and inositide kinases and APHs includes sensitivity to protein kinases inhibitors, though thus far these are neither sufficiently specific nor biologically active to demonstrate that these enzymes are viable targets for antibiotic potentiating agents. Similarly, highly potent bi-substrate inhibitors such as compound **11** (K_i of 76 nM versus AAC(6′)-Ii) of AAC(6′) have been reported but no biological activity was reported[96]; the potency of these inhibitors is encouraging, however.

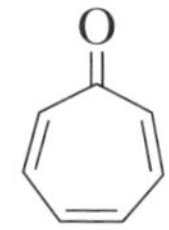

10

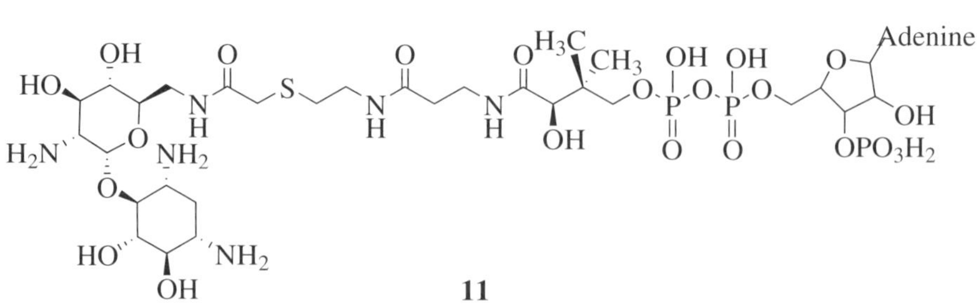

11

A challenge in the design of aminoglycoside potentiating agents is the large number of resistance mechanisms that exist in the clinic. Regional chemotherapeutic practice[71] and the fact that the distribution of resistance genes is not ubiquitous (e.g., certain resistance genes such as *aac(6′)-Ii* in *E. faecium* are associated with only one species) means that the type of aminoglycoside resistance can often be predicted and the genes encountered in relevant organisms are in fact restricted despite the daunting data in Table 3.2. Nonetheless, the ideal small-molecule inhibitor of aminoglycoside resistance would have activity against multiple enzymes and classes (AAC, APH, ANT).

One common structural element in all known aminoglycoside resistance proteins is a highly negatively charged binding site to accommodate the positively charged antibiotic. A series of cationic peptides were therefore screened against a number of aminoglycoside-inactivating enzymes, and peptides with inhibitor activity against several enzymes from different classes were identified, demonstrating that it is possible to inhibit multiple enzymes with one molecule.[97] These results are encouraging and suggest that a pan-enzyme inhibition strategy is achievable by focusing on the aminoglycoside-binding pocket.

3.5.2. "Resistance-Proof" Aminoglycosides

While the number of aminoglycoside antibiotic resistance strategies and associated genes means that truly "resistance-proof" aminoglycosides are likely to remain a fiction, there is great precedent for identifying aminoglycosides with improved activity against resistance strains, and recent semisynthetic approaches have significantly expanded the chemical diversity of the class with opportunities to explore new activities. For example, the natural products tobramycin (**2**) and gentamicin (**1**) lack 3′-hydroxyl groups and are thus not susceptible to APH(3′) enzymes. The semisynthetic compound amikacin (4-amino-2-hydroxybutrylamide derivative of kanamycin (**4**)), which was inspired by the natural product butirosin, has improved activity against aminoglycoside resistant-strains.

Recent efforts to expand chemical diversity have been quite successful. For example, a series of neamine dimmers such as **12** were shown to maintain good antibiotic activity and inhibited APH(2″) activity.[13] The Chang group has created a new class of semisynthetic derivatives of neamine, termed the pyranmycins (e.g., **13, 14**), with excellent antibiotic activity including against resistant strains.[14,98] This same group has also prepared libraries of kanamycin derivatives, many with excellent bioactivity, demonstrating the ability to greatly expand the chemical space of these compounds.[16] The Mobashery lab has used structure-based drug design to rationally designed semisynthetic antibiotic such as **15** using the crystal structure of the 30S ribosomal subunit as a guide.[12] Similarly, Bastida and colleagues compared the binding of aminoglycosides to resistance enzymes and 16S rRNA *in silico* to design conformationally restricted aminoglycosides that would favor rRNA binding over resistance enzymes such as compound **16**, which retained antibiotic activity even in the presence of resistance genes.[10]

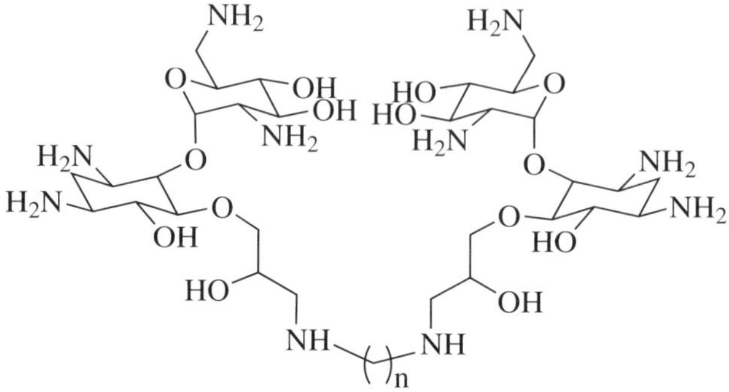

12

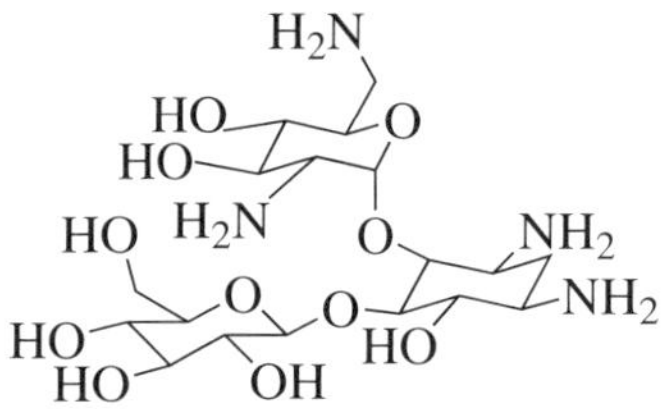

13

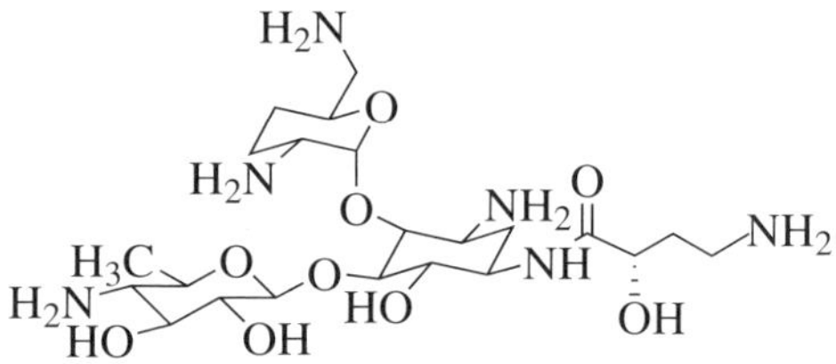

14

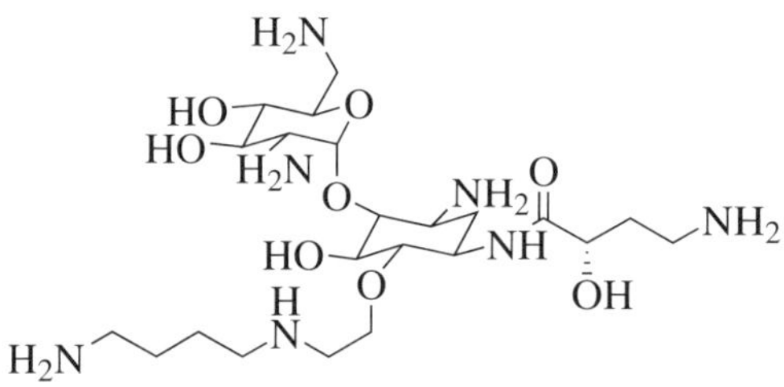

15

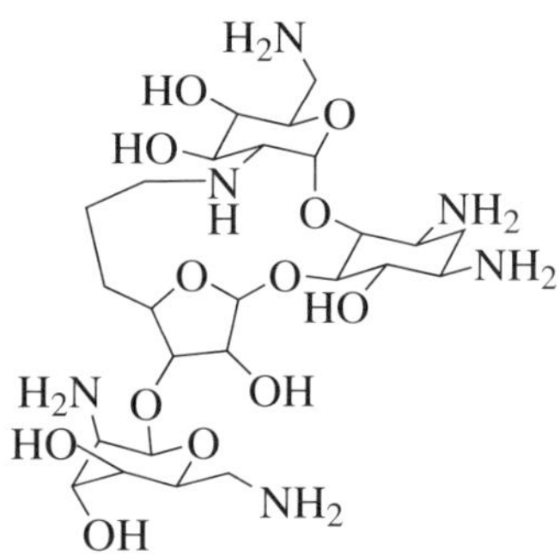

16

These selected studies demonstrate that carefully designed aminoglycosides can maintain antibiotic activity despite the presence of resistance elements that would inactivate compounds in current clinical use.

3.6. CONCLUSIONS

Aminoglycoside resistance has been a powerful impediment to the continued use of these potent antibacterial agents. The myriad of strategies that confer resistance including target modification (by mutation and gene acquisition), efflux, and a gamut of modifying enzymes makes understanding and overcoming resistance a significant challenge. However, new strategies in expanding aminoglycoside bioactive chemical space, guided by molecular knowledge of the ribosomal target and the major resistance proteins, has already resulted in novel compounds with improved activity. There is good reason to be optimistic that a new generation of this important class of antibiotics will emerge as serious contenders for clinical attention in the next few years.

ACKNOWLEDGMENTS

Research in GDW's lab on aminoglycoside antibiotics has been funded by the Canadian Institutes of Health Research and by a Canada Research Chair award.

REFERENCES

1. Weinstein, M. J.; Luedemann, G. M.; Oden, E. M.; Wagman, G. H.; Rosselet, J. P.; Marquez, J. A.; Coniglio, C. T.; Charney, W.; Herzog, H. L.; Black, J. *J. Med. Chem.* **1963**, *6*, 463.
2. Carter, A. P.; Clemons, W. M.; Brodersen, D. E.; Morgan-Warren, R. J.; Wimberly, B. T.; Ramakrishnan, V. *Nature* **2000**, *407*, 340.
3. Davies, J.; Gorini, L.; Davis, B. D. *Mol. Pharmacol.* **1965**, *1*, 93.
4. Davis, B. D. *Microbiol. Rev.* **1987**, *51*, 341.
5. Umezawa, H.; Okanishi, M.; Kondo, S.; Hamana, K.; Utahara, R.; Maeda, K.; Mitsuhashi, S. *Science* **1967**, *157*, 1559.
6. Doi, O.; Miyamoto, M.; Tanaka, N.; Umezawa, H. *Appl. Microbiol.* **1968**, *16*, 1282.
7. Francois, B.; Russell, R. J.; Murray, J. B.; Aboul-ela, F.; Masquida, B.; Vicens, Q.; Westhof, E. *Nucleic Acids Res.* **2005**, *33*, 5677.
8. Vicens, Q.; Westhof, E. *Chem. Biol.* **2002**, *9*, 747.
9. Vicens, Q.; Westhof, E. *J. Mol. Biol.* **2003**, *326*, 1175.
10. Bastida, A.; Hidalgo, A.; Chiara, J. L.; Torrado, M.; Corzana, F.; Perez-Canadillas, J. M.; Groves, P.; Garcia-Junceda, E.; Gonzalez, C.; Jimenez-Barbero, J.; Asensio, J. L. *J. Am. Chem. Soc.* **2006**, *128*, 100.
11. Kim, C.; Haddad, J.; Vakulenko, S. B.; Meroueh, S. O.; Wu, Y.; Yan, H.; Mobashery, S. *Biochemistry* **2004**, *43*, 2373.
12. Russell, R. J.; Murray, J. B.; Lentzen, G.; Haddad, J.; Mobashery, S. *J. Am. Chem. Soc.* **2003**, *125*, 3410.
13. Sucheck, S. J.; Wong, A. L.; Koeller, K. M.; Boehr, D. D.; Draker, K.-A.; Sears, P.; Wright, G. D.; Wong, C.-H. *J. Am. Chem. Soc.* **2000**, *122*, 5230.
14. Chang, C. W.; Hui, Y.; Elchert, B.; Wang, J.; Li, J.; Rai, R. *Org Lett* **2002**, *4*, 4603.
15. Elchert, B.; Li, J.; Wang, J.; Hui, Y.; Rai, R.; Ptak, R.; Ward, P.; Takemoto, J. Y.; Bensaci, M.; Chang, C. W. *J Org Chem* **2004**, *69*, 1513.

16. Wang, J.; Li, J.; Chen, H. N.; Chang, H.; Tanifum, C. T.; Liu, H. H.; Czyryca, P. G.; Chang, C. W. *J Med Chem* **2005**, *48*, 6271.
17. Davies, J. E. In *Antibiotics in Laboratory Medicine*, 3rd edition; Lorian, V., Ed.; Baltimore: Williams & Wilkins, 1991; pp. 691–713.
18. Umezawa, H.; Kondo, S. In *Aminoglycoside Antibiotics*; Umezawa, H., Hooper, I. R., Eds.; Berlin: Springer-Verlag, 1982; pp. 267–292.
19. Wright, G. D.; Berghuis, A. M.; Mobashery, S. In *Resolving the Antibiotic Paradox: Progress in Drug Design and Resistance*; Rosen, B. P., Mobashery, S., Eds.; New York: Plenum Press, 1998; pp. 27–69.
20. Mingeot-Leclercq, M. P.; Glupczynski, Y.; Tulkens, P. M. *Antimicrob Agents Chemother* **1999**, *43*, 727.
21. Vakulenko, S. B.; Mobashery, S. *Clin Microbiol Rev* **2003**, *16*, 430.
22. Moazed, D.; Noller, H. F. *Nature* **1987**, *27*, 389.
23. De Stasio, E. A.; Moazed, D.; Noller, H. F.; Dahlberg, A. E. *EMBO J.* **1989**, *8*, 1213.
24. Prammananan, T.; Sander, P.; Brown, B. A.; Frischkorn, K.; Onyi, G. O.; Zhang, Y.; Bottger, E. C.; Wallace, R. J., Jr. *J. Infect. Dis.* **1998**, *177*, 1573.
25. Recht, M. I.; Douthwaite, S.; Puglisi, J. D. *EMBO J.* **1999**, *18*, 3133.
26. Basso, L. A.; Blanchard, J. S. *Adv. Exp. Med. Biol.* **1998**, *456*, 115.
27. Springer, B.; Kidan, Y. G.; Prammananan, T.; Ellrott, K.; Bottger, E. C.; Sander, P. *Antimicrob. Agents Chemother.* **2001**, *45*, 2877.
28. Gregory, S. T.; Carr, J. F.; Dahlberg, A. E. *J Bacteriol* **2005**, *187*, 2200.
29. Bjorkman, J.; Samuelsson, P.; Andersson, D. I.; Hughes, D. *Mol. Microbiol.* **1999**, *31*, 53.
30. Vasiljevic, B.; Cundliffe, E. *J. Cell. Biochem.* **1990**, *Suppl. 14A*.
31. Skeggs, P. A.; Holmes, D. J.; Cundliffe, E. *J. Gen. Microbiol.* **1987**, *133*, 915.
32. Beauclerk, A. A.; Cundliffe, E. *J. Mol. Biol.* **1987**, *93*, 661.
33. Yokoyama, K.; Doi, Y.; Yamane, K.; Kurokawa, H.; Shibata, N.; Shibayama, K.; Yagi, T.; Kato, H.; Arakawa, Y. *Lancet* **2003**, *362*, 1888.
34. Doi, Y.; Yokoyama, K.; Yamane, K.; Wachino, J.; Shibata, N.; Yagi, T.; Shibayama, K.; Kato, H.; Arakawa, Y. *Antimicrob. Agents Chemother.* **2004**, *48*, 491.
35. Galimand, M.; Courvalin, P.; Lambert, T. *Antimicrob. Agents Chemother.* **2003**, *47*, 2565.
36. Gonzalez-Zorn, B.; Teshager, T.; Casas, M.; Porrero, M. C.; Moreno, M. A.; Courvalin, P.; Dominguez, L. *Emerg. Infect. Dis.* **2005**, *11*, 954.
37. Galimand, M.; Sabtcheva, S.; Courvalin, P.; Lambert, T. *Antimicrob. Agents Chemother.* **2005**, *49*, 2949.
38. Gonzalez-Zorn, B.; Catalan, A.; Escudero, J. A.; Dominguez, L.; Teshager, T.; Porrero, C.; Moreno, M. A. *J. Antimicrob. Chemother.* **2005**, *56*, 583.
39. Yamane, K.; Doi, Y.; Yokoyama, K.; Yagi, T.; Kurokawa, H.; Shibata, N.; Shibayama, K.; Kato, H.; Arakawa, Y. *Antimicrob Agents Chemother.* **2004**, *48*, 2069.
40. Bryan, L. E.; Kwan, S. *Antimicrob. Agents Chemother.* **1983**, *23*, 835.
41. Muir, M. E.; Hanwell, D. R.; Wallace, B. J. *Biochim. Biophys. Acta* **1981**, *638*, 234.
42. Arrow, A. S.; Taber, H. W. *Antimicrob. Agents Chemother.* **1986**, *29*, 141.
43. McEnroe, A. S.; Taber, H. W. *Antimicrob. Agents Chemother.* **1984**, *26*, 507.

44. Young, M. L.; Bains, M.; Bell, A.; Hancock, R. E. *Antimicrob. Agents Chemother.* **1992**, *36*, 2566.
45. Bryan, L. E.; O'Hara, K.; Wong, S. *Antimicrob. Agents Chemother.* **1984**, *26*, 250.
46. Shearer, B. G.; Legakis, N. J. *J. Infect. Dis.* **1985**, *152*, 351.
47. Poole, K. *J Antimicrob. Chemother.* **2005**, *56*, 20.
48. Poole, K. *Clin. Microbiol. Infect.* **2004**, *10*, 12.
49. Moore, R. A.; DeShazer, D.; Reckseidler, S.; Weissman, A.; Woods, D. E. *Antimicrob. Agents Chemother.* **1999**, *43*, 465.
50. Rosenberg, E. Y.; Ma, D.; Nikaido, H. *J. Bacteriol.* **2000**, *182*, 1754.
51. Edgar, R.; Bibi, E. *J. Bacteriol.* **1997**, *179*, 2274.
52. Chan, Y. Y.; Tan, T. M.; Ong, Y. M.; Chua, K. L. *Antimicrob. Agents Chemother.* **2004**, *48*, 1128.
53. Sobel, M. L.; McKay, G. A.; Poole, K. *Antimicrob. Agents Chemother.* **2003**, *47*, 3202.
54. Li, X. Z.; Nikaido, H.; Poole, K. *Antimicrob. Agents Chemother.* **1995**, *39*, 1948.
55. Li, X. Z.; Poole, K.; Nikaido, H. *Antimicrob. Agents Chemother.* **2003**, *47*, 27.
56. Poelarends, G. J.; Mazurkiewicz, P.; Putman, M.; Cool, R. H.; Veen, H. W.; Konings, W. N. *Drug Resist. Update.* **2000**, *3*, 330.
57. Marchand, I.; Damier-Piolle, L.; Courvalin, P.; Lambert, T. *Antimicrob. Agents Chemother.* **2004**, *48*, 3298.
58. Li, X. Z.; Zhang, L.; Poole, K. *Antimicrob. Agents Chemother.* **2002**, *46*, 333.
59. Jo, J. T.; Brinkman, F. S.; Hancock, R. E. *Antimicrob. Agents Chemother.* **2003**, *47*, 1101.
60. Elkins, C. A.; Nikaido, H. *J. Bacteriol.* **2003**, *185*, 5349.
61. Shaw, K. J.; Rather, P. N.; Hare, R. S.; Miller, G. H. *Microbiol. Rev.* **1993**, *57*, 138.
62. Costa, Y.; Galimand, M.; Leclercq, R.; Duval, J.; Courvalin, P. *Antimicrob. Agents Chemother.* **1993**, *37*, 1896.
63. Miller, G. H.; Sabatelli, F. J.; Hare, R. S.; Glupczynski, Y.; Mackey, P.; Shlaes, D.; Shimizu, K.; Shaw, K. J.; Groups, a. A. R. S. *Clin. Infect. Dis.* **1997**, *24*, S46.
64. Over, U.; Gur, D.; Unal, S.; Miller, G. H. *Clin. Microbiol. Infect.* **2001**, *7*, 470.
65. Wolf, E.; Vassilev, A.; Makino, Y.; Sali, A.; Nakatani, Y.; Burley, S. K. *Cell* **1998**, *94*, 439.
66. Wybenga-Groot, L.; Draker, K. a.; Wright, G. D.; Berghuis, A. M. *Structure* **1999**, *7*, 497.
67. Vetting, M. W.; Magnet, S.; Nieves, E.; Roderick, S. L.; Blanchard, J. S. *Chem. Biol.* **2004**, *11*, 565.
68. Vetting, M. W.; Hegde, S. S.; Javid-Majd, F.; Blanchard, J. S.; Roderick, S. L. *Nat. Struct. Biol.* **2002**, *9*, 653.
69. Vetting, M. W.; de Carvalho LP, S. d. C.; Yu, M.; Hegde, S. S.; Magnet, S.; Roderick, S. L.; Blanchard, J. S. *Arch. Biochem. Biophys.* **2005**, *433*, 212.
70. Robicsek, A.; Strahilevitz, J.; Jacoby, G. A.; Macielag, M.; Abbanat, D.; Park, C. H.; Bush, K.; Hooper, D. C. *Nat. Med.* **2006**, *12*, 83.
71. Miller, G. H.; Sabatelli, F. J.; Naples, L.; Hare, R. S.; Shaw, K. J.; *J. Chemother.* **1995**, *7 Suppl. 2*, 17.
72. Gates, C. A.; Northrop, D. B. *Biochemistry* **1988**, *27*, 3820.

73. Gates, C. A.; Northrop, D. B. *Biochemistry* **1988**, *27*, 3826.
74. Gates, C. A.; Northrop, D. B. *Biochemistry* **1988**, *27*, 3834.
75. Ekman, D. R.; DiGiammarino, E. L.; Wright, E.; Witter, E. D.; Serpersu, E. H. *Biochemistry* **2001**, *40*, 7017.
76. Pedersen, L. C.; Benning, M. M.; Holden, H. M. *Biochemsitry* **1995**, *34*, 13305.
77. Holm, L.; Sander, C. *Trends Biol. Chem.* **1995**, *20*, 345.
78. Hon, W. C.; McKay, G. A.; Thompson, P. R.; Sweet, R. M.; Yang, D. S. C.; Wright, G. D.; Berghuis, A. M. *Cell* **1997**, *89*, 887.
79. Fong, D. H.; Berghuis, A. M. *EMBO J* **2002**, *21*, 2323.
80. Burk, D. L.; Hon, W. C.; Leung, A. K.; Berghuis, A. M. *Biochemistry* **2001**, *40*, 8756.
81. McKay, G. A.; Wright, G. D. *Biochemistry* **1996**, *35*.
82. Thompson, P. R.; Hughes, D. W.; Wright, G. D. *Chem. Biol.* **1996**, *3*, 747.
83. Thompson, P. R.; Schwartzenhauer, J.; Hughes, D. W.; Berghuis, A. M.; Wright, G. D. *J. Biol. Chem.* **1999**, *274*, 30697.
84. Boehr, D. D.; Thompson, P. R.; Wright, G. D. *J. Biol. Chem.* **2001**, *276*, 23929.
85. Boehr, D. D.; Farley, A. R.; Wright, G. D.; Cox, J. R. *Chem. Biol.* **2002**, *9*, 1209.
86. Thompson, P. R.; Boehr, D. D.; Berghuis, A. M.; Wright, G. D. *Biochemistry* **2002**, *41*, 7001.
87. Boehr, D. D.; Farley, A. R.; LaRonde, F. J.; Murdock, T. R.; Wright, G. D.; Cox, J. R. *Biochemistry* **2005**, *44*, 12445.
88. Nurizzo, D.; Shewry, S. C.; Perlin, M. H.; Brown, S. A.; Dholakia, J. N.; Fuchs, R. L.; Deva, T.; Baker, E. N.; Smith, C. A. *J. Mol. Biol.* **2003**, *327*, 491.
89. Daigle, D. M.; McKay, G. A.; Thompson, P. R.; Wright, G. D. *Chem. Biol.* **1998**, *6.*, 11.
90. Daigle, D. M.; McKay, G. A.; Wright, G. D. *J. Biol. Chem.* **1997**, *272*, 24755.
91. Boehr, D. D.; Lane, W. S.; Wright, G. D. *Chem Biol* **2001**, *8*, 791.
92. Petrickova, K.; Petricek, M. *Microbiology* **2003**, *149*, 1609.
93. Umeyama, T.; Lee, P. C.; Horinouchi, S. *Appl. Microbiol. Biotechnol.* **2002**, *59*, 419.
94. Buynak, J. D. *Biochem. Pharmacol.* **2006**. 71, 930.
95. Allen, N. E., Jr.; Alborn, W. E. A., Jr.; J. N. H.; Kirst, H. A. *Antimicrob. Agents Chemother.* **1982**, *22*, 824.
96. Gao, F.; Yan, X.; Baettig, O. M.; Berghuis, A. M.; Auclair, K. *Angew. Chem. Int. Ed. Engl* **2005**, *44*, 6859.
97. Boehr, D. D.; Draker, K.; Koteva, K.; Bains, M.; Hancock, R. E.; Wright, G. D. *Chem. Biol.* **2003**, *10*, 189.
98. Rai, R.; Chen, H. N.; Czyryca, P. G.; Li, J.; Chang, C. W. *Org. Lett.* **2006**, *8*, 887.

4

DESIGN, CHEMICAL SYNTHESIS, AND ANTIBACTERIAL ACTIVITY OF KANAMYCIN AND NEOMYCIN CLASS AMINOGLYCOSIDE ANTIBIOTICS

JINHUA WANG AND CHENG-WEI TOM CHANG
Department of Chemistry and Biochemistry, Utah State University, Logan, UT 84322

Aminoglycoside Antibiotics: From Chemical Biology to Drug Discovery, Second Edition, by Dev P. Arya

4.1. INTRODUCTION

Ever since the discovery of aminoglycoside antibiotics, their challenging synthesis has attracted much attention. Pioneered by Umezawa and others in the 1970s, a vast amount of aminoglycoside derivatives along with various modification methods have been created.[1–3] Due to the false belief that infectious diseases can be defeated by available antibiotics, the development of new antibiotics has dwindled for more than a decade. With the advancements in studies of resistance mechanisms[4–6] and structural information from the binding of aminoglycosides with the target, the A-site decoding region of 16S rRNA,[7–10] new strategies have been developed that aim to revive antibacterial activity against aminoglycoside-resistant bacteria.

Aminoglycosides are a group of structurally diverse antibiotics consisting of various numbers of normal and unusual sugars (Figure 4.1). A wide range of synthetic techniques have been documented in the literature. However, to provide a brief coverage, we focus only on the general chemical synthetic strategies and several related examples from the neomycin and kanamycin class of aminoglycosides. In addition, several designs that are effective against resistant bacteria will also be covered.

4.2. STRATEGIES IN MODIFICATION OF AMINOGLYCOSIDES

Neomycin and kanamycin are two of the most-studied aminoglycoside antibiotics. Neomycin belongs to a group of aminoglycosides containing a 4,5-disubstituted 2-deoxystreptamine core (ring II), while kanamycin contains a 4,6-disubstituted 2-deoxystreptamine core. In general, there are two types of approaches reported for syntheses of new aminoglycoside antibiotics (Figure 4.2). The first one is

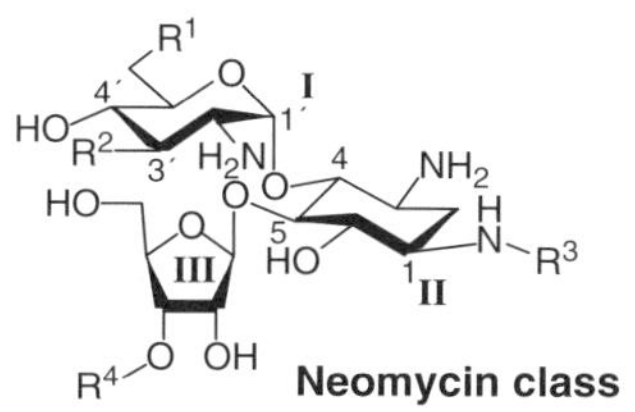

	R^1	R^2	R^3	R^4
Neomycin B	NH_2	OH	H	DAG
Paromomycin I	OH	OH	H	DAG
Lividomycin B	OH	H	H	DAG
Butirosin B	NH_2	OH	AHB	H
Ribostamycin	NH_2	OH	H	H

DAG: 2,6-diamino-2,6-dideoxy–α-L-glucopyranosyl

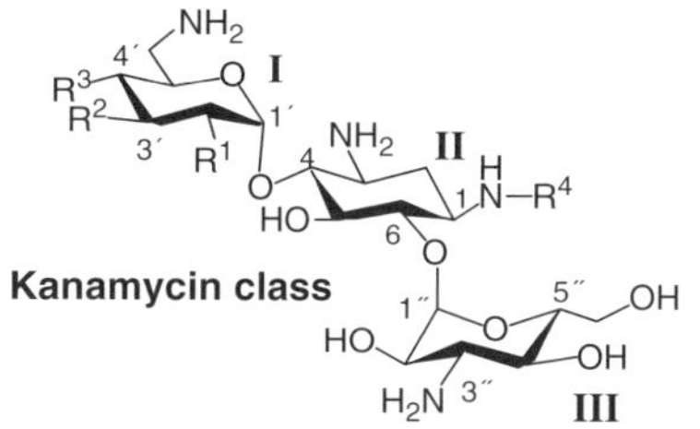

	R^1	R^2	R^3	R^4
Kamanycin A	OH	OH	OH	H
Kamanycin B	NH_2	OH	OH	H
Dibekacin	NH_2	H	H	H
Tobramycin	NH_2	H	OH	H
Amikacin	OH	OH	OH	AHB
Arbekacin	NH_2	H	H	AHB

AHB:(S)-4-amino-2-hydroxybutyryl

Figure 4.1. Structures of neomycin and kanamycin classes of aminoglycosides.

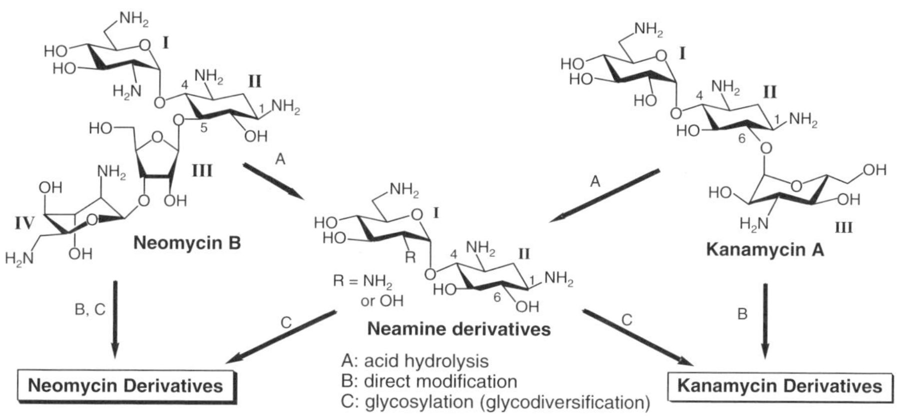

Figure 4.2. General strategies for the synthesis of neomycin and kanamycin classes of aminoglycosides.

direct modification of existing aminoglycosides. The second approach is to apply glycosylation strategies on the selected cores, such as 2-deoxystreptamine and neamine (rings I and II), and create libraries of new aminoglycosides. A tremendous amount of effort in the study of aminoglycosides has produced many fruitful results. In order to provide a concise review that covers up-to-date advancements, we will focus mostly on works that have appeared since late 1990s.

For direct chemical modifications of existing aminoglycosides, several criteria need to be considered first. These include the available source of aminoglycosides, appropriate protecting groups for the amino groups, and regioselective differentiation of these amino groups. Neomycin B and kanamycin A are two of the pertinent low-cost aminoglycosides to be employed as the starting materials for modifications. Although commercially available neomycin B is usually mixed with 10–20% neomycin A and C while kanamycin A is commonly mixed with 10–20% kanamycin B, these minor components do not pose significant difficulty in the chemical synthesis and characterization of the synthesized compounds.

Two types of amino group protection/masking have been commonly employed. The first one is to use carbamate types of protecting groups. The second strategy is to mask amino groups into azido groups. Using the carbamate type of protecting groups is more cost effective. However, the poor solubility of the polycarbamate-containing compounds makes purification and characterization of these compounds very difficult. Masking the amino groups of aminoglycosides as azido groups has the advantage of better solubility in organic media, therefore making the modification of azidoaminoglycosides compatible with the traditional synthetic techniques.

Regioselective differentiation of amino groups is the most challenging issue in synthesizing aminoglycoside derivatives. Fortunately, the regioselective manipulation of the amino groups on aminoglycosides has been extensively studied, especially in neomycin, kanamycin, and neamine. In general, these methods utilize metal-mediated chelation or cyclization between the selected carbamate-protected amino group and its vicinal hydroxyl group. However, selectivity of these two strategies closely depends on the scaffolds of parent aminoglycosides.

It has been shown that the stereochemistry of the glycosidic bond to which the carbohydrate component is attached at the neamine core is essential for antibacterial activity.[11] The neomycin class aminoglycoside consists of a neamine core and a β-linked carbohydrate component attached at the O-5 position, while the kanamycin class aminoglycoside consists of a neamine core with α-linked carbohydrate component attached at the O-6 position. Since neamine is the pivotal component of both neomycin and kanamycin, a readily accessible library of unusual sugars will provide opportunity for the facile construction of both classes of aminoglycosides via glycosylation approach.

4.3. SYNTHESIS OF NEAMINE DERIVATIVES

4.3.1. Preparation of Neamine

Neamine is the core molecule that can be used for synthesizing aminoglycosides of both kanamycin and neomycin classes. Therefore, modification of neamine has been extensively studied. Acid hydrolysis of neomycin B, followed by purification using Amberlite IRA 410 (OH^-), is the most convenient method of obtaining neamine (Scheme 4.1).[12] Such a protocol can be used for as much as multi-hundred grams although the purity of neamine prepared in such quantity is only about 50–60%. However, further purification is doable using Amberlite CG50.

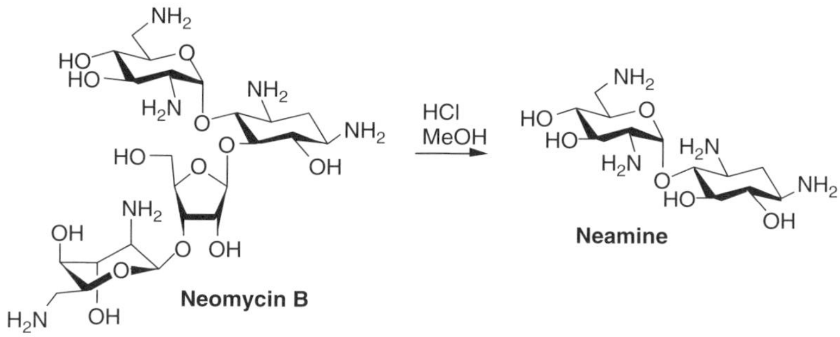

Scheme 4.1. Synthesis of neamine from acid hydrolysis of neomycin B.

4.3.2. Modification of Neamine: Use of Carbamate Type of Protecting Groups

Since aminoglycosides contain multiple amino groups, the carbamate type of protecting groups—such as MeO_2C-, EtO_2C-(Cbe), PhO_2C-, BnO_2C-(Cbz, Z), and $t$$BuO_2C$- (Boc)—were commonly used for direct synthetic modifications of aminoglycosides, especially in early works. After being incorporated on the aminoglycoside, these protecting groups can be selectively modified with their vicinal hydroxyl groups, thereby leading to more versatile structural modifications. For example, by using appropriate bases, carbamate protecting groups can be transformed into cyclic urea or carbamate with the neighboring function groups (Scheme 4.2).[13]

4.3.3. Modification of Neamine: Regioselective Protection of the Amino Groups Using Metal Chelation

Differentiation of amino groups by using transition metals, such as Cu(II) and Zn (II), has been reported by Umezawa and co-workers[14] and Nagabhushan et al.[15] Rationale for the regioselectivity has been provided, however, this method relies

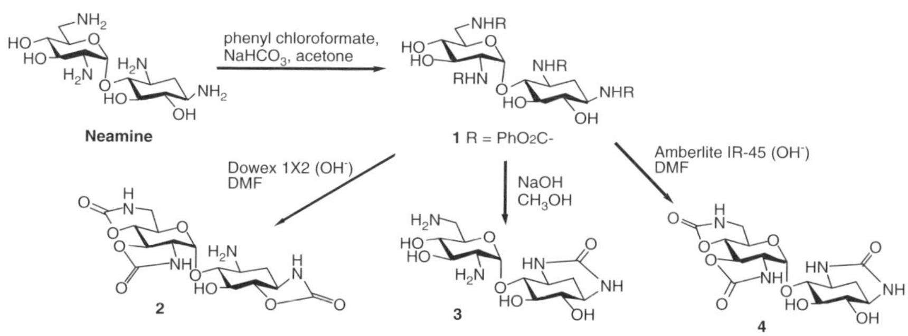

Scheme 4.2. Regioselective modification of neamine.

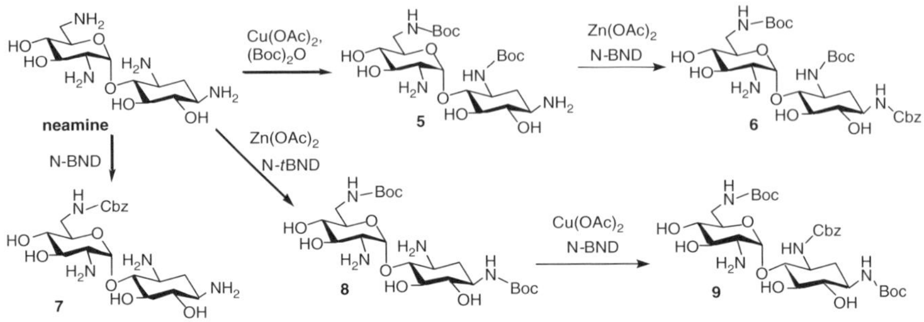

N-BND: *N*-benzyloxycarbonyloxy-5-norbornene-endo-2,3-dicarboximide
N-*t*BND: *N*-butoxycarbonyloxy-5-norbornene-endo-2,3-dicarboximide

Scheme 4.3. Selective exposure of amino groups on neamine.

on the specific configurations of hydroxy and amino groups. Therefore, it was commonly applied to kanamycin A and neamine. For example, all four amino groups on neamine can be selectively protected or revealed using the metal-chelating method (Scheme 4.3).[16]

4.3.4. Modification of Neamine: Regioselective Protection of the Hydroxy Groups

The two pairs of *trans*-diols (O-3′/O-4′ and O-5/O-6) on neamine can be selectively protected by protecting groups, such as cyclohexylidene (Scheme 4.4).[17] The O-3′/O-4′ diol is less hindered than the O-5/O-6 diol. Therefore, the

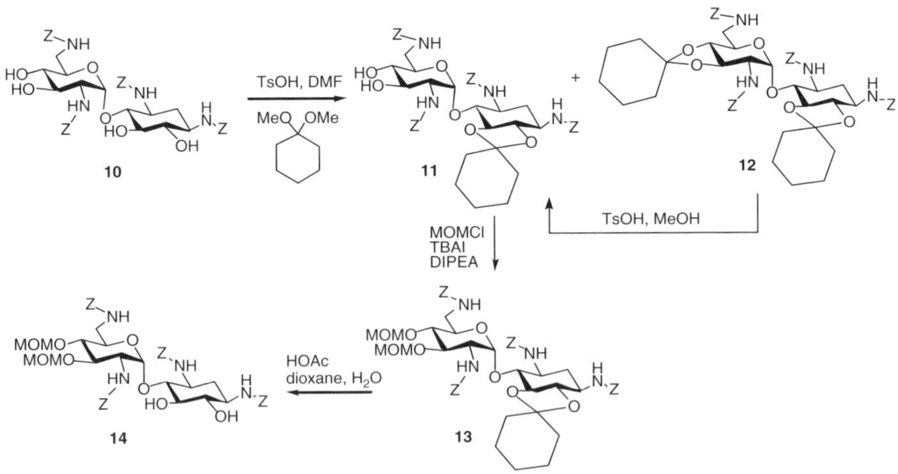

Scheme 4.4. Regioselective protection of hydroxy groups on neamine.

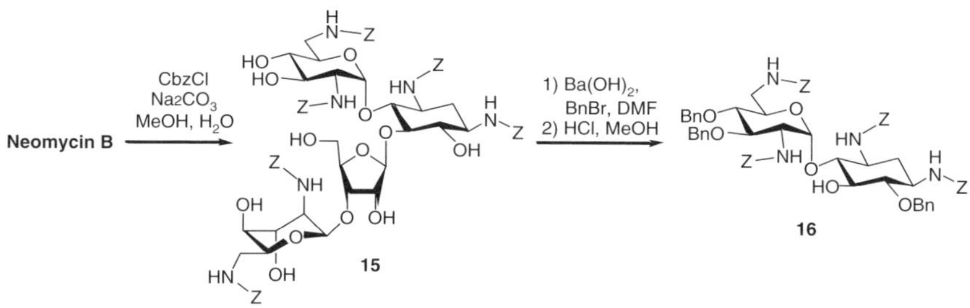

Scheme 4.5. Synthesis of 3′,4′,6-tri-*O*-protected neamine.

protection group at the O-3′/O-4′ diol is prone to acid-catalyzed deprotection. As a result, in the appropriate acidic conditions, the formation of the O-5/O-6 diol protected neamine, **11**, will be the major product, which leads to the preparation of neamine with free O-5/O-6 diol. The O-5 hydroxy group is more sterically hindered than O-6, allowing the synthesis of 3′,4′,6-tri-*O*-protected neamine.

Alternatively, a 3′,4′,6-tri-*O*-protected neamine, such as **16**, can be prepared from hydrolysis of protected neomycin B (Scheme 4.5).[18]

4.3.5. Modification of Neamine: Transformation of Neamine into Tetraazidoneamine

Wong and co-workers[19] have utilized TfN_3 for the conversion of neamine into tetraazidoneamine, **17** (Scheme 4.6). Unlike carbamate, the azido group has much superior solubility in organic media, which is advantageous in the purification

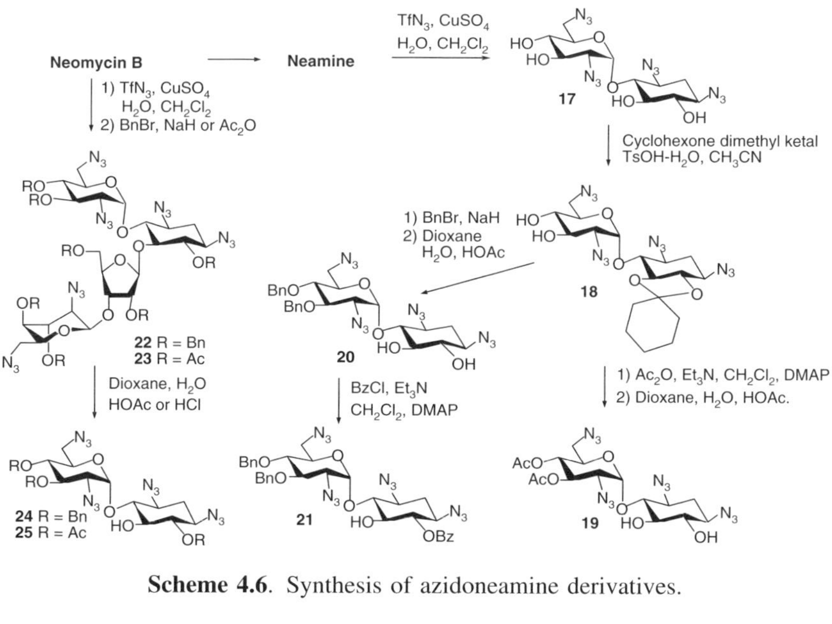

Scheme 4.6. Synthesis of azidoneamine derivatives.

and characterization of poly-azido compound. The preparation of 3′,4′,6-tri-*O*-protected tetrazidoneamines (**21, 24**, and **25**) and 3′,4′-di-*O*-protected tetrazidoneamines (**19** and **20**) can be achieved in a similar fashion as previously described, leading to the synthesis of neomycin and kanamycin analogs, respectively, that will be discussed later.[20,21] Compound **18** will allow modification of 3′,4′-diol with important biological consequence, which will also be discussed later.

4.3.6. Modification of Neamine: Neamine Derivatives with Manipulated Amino Groups

Wong and co-workers[20] have also reported the synthesis of neamine derivatives bearing modified ring I with various numbers of amino groups at different positions (Figure 4.3). The key step is the regioselective protection of the triol on 2-deoxystreptamine (ring II), which can be accomplished via chiral protecting reagent, **33**, or lipase-catalyzed deacetylation (Scheme 4.7). The first method, however, provides the undesired regioisomer.

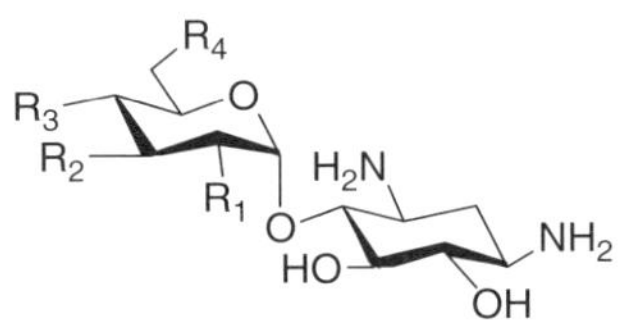

	R_1	R_2	R_3	R_4
26	OH	OH	OH	OH
27	NH_2	OH	OH	OH
28	OH	NH_2	OH	OH
29	OH	OH	NH_2	OH
30	OH	OH	OH	NH_2

Figure 4.3. Neamine derivatives.

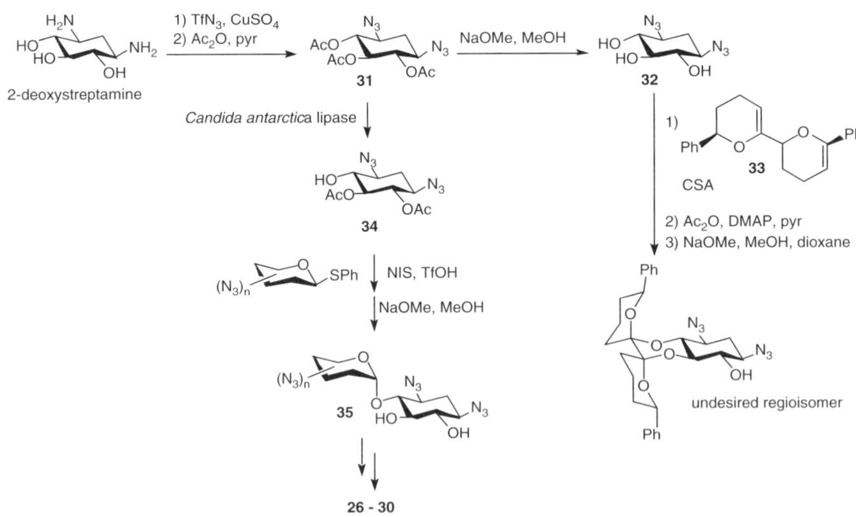

Scheme 4.7. Synthesis of neamine derivatives.

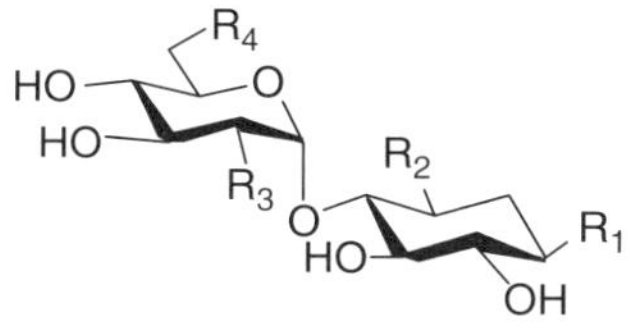

	R_1	R_2	R_3	R_4
36	H	NH_2	NH_2	NH_2
37	NH_2	H	NH_2	NH_2
38	NH_2	NH_2	H	NH_2
39	NH_2	NH_2	NH_2	H

Figure 4.4. Neamine derivatives with deamination.

Neamine analogs, in which four amino groups were individually replaced with hydrogen (deamination), were reported by Mobashery and co-workers[16] (Figure 4.4). Regioselective differentiation of amino groups using metal-chelating method, followed by deamination, offered four neamine derivatives bearing deamination: The NH_2 groups were replaced by H (Scheme 4.8). With the exception of **39**, all these neamine derivatives manifested better antibacterial activity against susceptible and resistant strains of *Escherichia coli* (Table 4.1).

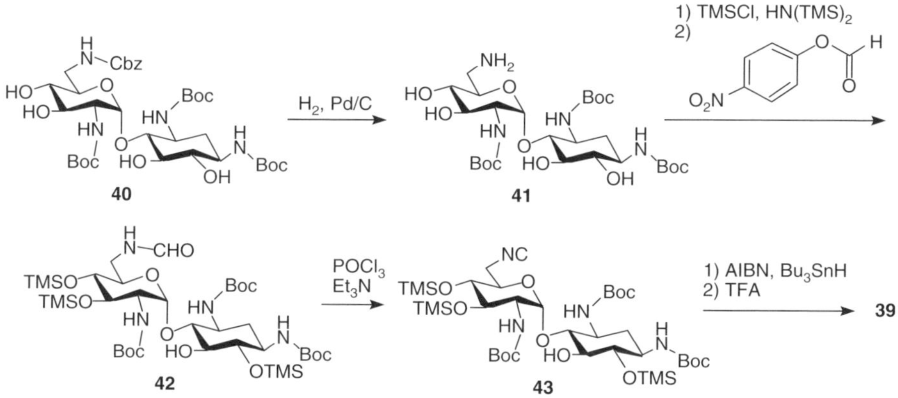

Scheme 4.8. Synthesis of deaminated neamine derivatives.

TABLE 4.1. Minimum Inhibitory Concentration (MIC) of Neamine Derivatives

Compounds	*E. coli*	*E. coli* (APH(3′)-Ia)	*E. coli* (APH(3′)-IIa)
Neamine	3.9	15.5	15.5
36	1.6	1.5	1.5
37	3.1	4.4	4.4
38	0.8	0.7	1.4
39	14.7	14.7	14.7

Unit, mM; APH, aminoglycoside phosphotransferase.

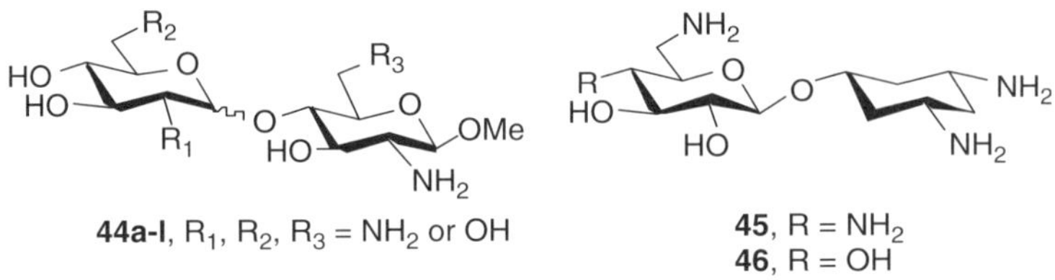

Figure 4.5. Structures of disaccharide-based aminoglycosides.

4.3.7. Neamine Mimics

Designs involving disaccharide have also been attempted (Figure 4.5).[22,23] However, disaccharide-based aminoglycosides are less active in general. For example, the designs of **45** and **46** that mimic rings II and III of pyranmycin were found to be inactive.

4.4. SYNTHESIS OF KANAMYCIN CLASS AMINOGLYCOSIDES

4.4.1. Synthetic Strategies in Direct Modification from Kanamycin

Modifications directly from kanamycin A were described in one of the earliest works engaged in the synthesis of aminoglycoside derivatives. Combined with the strategies of amino group differentiation, various designs aimed to increase the activity, especially against aminoglycoside-resistant bacteria, have been developed. For example, by using a Cbz protecting group, cyclic carbamate formation can be carried out by appropriate base (Scheme 4.9).[24] Compound **53**,

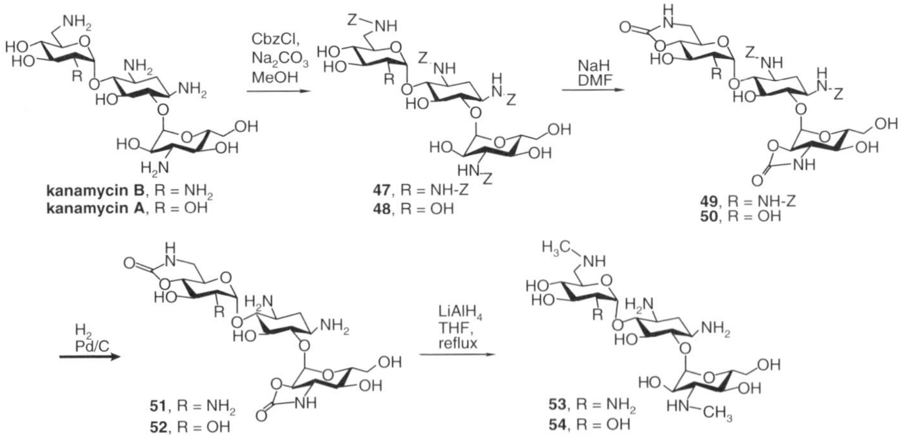

Scheme 4.9. Synthesis of kanamycin derivatives via direct modification.

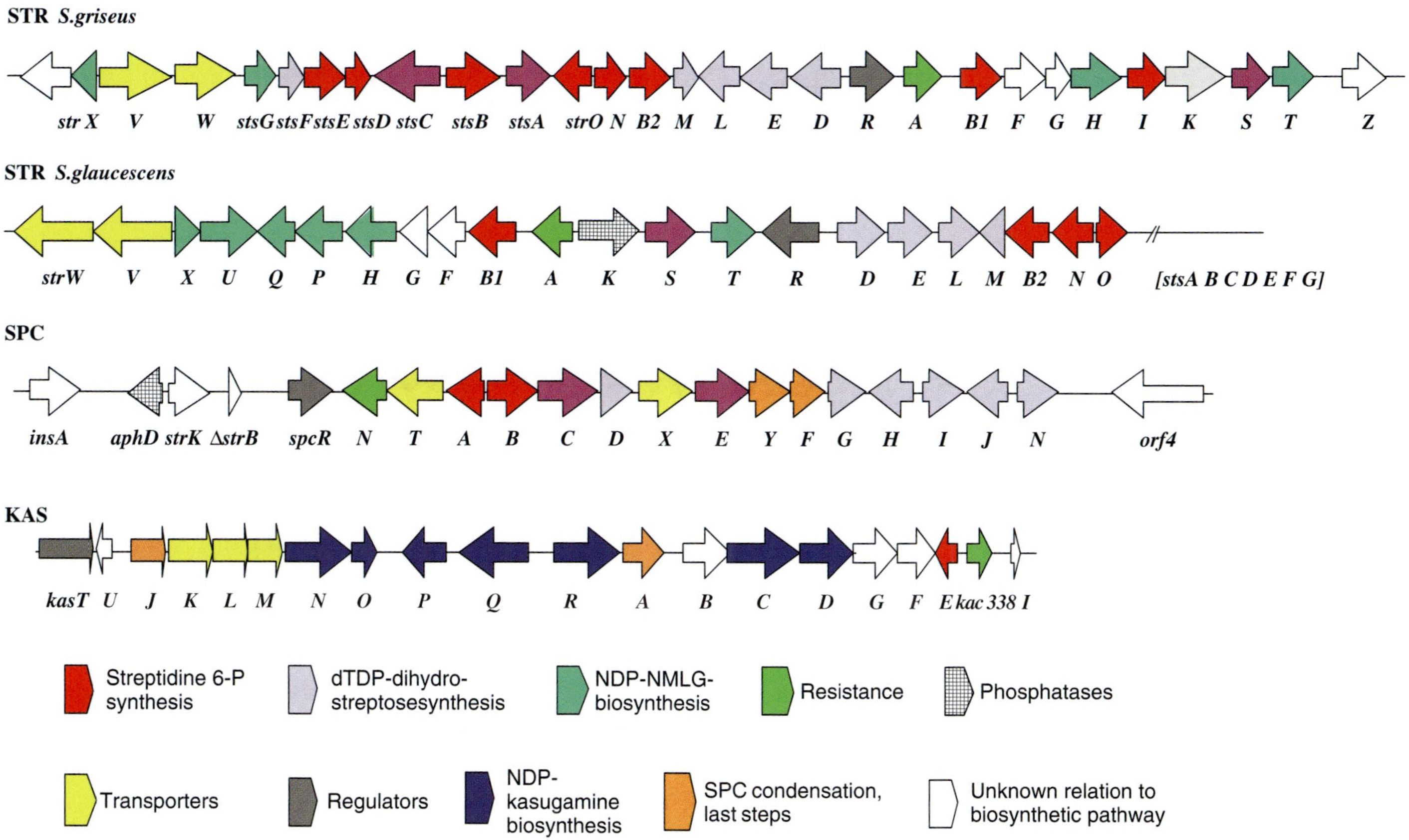

Figure 2.2. Biosynthetic gene clusters for STR-related AGAs. Functionally related groups of genes are differentiated by the color code given below. NMLG, *N*-methyl-L-glucosamine. Related groups of genes are differentiated by the color code given above.

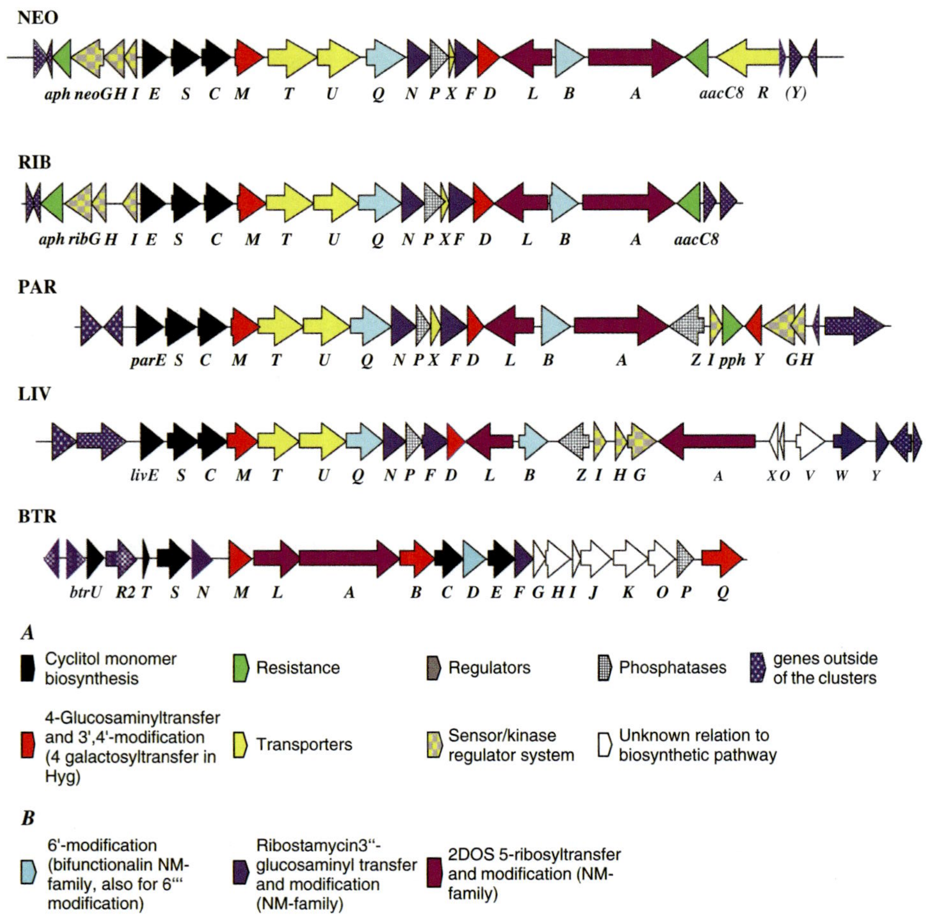

Figure 2.9. Biosynthetic gene clusters for NEO-family AGAs. Functionally related groups of genes are differentiated by the color code given below: (A) color code used in all genetic maps for 2DOS and 2DOS-related AGAs; (B) color code for pathway-specific functions in the NEO family.

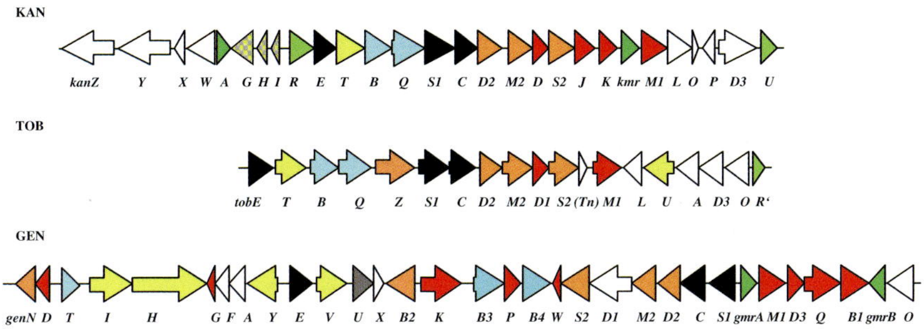

Figure 2.15. Biosynthetic gene clusters for KAN/GEN-related AGAs. 2DOS 6-glycosyltransfer and modification-specific genes, all other colors see Figure 2.9.

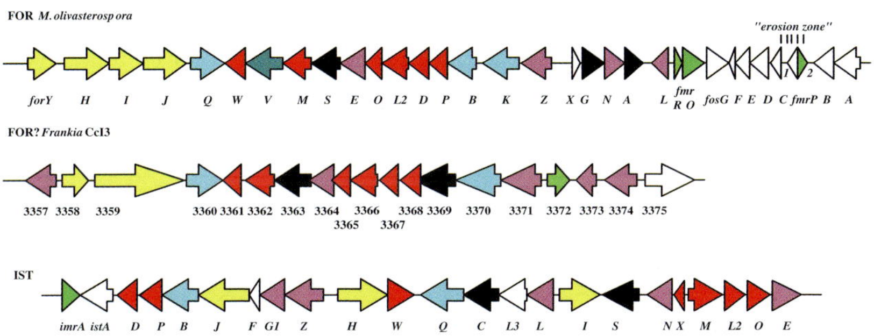

Figure 2.22. Biosynthetic gene clusters for fortimicin (FOR), the putative FOR cluster from *Frankia* sp. CcI3 (*fra* cluster, cf. Table 18) and istamycin (IST) AGAs. Late cyclitol modification, for all other colors see Figure 2.9. See color plates.

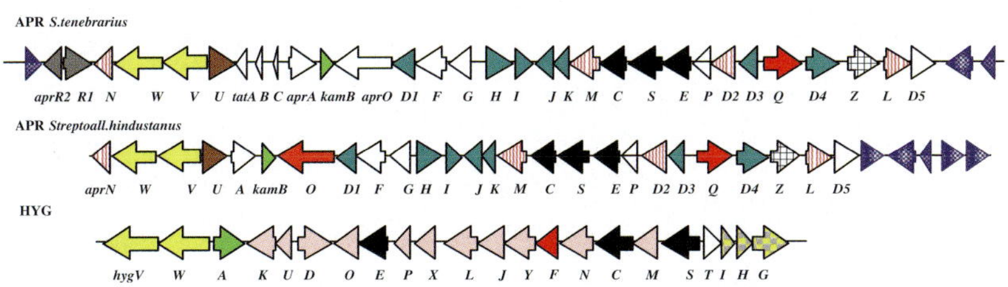

Figure 2.25. Biosynthetic gene clusters for APR and HYG-B. octose synthase, transfer, modification (APR), 8′Glycosyltransfer/modification (APR), HYG-specific genes; all other colors see Figure 2.9.

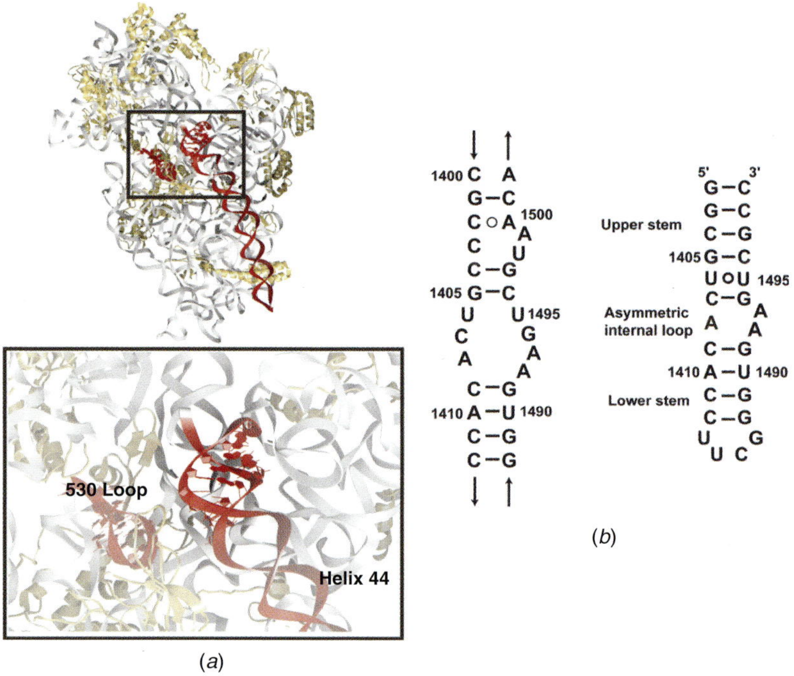

Figure 5.2. Design of a model oligonucleotide for the decoding site. See text for full caption.

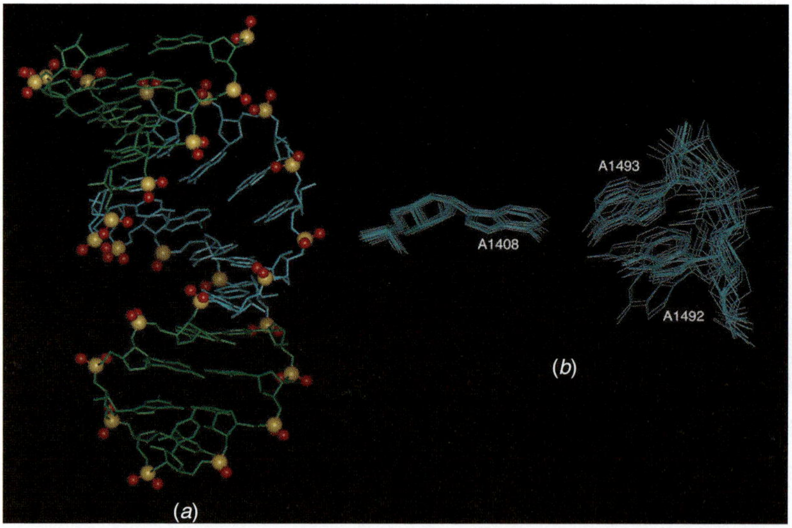

Figure 5.3. NMR structure of the A-site oligonucleotide. See text for full caption.

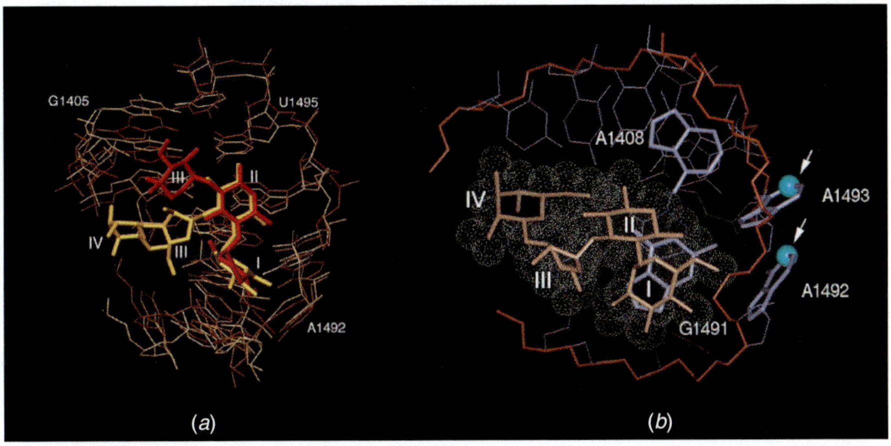

Figure 5.5. NMR structures of paromomycin and gentamicin C1a in complex with the A-site oligonucleotide. See text for full caption.

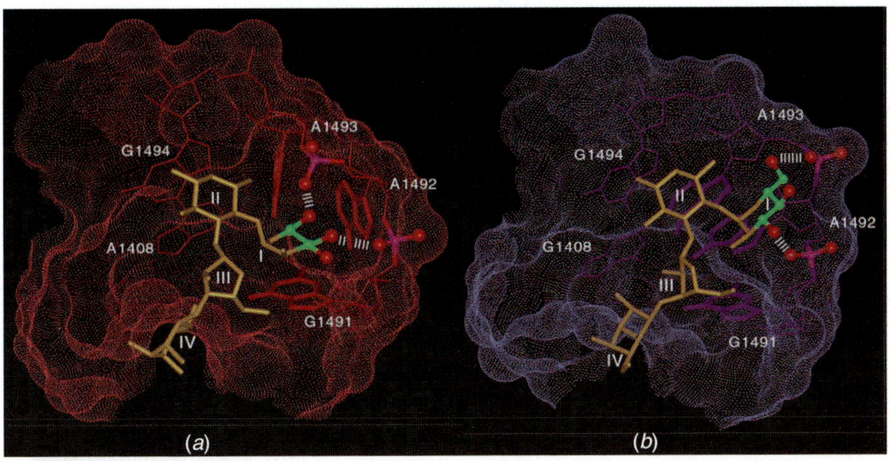

Figure 5.7. Structural origins of aminoglycoside specificity. See text for full caption.

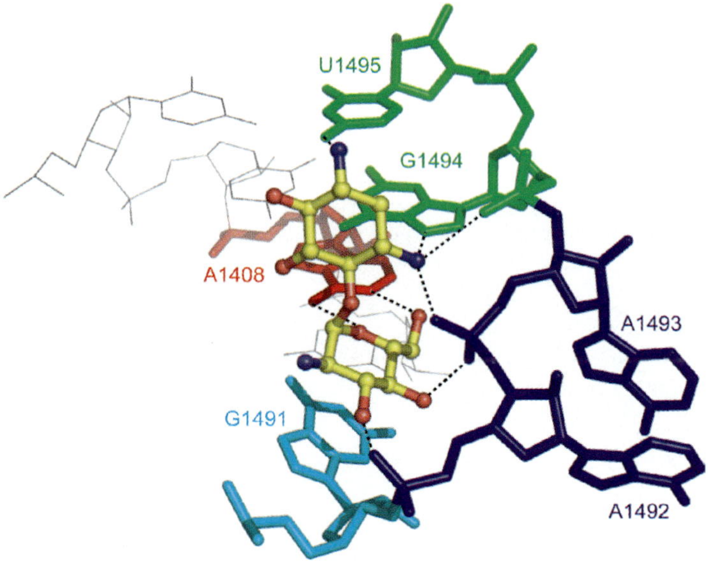

Figure 6.3. Conserved interactions between the prokaryotic A-site and the neamine core of 4,5- and 4,6-disubstituted aminoglycosides as deduced from crystal structures.[1,3,6,10,12–20]

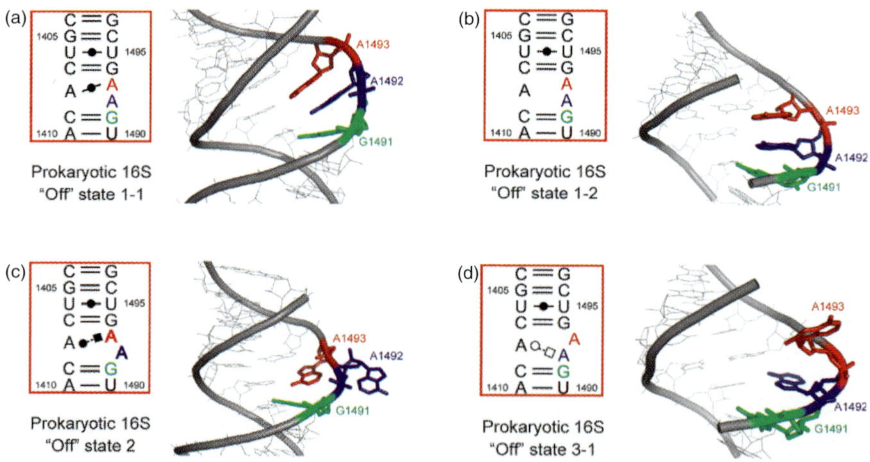

Figure 6.5. The observed structures of the resting "off" state of the prokaryotic (a (references[24]), b (reference[25]), c (reference[6]), d (reference[25]), e (reference[9]), f (reference[3]), and the eukaryotic (g (reference[5]), h (reference[26])) cytoplasmic A-site.

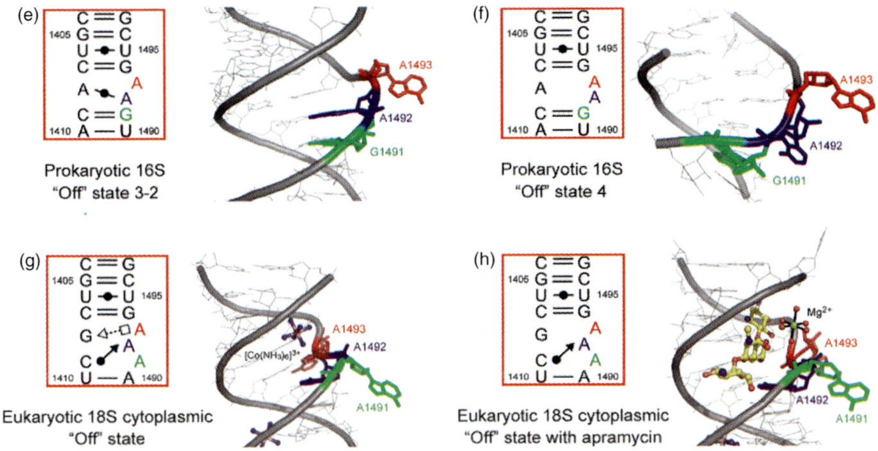

Figure 6.5. (*continued*)

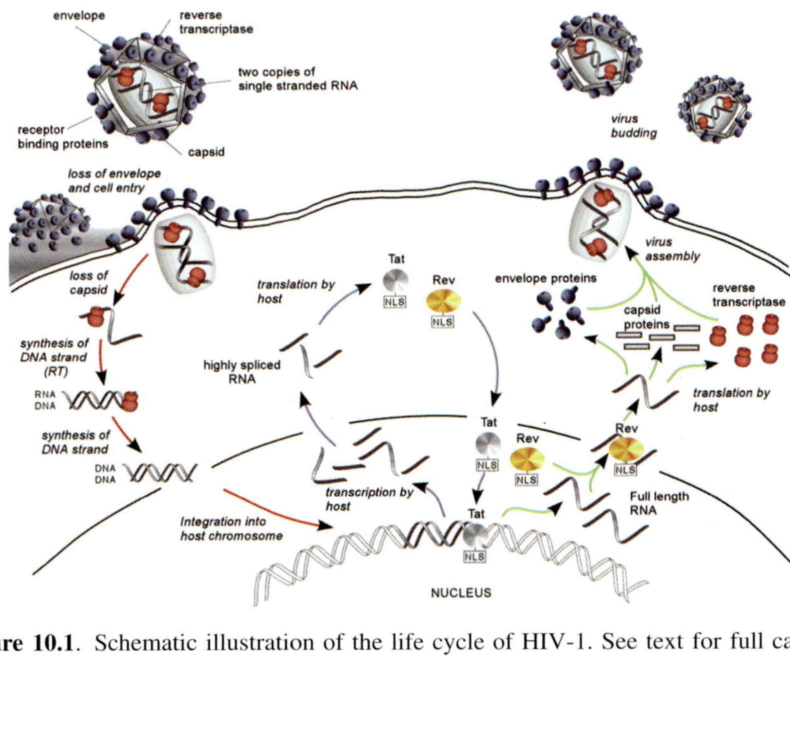

Figure 10.1. Schematic illustration of the life cycle of HIV-1. See text for full caption.

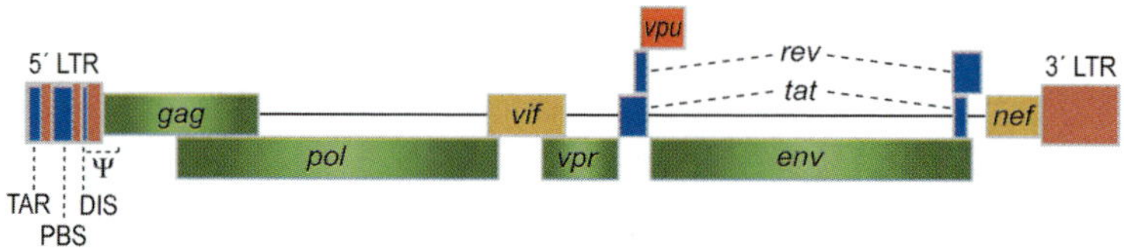

Figure 10.3. Schematic view of the HIV-1 genome, showing gene structure and splicing patterns, as well as important RNA sites localized in the 5′-long terminal repeat (LTR) and the ψ-site.

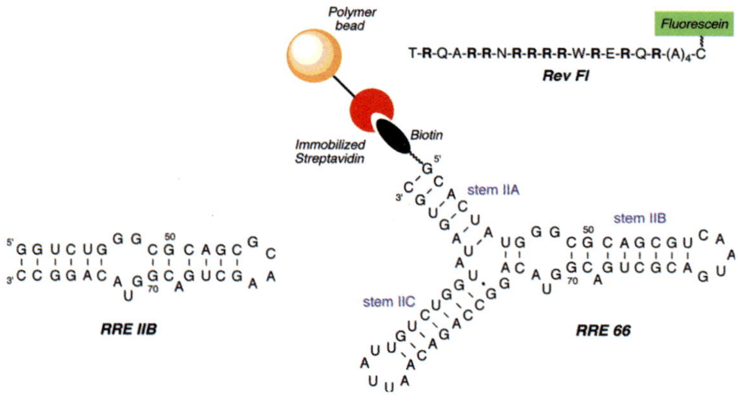

Figure 10.8. *Left*: The sequence of RRE IIB, a fragment used to study RRE–ligand binding. *Right*: The components of our solid-phase assay that utilize the native RRE sequence.

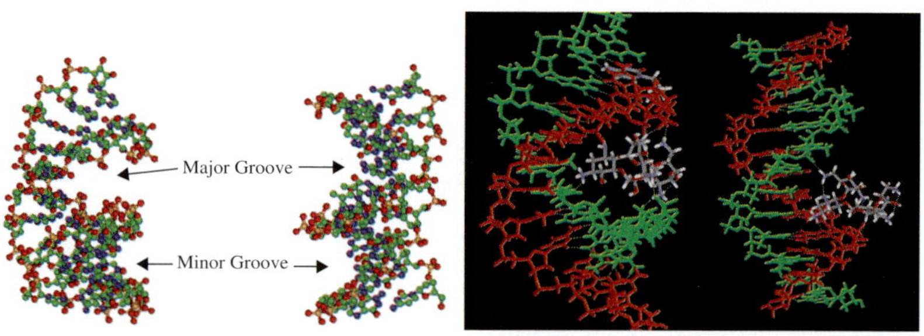

Figure 11.2. Globular conformations of an A-type duplex (left) and a B-type duplex (right), generally seen for RNA·RNA and DNA·DNA duplexes, respectively. See text for full caption.

TABLE 4.2. Minimum Inhibitory Concentration (MIC) of Neamine Derivatives

Bacterial Strains	Kan A	Kan B	**53**	**54**
S. aureus	1	0.5	4	>125
S. aureus (APH(3′)-IV)	>125	>125	8	>125
E. coli	4	2	16	>125
E. coli (APH(3′)-I)	>125	>125	63	>125
E. coli (ANT(2″))	125	32	63	>125
E. coli (AAC(6′))	>125	32	63	125
P. aeruginosa	63	63	8	>125
P. aeruginosa (APH(3′) + AAC(3)-I)	32	32	8	>125
P. aeruginosa (APH(3′)-IV)	>125	>125	8	>125
P. aeruginosa (APH(3′)-I+ AAC(6′))	>125	>125	8	>125
P. aeruginosa (perm. mutant)	>125	>125	16	>125

Unit, μg/mL; Kan, kanamycin; *S. aureus, Staphylococcus aureus; P. aeruginosa, Pseudomonas aeruginosa;* AAC, aminoglycoside acetyltransferase; ANT, aminoglycoside nucleotidyltransferase; Perm. mutant, mutant strain impermeable to kanamycin A and B.

6′,3″-di-*N*-methyl kanamycin B, displayed activity against various resistant strains. In the design of **54**, 6′,3″-di-*N*-methyl kanamycin A, replacing the 2′-NH_2 with 2′-OH, caused drastic decrease in activity.

4.4.2. Synthesis of Kanamycin Derivatives via Glycosylation

Kanamycin can be prepared via glycosylation of the O-6 hydroxy group of neamine derivatives or 2,5-dideoxystreptamine. A regioselective glycosylation on the O-6 hydroxy group of a 3′,4′-di-*O*-protected neamine can be readily achieved because O-5 OH is sterically hindered. The challenge is, however, to form the desired α-glycosidic bond since those with a β-glycosidic bond exhibit much weaker antibacterial activity.

For example, glycosylation using 2,3,4,6-tetra-*O*-acetylglucopyranosyl bromide provided the β-anomer as the major product due to the neighboring group assistance (Scheme 4.10).[11] However, the preferred α-anomer can be separated from the less active β-anomer, and assayed individually (Table 4.3).

Seeberger and co-workers[25] have reported synthesis of interesting C_2-symmetrical kanamycin derivatives, such as **65**, from glycosylation of diazido-2,5-dideoxystreptamine, **61** (Scheme 4.11).

4.4.3. Synthesis of Kanamycin B Analogs via Glycodiversification Strategy

Based on the method of glycosylation, our group has developed a facile library synthesis of kanamycin B analogs (Scheme 4.12).[21] The strategy, which is termed glycodiversification, is to "swap" the originally attached carbohydrate components with the synthetic carbohydrates followed by further structural modifications. With the attachment of various carbohydrates bearing designed modifications for

TABLE 4.3. Zone of Inhibition of Kanamycin Derivatives

Compounds	*S. aureus*	*B. subtillis*	*E. coli K-12*	*M. smegmatis*
Neamine	19.5	31.5	28.9	26.5
59	18.5	30.4	31.3	18.9
60	0	12.0	16.4	0

Unit, mm; *"B. subtillis," Bacillus subtillis; M. smegmatis, Mycobacterium smegmatis.*

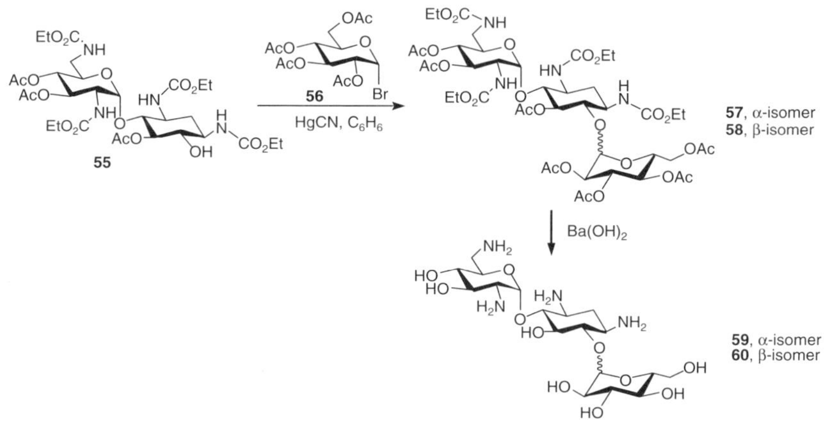

Scheme 4.10. Synthesis of kanamycin derivatives via glycosylation.

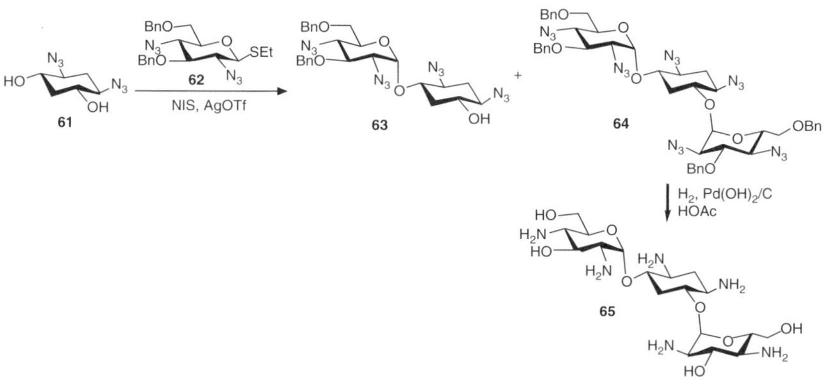

Scheme 4.11. Synthesis of kanamycin derivatives via glycosylation.

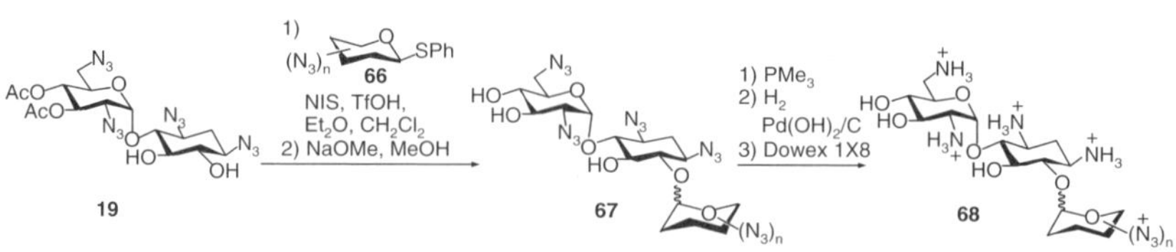

Scheme 4.12. Synthesis of kanamycin derivatives via glycosylation.

probing the efficacy of various structural modifications, a rationale-based drug design can be employed.

To expedite the purification process, we use the azido group as the surrogate of the amino group. The glycosylation donors that have benzyl or azido groups at the C-2 position favor the formation of an α-glycosidic bond under the influence of anomeric and solvent effects. The neamine acceptor, **19**, undergoes regiospecific glycosylation at O-6 position, resulting in the desired 4,6-disubstituted 2-deoxystreptamine motif. A wide range of structural features can be introduced following the same strategy.

4.5. SYNTHESIS OF NEOMYCIN CLASS AMINOGLYCOSIDES

4.5.1. Synthetic Strategies in Direct Modification from Neomycin and Paromonmycin

Extensive efforts have yielded numerous neomycin derivatives.[1,3] Recent advancements in structural studies involving the aminoglycoside-bound rRNA molecules have provided new concepts for the modification of aminoglycoside.[7–10] Based on the X-ray structure of neomycin B bound to the APH(3′) combined with ADP, 5″-OH was found to be one of the two hydroxy groups not employed for binding in the AME.[10] Therefore, design of introducing modification via glycosylation at O-5″ was proposed leading to the synthesis of 5″-glycosylated neomycin derivatives (Figure 4.6).[26,27] Among these compounds, **72** displays improved antibacterial activity against resistant strain (APH(3′)) as compared to the parent aminoglycoside, neomycin B (Table 4.4).

Recently, two research groups have independently synthesized the conformationally constrained neomycin derivatives, **74**, with the 2′-NH_2 group linked to the C-5″ (Scheme 4.13).[28,29] The information from X-ray crystallography has revealed a drastic conformational "flipping" of the 2-deoxystreptamine component between the binding sites of rRNA and ANT(4′).[30] Thus, it is proposed that an intramolecular link may obstruct the needed conformational change for ANT(4′)-catalyzed modification. Compound **74** shows lower activity against

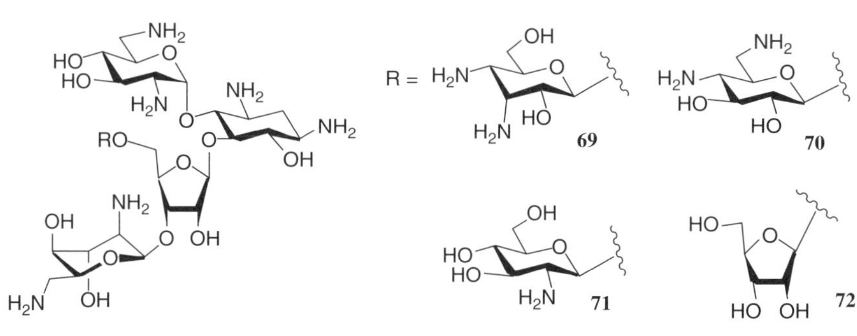

Figure 4.6. Structures of 5″-Glycosylated Neomycin.

TABLE 4.4. MIC of 5″-Glycosylated Neomycin Derivatives

Compounds	Kan	Neo	**69**	**70**	**71**	**72**
E. coli (ATCC 25922)	—	8–10	95	40–50	25–30	10–11
E. coli (APH(3′))	260–270	50–60	>200	>200	>200	35–45
S. epidermidis (ATCC 12228)	—	0.3–0.4	5.5–7	1.5–1.8	1.4–1.8	0.2–0.4
B. subtilis (ATCC 6633)	—	0.8–0.9	8.5–10	3.5–4	1.4–1.8	0.6–0.8
Salmonella virchow (APH(3′))	500–570	200–250	>1250	>1250	>1250	75–125
P. aeruginosa (ATCC 27853)	450–500	55–60	110–130	30–35	40–50	60–65

Unit, μg/mL; Neo, neomycin B.

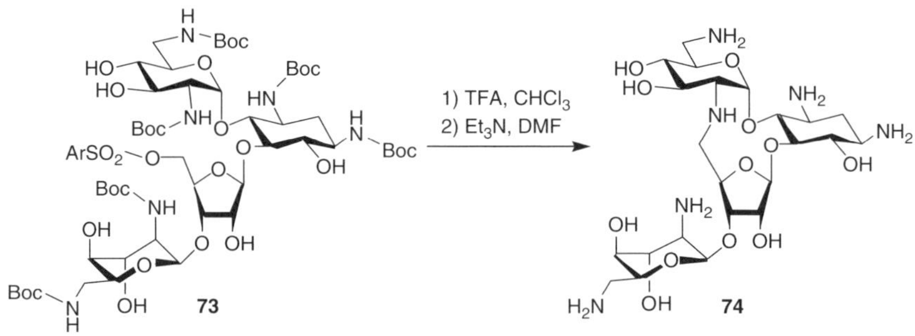

Scheme 4.13. Conformational constrained neomycin.

susceptible strains of bacteria; however, it maintains its activity against resistant bacteria expressing ANT(4′) while the activity of neomycin decreases drastically (Table 4.5).

4.5.2. Synthesis of Neomycin Derivatives via Glycosylation

After selective cyclic carbamate formation, the 3′-OH can be selectively protected using THP (Scheme 4.14).[18] The 5-OH was then glycosylated with a thiofuranose, creating a ribostamycin analog, **78**. However, compound **78** was

TABLE 4.5. MIC of Conformational Constrained Neomycin

Compounds	Neo B	**74**
S. epidermis	2	10
B. cereus	1	5
E. coli (BL21)	0.5	10
Alcaligenes faecalis	1	20
E. coli (DH5α)	3	20
E. coli (DH5α) (ANT(4′))	60	20

Unit, μg/mL.

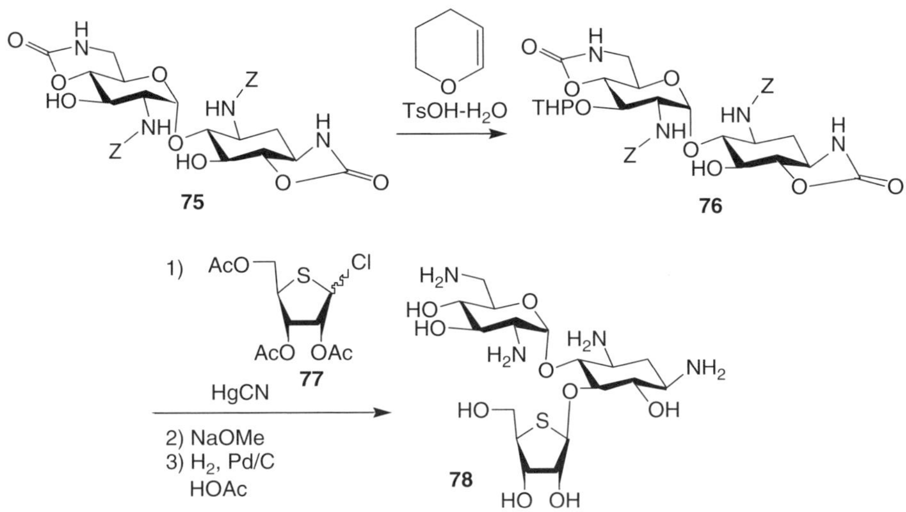

Scheme 4.14. Synthesis of kanamycin derivatives via glycosylation.

found to be inactive against various strains of bacteria in contrast to ribostamycin. The result may be explained by the disruption of the H-bond between 2′-NH_2 and 5″-O, which has been suggested to be pivotal for orienting the optimal conformation of ring I.[7]

Ribostamycin analogs have also been synthesized via glycosylation of tetraazidoneamine (Scheme 4.15).[31] Aminoalkyl side chains were employed in place of ring IV of neomycin B. Two such ribostamycin analogs, **83** and **84**, were found to have activity similar to that of neomycin B but better than that of ribostamycin (Table 4.6). The results proved that the aminoalkyl group can be a replacement for ring IV. However, **83** and **84** were inactive against *P. aeruginosa*, which has intrinsic aminoglycoside resistance.

4.5.3. Synthesis of Pyranmycin via Glycodiversification Strategy

Pyranmycin denotes a group of synthetic aminoglycoside antibiotics containing a 4,5-disubstituted 2-deoxystreptamine core, which is the character of neomycin class antibiotics.[22,32–34] However, it differs from neomycin by containing a pyranose in place of furanose at the O-5 position of neamine. The glycosidic bond of a furanose (five-membered ring) is more acid-labile than that of a pyranose (six-membered ring).[35,36] Therefore, it is expected that the replacement of the furanose ring (III) with an aminopyranose will increase the much-needed acid stability and, hopefully, reduce the cytotoxicity due to the lower administered dosage needed for achieving the therapeutically effective concentration of antibiotics.

Pyranmycin which contains both D-pyranose and L-pyranose at ring III, has been synthesized (Scheme 4.16). The β-glycosidic bond between rings II and

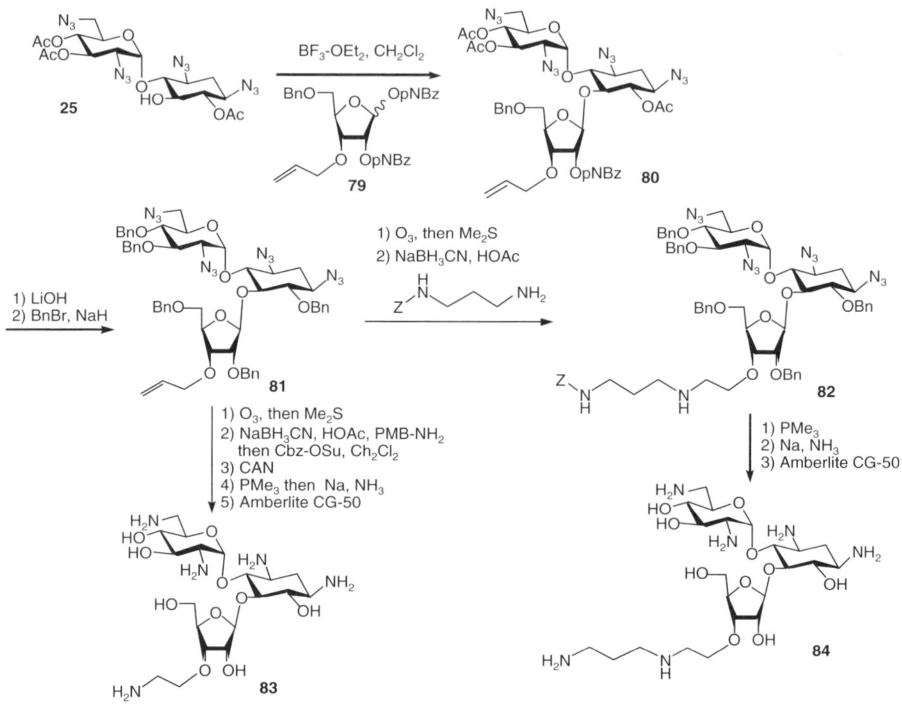

Scheme 4.15. Synthesis of ribostamycin derivatives via glycosylation.

TABLE 4.6. Antibacterial Assay of Ribostamycin Derivatives

Compounds	*S. aureus* (ATCC 25923)	*E. coli* (ATCC 25922)		*P. aeruginosa* (ATCC 27853)
	Inhibition zone	Inhibition zone	MIC	Inhibition zone
Ribostamycin	14.5	16.5	8	Inactive
Neomycin B	21.5	20.5	1.5	9.5
Paromomycin	19.5	18	5.5	Inactive
83	18.5	18.5	2.3	Inactive
84	21	19	1.4	Inactive

Unit, mm for inhibition zone, μg/mL for MIC.

III was directed by neighboring acetyl group of glycosyl trichloroacetimidate donors. The antibacterial assay result reveals three leads, **TC005, TC006**, and **TC010**. The presence of 6″-CH_3 was found to be an important factor in increasing the antibacterial activity for D-pyranose, while 6″-CH_2OH is important for L-pyranose. Nevertheless, adding a side chain on ring III did not increase the antibacterial activity.[22]

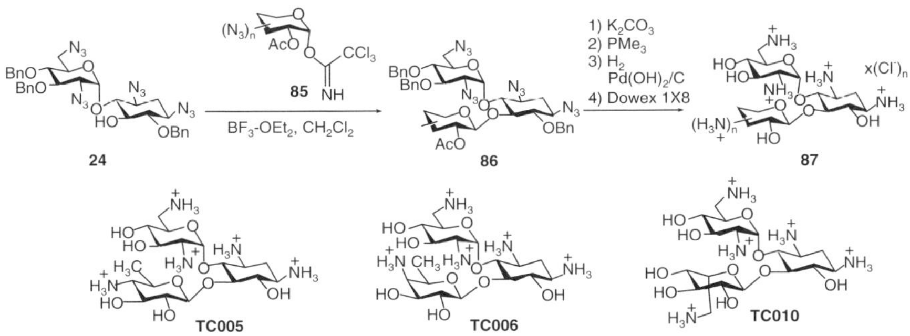

Scheme 4.16. Synthesis of designed pyranmycin.

4.6. DESIGNS OF AMINOGLYCOSIDES AGAINST RESISTANT BACTERIA

Overexpression of aminoglycoside modifying enzymes (AME) from resistant bacteria is the most commonly encountered mode of resistance.[4,5,37] Various aminoglycoside-modifying enzymes have been identified that catalyze a wide range of modifications including acetylation, phosphorylation, and adenylation, which prevent the modified aminoglycosides from binding to the targeted site of rRNA and, thus, enable the bacteria to acquire resistance. These enzymes are grouped as aminoglycoside phosphotransferases (APHs), aminoglycoside acetyltransferases (AACs), and aminoglycoside nucleotidyltransferases (ANTs). However, more than 50 different isoforms of these enzymes have been clinically isolated with subtle to significant differences in their capability of modifying various aminoglycosides.

Over the course of more than 30 years, many aminoglycoside derivatives have been developed with the goal of evading the action of AME, thus regaining the activity against aminoglycoside-resistant bacteria. Examples include methylation at the amino group where acetylation may take place, substitution of hydroxy group with fluoride to avoid the action of APHs or ANTs, and deoxygenation of hydroxy group(s) at the site of modification(s). Nevertheless, in order to provide a narrative among these works, we will focus on the recent synthetic modifications that are effective against a broader range of resistant bacteria with different AME.

4.6.1. 3′,4′-Dideoxygenation

One of the most prevalent modifying enzymes is APH(3′), which catalyzes phosphorylation at the 3′-OH of both neomycin and kanamycin classes of aminoglycosides, rendering the phosphorylated adduct incapable of binding toward the ribosomal target (Figure 4.7). Dideoxygenation at 3′ and 4′ positions has been proved to be effective against APH(3′) as reported by Umezawa and others.[38–47] The concept has led to the syntheses and discovery of tobramycin,[48,49] arbekacin,[50] and other similar aminoglycosides.

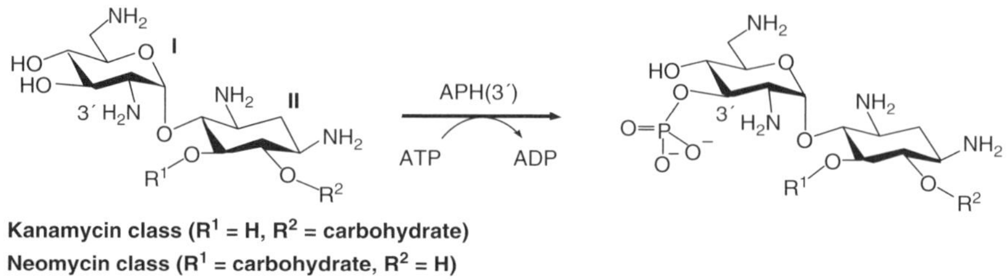

Figure 4.7. Phosphorylation of kanamycin and neomycin classes of aminoglycosides by APH(3′).

The reported methods for 3′,4′-dideoxygenation on aminoglycosides often involve converting a *trans*-diol into dimesylate, followed by zinc-mediated elimination.[51,52] In an improved synthesis of 3′,4′-dideoxykanamycin B (arbekacin), the *trans*-3′,4′-diol were converted into benzylsulfonates and then followed by elimination using NaI (Scheme 4.17).[46] It is a convenient way for the dideoxygenation of carbamate-protected aminoglycosides. Nevertheless, the condition may not be compatible with the presence of azido groups. Most of the syntheses begin with the kanamycin scaffold. There are very few examples of deoxygenation on neomycin class antibiotics.[47] In addition, the reported syntheses of both classes of antibiotics usually derive from kanamycin or neomycin, which limits the options for introducing novel structural motifs at other desirable places of aminoglycosides.

In light of the existing deficiencies in the development of aminoglycosides with 3′-deoxygenation, our group has developed an improved synthesis for the preparation of 3′,4′-dideoxy-1,3,2′,6′-tetraazidoneamine, **92** (Scheme 4.18). The dideoxygenation can be carried out in the presence of azido groups. Utilizing **92**

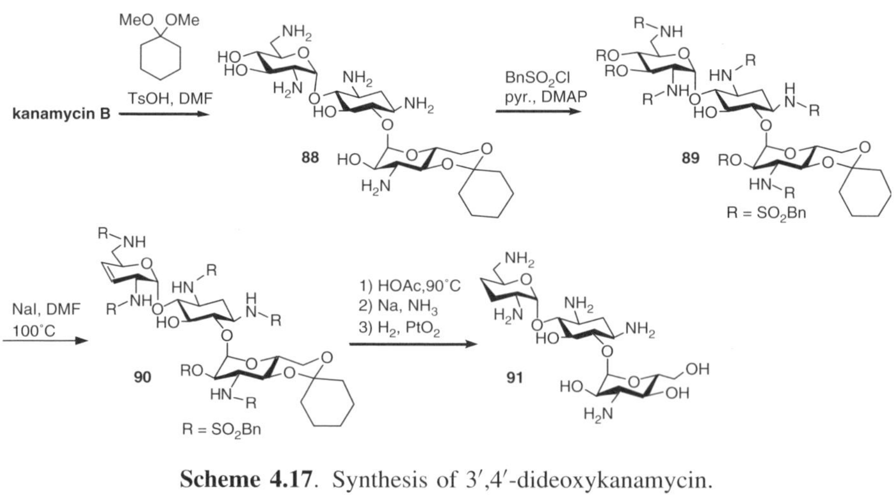

Scheme 4.17. Synthesis of 3′,4′-dideoxykanamycin.

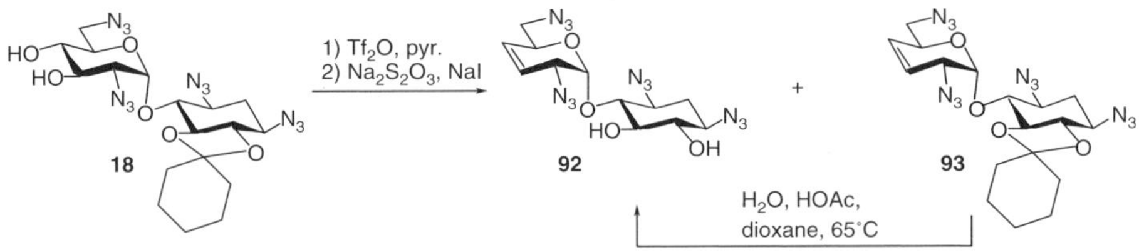

Scheme 4.18. Synthesis of 3′,4′-dideoxyneamine acceptor.

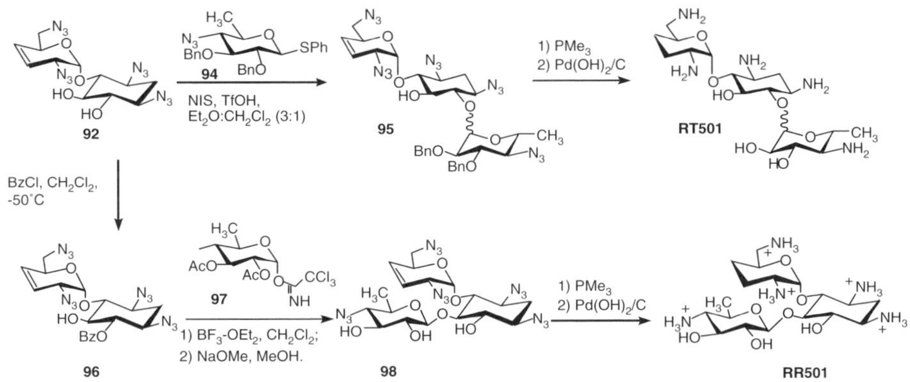

Scheme 4.19. Synthesis of 3′,4′-dideoxy kanamycin B analog and pyranmycin.

as the common core, the syntheses of 3′,4′-dideoxygenation products of both classes of aminoglycosides have been accomplished (Scheme 4.19).[53] Direct glycosylation of **92** with thioglycoside gives 3′,4′-dideoxygenated kanamycin analogs, while regioselective protection of O-6 hydroxyl group of **92**, followed by glycosylation with trichloroacetimidate donors, gives 3′,4′-dideoxy pyranmycin adduct.

The synthesized 3′,4′-dideoxyaminoglycosides were assayed against aminoglycoside susceptible and resistant strains of *E. coli* (Table 4.7). As expected, both **RR501** and **RT501** are active against resistant bacteria equipped with APH(3′). They are, however, less active against bacteria equipped with AAC(6′)/APH(2″)″.

4.6.2. N-1 Modification

Attaching functionalities at the N-1 position of the 2-deoxystreptamine among kanamycin or neomycin class antibiotics, is one of the other most effective methods of reviving the activity against aminoglycoside resistant bacteria. This strategy has led to the development of semisynthetic amikacin that has an (*S*)-4-amino-2-hydroxybutyryl (AHB) group at N-1 position.

The synthesis of kanamycin and neomycin class aminoglycosides with N-1 modification can be achieved via enzymatic[47,54] or chemical methods (for

TABLE 4.7. MIC of Synthesized 3′,4′-Dideoxyaminoglycosides

Compounds	Strains		
	E. coli (TG1)	*E. coli* (TG1) (AAC(6′)/APH(2″))	*E. coli* (TG1) (APH(3′)-I)
Amikacin	1	1	0.5
Kanamycin B	4	Inactive	32
Ribostamycin	2	16	Inactive
Butirosin	0.5	0.5	0.5
RR501	8	4	4
RT501	8	Inactive	4

Unit, μg/mL.

example, metal chelation methods).[14,39,55–61] A wealth of compounds bearing an AHB group at N-1 have been reported. For example, the synthesis of neomycin class with AHB at N-1 has been accomplished using regioselective carbamate cyclization followed by hydrolysis of cyclized carbamate (Figure 4.8).[55,61] Two compounds, **99** and **101**, with ring IV attached to the O-2″ were also prepared. However, these two compounds were much less active than those with ring IV attached to O-3″ (Table 4.8). Having an AHB group at N-1 in the normal neomycin scaffold did increase the activity even against a resistant strain of *E. coli*. Finally, all these four analogs were ineffective against *P. aeruginosa*, which exemplified the challenge in finding an active aminoglycoside against this formidable pathogen.

Direct chemical modification of butirosin (ribostamycin analog with AHB at N-1) has led to the synthesis of several butirosin derivatives (Figure 4.9).[47] These butirosin analogs manifested superior activity than butirosin A, especially

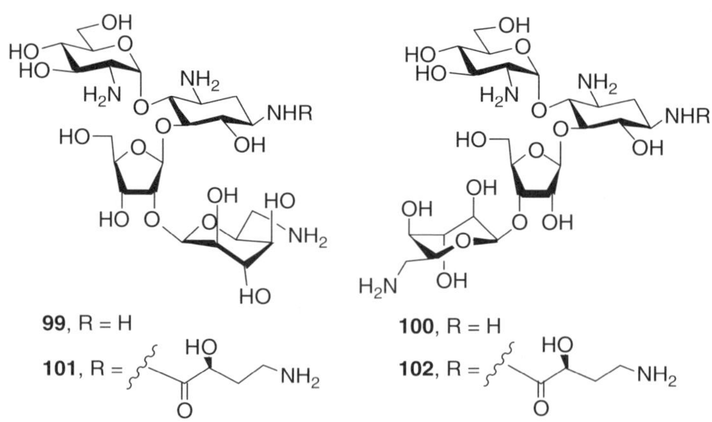

Figure 4.8. Neomycin class of aminoglycosides with N-1 AHB.

TABLE 4.8. MIC of Neomycin Class Aminoglycosides with N-1 AHB

Bacterial Strains	Par	**99**	**100**	**101**	**102**
S. aureus	0.39	6.25	3.12	100	3.12
B. subtilis	<0.2	0.78	<0.2	3.12	<0.2
E. coli K-12	3.12	50	12.5	>100	6.25
E. coli (APH(3′)-APH(5″))	>100	>100	>100	>100	6.25
P. aeruginosa A3	12.5	100	50	>100	100
M. smegmatis	0.39	3.12	6.25	100	1.56

Unit, μg/mL; Par, paromomycin.

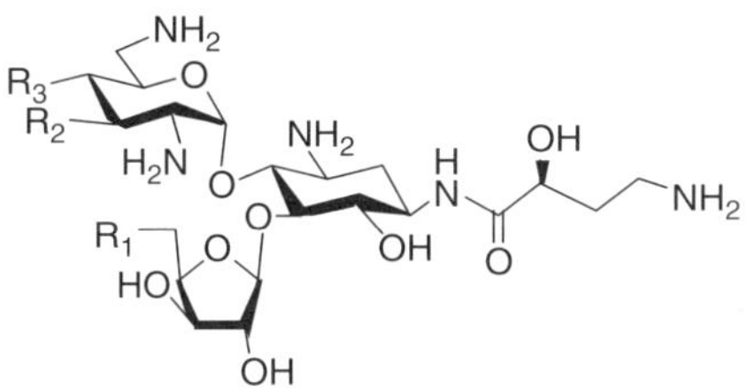

	R_1	R_2	R_3
Butirosin A	OH	OH	OH
103	NH_2	OH	OH
104	NH_2	H	H
105	OH	H	OH
106	OH	OH	H
107	NH_2	H	OH
108	NH_2	OH	H

Figure 4.9. Ribostamycin analogs with N-1 AHB.

TABLE 4.9. MIC of Synthesized Ribostamycin with N-1 AHB (Butirosin)

Bacterial strains	But A	Gen	**103**	**104**	**105**	**106**	**107**	**108**
S. aureus (APH)	50	3.1	25	3.1	6.3	—	3.1	6.3
E. coli	4.4	1.6	6.2	2.2	—	—	—	—
E. coli (APH)	200	200	200	12.5	3.1	—	25	25
P. aeruginosa	200	25	200	6.3	25	—	12.5	25
P. aeruginosa GNT (resistant strain)	31.3	126	10.0	3.9	12.5	—	6.3	6.3

Unit, μg/mL; But, butirosin; Gen, gentamicin.

for compound **104** against *P. aeruginosa* (Table 4.9). Compounds with deoxygenation at C-3′ maintain their activity against bacteria equipped with APH while butirosin, which still has its 3′-OH, is ineffective.

Metal-chelating method has also been commonly employed for the regioselective incorporation of AHB at N-1 of kanamycin class aminoglycoside (Scheme 4.20).[39] Other kanamycin derivatives have been prepared in similar fashion (Figure 4.10).[57]

The trend of antibacterial activity of these kanamycin derivatives with N-1 AHB (amikacin analogs) is similar to amikacin regardless of the presence of 3′-deoxygenation (Table 4.10). The design of amikacin is, however, less effective against AAC(6′). One interesting finding is the difference between 3′,

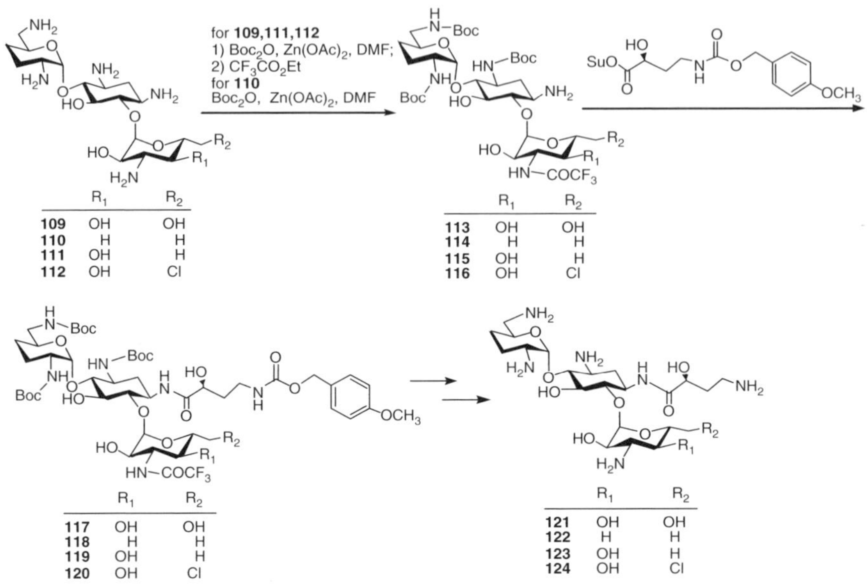

Scheme 4.20. Synthesis of kanamycin class of aminoglycosides with N-1 AHB.

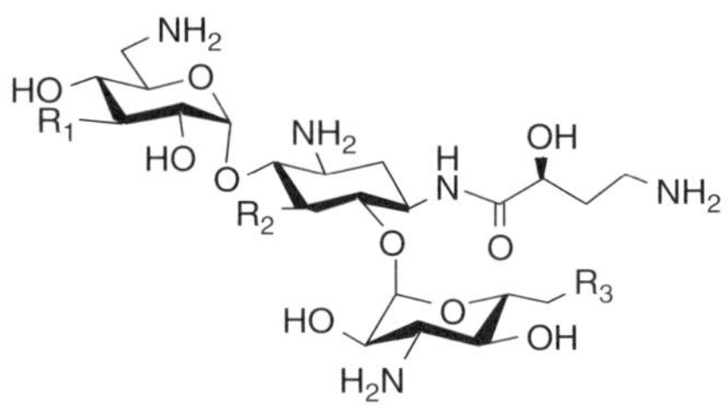

	R_1	R_2	R_3
amikacin	OH	OH	OH
125	H	OH	OH
126	H	OH	H
127	H	H	OH
128	H	H	H
129	OH	H	OH

Figure 4.10. Kanamycin class of aminoglycosides with N-1 AHB.

4′-dideoxygenation and 3′-deoxygenation. The former design seems to be ineffective against bacteria with ANT(2″), while the latter retains antibacterial activity. In addition, 3′-deoxyamikacin appears to be more active against bacteria equipped with APH(3′)-I than APH(3′)-III. Compound **124** bearing 6″-Cl was found to be equally active; however, it was reported to have higher toxicity.

Direct modification has also been applied to the synthesis of fluoro-substituted amikacin (Scheme 4.21).[58] Utilizing the metal-chelating protocol and the steric hindrance of 5-OH, fluorination was carried out selectively at the 5-OH. Two compounds (132 and 133) were found to have similar trends of antibacterial activity (Table 4.11). However, the activities of amikacin, **132** and **133**, against *E. coli* (ANT(2″)) were found to be much higher than those reported in Table 4.10 regardless of the sites of deoxygenation. We believe that the results from Table 4.11 may better explain the trend of structure activity relationship

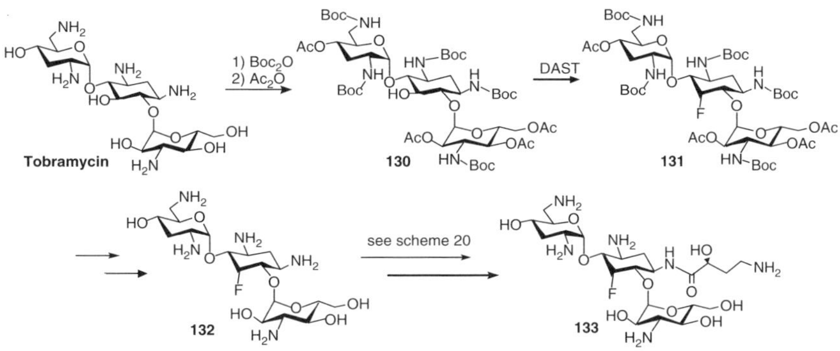

Scheme 4.21. Synthesis of fluorinated amikacin.

TABLE 4.10. MIC of Synthesized Kanamycin Derivatives with N-1 AHB

Bacterial Strains	Amk	**121**	**122**	**123**	**124**	**125**	**126**	**127**	**128**	**129**
S. aureus	1.56	3.13	3.13	3.13	0.78	<0.2	<0.2	0.39	0.39	0.78
S. aureus (ANT(4′))	1.56	6.25	12.5	6.25	0.78	1.56	3.13	6.25	12.5	6.25
E. coli	0.78	12.5	>100	25	3.13	0.78	0.39	1.56	0.78	1.56
E. coli (AAC(6′))	100	>100	>100	>100	>100	>100	100	>100	>100	>100
E. coli (APH(3′)-I)	1.56	12.5	12.5	6.25	3.13	0.78	0.78	1.56	0.78	3.13
E. coli (ANT(2″))	3.13	>100	>100	>100	>100	1.56	1.56	3.13	3.13	6.25
P. aeruginosa	3.13	12.5	6.25	3.13	3.13	0.78	1.56	1.56	3.13	3.13
P. aeruginosa (APH(3′)-I)	6.25	12.5	25	12.5	50	12.5	12.5	12.5	25	12.5
P. aeruginosa (AAC(6′))	>100	>100	>100	>100	>100	>100	100	>100	>100	>100
P. aeruginosa (AAC(3))	12.5	25	>100	25	12.5	12.5	12.5	25	50	12.5
P. aeruginosa (APH(3′)-III)	12.5	>100	>100	>100	50	12.5	100	25	100	25

Unit, μg/mL; Amk, amikacin.

TABLE 4.11. MIC of Fluoro-Substituted Amikacin

Bacterial Strains	Abk	amk	**132**	**133**
S. aureus	<0.2	<0.2	<0.2	<0.2
S. aureus (ANT(4′))	0.78	25	50	6.25
E. coli	0.39	<0.2	0.39	0.39
E. coli (AAC(6′))	>100	50	50	12.5
E. coli (APH(3′)-I)	0.78	0.78	0.39	0.78
E. coli (ANT(2″))	0.39	6.25	0.78	1.56
P. aeruginosa	<0.2	0.78	0.39	<0.2
P. aeruginosa (APH(3′)-II)	1.56	3.12	1.56	0.78
P. aeruginosa (AAC(6′))	3.12	25	>100	6.25

Unit, μg/mL; Abk, arbekacin (3′,4′-dideoxykanamycin with N-1 AHB).

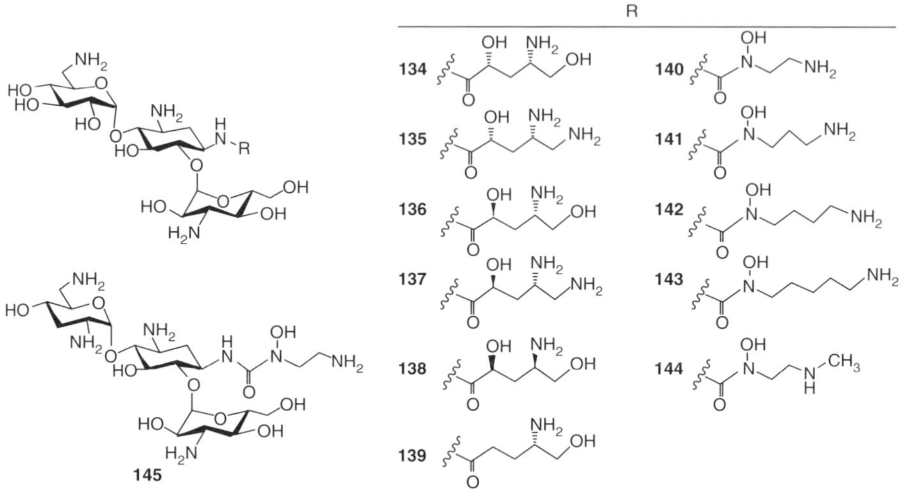

Figure 4.11. Kanamycin with different side chains at N-1.

(SAR) among modified aminoglycosides. The lower activity against AAC(6′), even with the presence of AHB group at N-1, was also observed in these fluoro-substituted amikacin analogs.

In a recent article, the metal-chelating method was used to provide a different functional group at the N-1 position (Figure 4.11).[62]

Among these amikacin-type derivatives, compounds **135, 136,** and **145** display superior activity, with **145** being the most active (Table 4.12). The presence of 2-OH in S-configuration on the side chain appears to be important. Very interestingly, compound **145**, the *N*-hydroxyurea analog, is even more active than amikacin while being one atom short of the normal AHB scaffold. Compounds **140** and **145** were also found to be active against *P. aeruginosa* (ATCC 27853) (IC_{50} 6.25 μg/mL). However, both compounds were inactive against resistant *P. aeruginosa* and methicillin-resistant *S. aureus* (MRSA) (ATCC 33591).

Other functionalities at the N-1 position, such as a 2-aminoethanesulfonyl group, have also been attempted (Figure 4.12).[60] The attachment of a 2-aminoethanesulfonyl side chain rendered similar activity against *E. coli*, which is expected from an amikacin-type aminoglycoside (Table 4.13). However, these compounds, **146–149**, appeared to be much less active against *P. aeruginosa* even equipped with the same type of AME (APH(3′)-I). The result could be attributed to the observed relative low permeability of *P. aeruginosa*.

Despite these prior endeavors, regioselective modification at the N-1 position remains a challenging task, especially for the azidoaminoglycosides. Wong and others have reported that the electron-withdrawing protecting groups will enhance the reactivity of their vicinal azido group toward the Staudinger reaction.[63] Such an electron-withdrawing effect can be correlated with chemical shifts of the corresponding protons (H-1, H-3, H-2′, and H-6′) adjacent to the azido groups.

TABLE 4.12. MIC of Kanamycin with Different Side Chains at N-1

Compounds	*S. aureus* (ATCC 13709)	*E. coli* (ATCC 25922)
Kanamycin A	1.2–2.5	2.5–5
Kanamycin B	0.3–0.6	1.2–2.5
Tobramycin	0.3–0.6	0.6–1.2
Amikacin	1.2–2.5	1.2–2.5
Paromomycin	1.2–2.5	2.5–5
134	>10	>10
135	1.2–2.5	5–10
136	1.2–2.5	2.5–5
137	>10	2.5–5
138	>10	5–10
139	>10	>10
140	2.5–5	2.5–5
141	>10	2.5–5
142	2.5–5	>10
143	>10	>10
144	2.5–5	>10
145	0.6–1.2	0.6–1.2

Unit, μM.

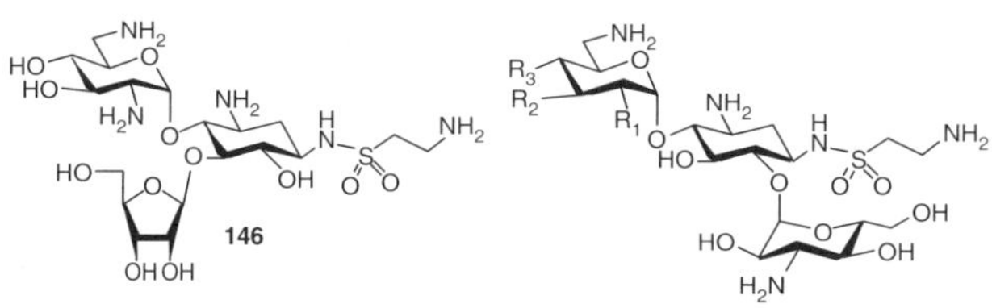

	R_1	R_2	R_3
147	OH	OH	OH
148	NH_2	OH	OH
149	NH_2	H	H

Figure 4.12. Aminoglycosides with 2-aminoethenesulfonyl side chains at N-1.

TABLE 4.13. MIC of Amikacin and Butirosin Analogs

Bacterial Strains	Kan B	**146**	**147**	**148**	**149**
S. aureus	0.39	6.25	6.25	<0.39	3.12
E. coli K-12	0.78	1.56	3.12	1.56	12.5
E. coli (APH(3′)-I)	Inactive	1.56	3.12	1.56	6.25
E. coli (ANT(2″))	12.5	3.12	6.25	3.12	50
P. aeruginosa	50	12.5	3.12	1.56	6.25
P. aeruginosa (APH(3′)-I)	100	>100	25	25	>100
P. aeruginosa (AAC(3))	>100	>100	25	25	>100

Unit, μg/mL.

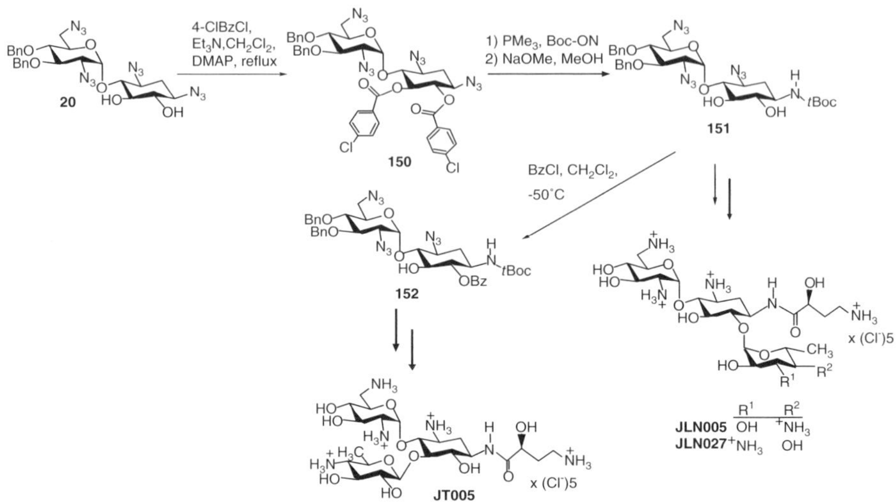

Scheme 4.22. Synthesis of pyranmycin and kanamycin B analogs with AHB group at N-1.

However, the N-2′ azido group appears to be more reactive than the N-1 azido group from the experimental and spectroscopic data.

Our group has successfully increased the reactivity of N-1 azido group by introducing di-(4-chlorobenzoyl) at the O-5 and O-6 positions, leading to the synthesis of the essential intermediate, **151** (Scheme 4.22).[64] Glycosylation at either O-5 or O-6 position followed by attaching the AHB side chain generates pyranmycin and kanamycin analogs with N-1 modification.

The constructed N-1-modified aminoglycosides were assayed against aminoglycoside susceptible and resistant strains of *E. coli*. All the synthesized aminoglycosides with the attachment of AHB group at N-1 regain the activity against both resistant strains of bacteria (Table 4.14). One of the synthesized kanamycin analogs, **JLN027**, is even more active against APH(3′)-I than the clinically used amikacin.

4.6.3. 3′-Deoxygenated Aminoglycoside: Synthesis of Tobramycin Derivatives

Recently, Hanessian has reported methods for direct modification of tobramycin.[65] Tobramycin (or nebramycin) differs from kanamycin B only in its 3′-deoxygenation, which makes tobramycin immune from the modification catalyzed by APH(3′). A library of tobramycin analogs bearing various functionalities at O-5 was prepared (Scheme 4.23). In general, these tobramycin analogs were less active than tobramycin (Table 4.15). However, compounds **157** and **159** showed activity against *P. aeruginosa* (ATCC 27853) (MIC = 12.5 μg/mL).

TABLE 4.14. MIC of Pyranmycin and Kanamycin B Analogs with AHB at N-1

	Strains		
Compound	*E.coli* TG1	*E.coli* TG1 (APH(3′)-I)	*E.coli* TG1 (AAC(6′)/APH(2″))
Butirosin	1	0.5	0.25
Ribostamycin	2	Inactive	8
JT005	4	4	4
Amikacin	1	0.5	1
Kanamycin	4	Inactive	Inactive
JLN005	4	2	2
JLN027	1	0.25	1

Unit, μg/mL.

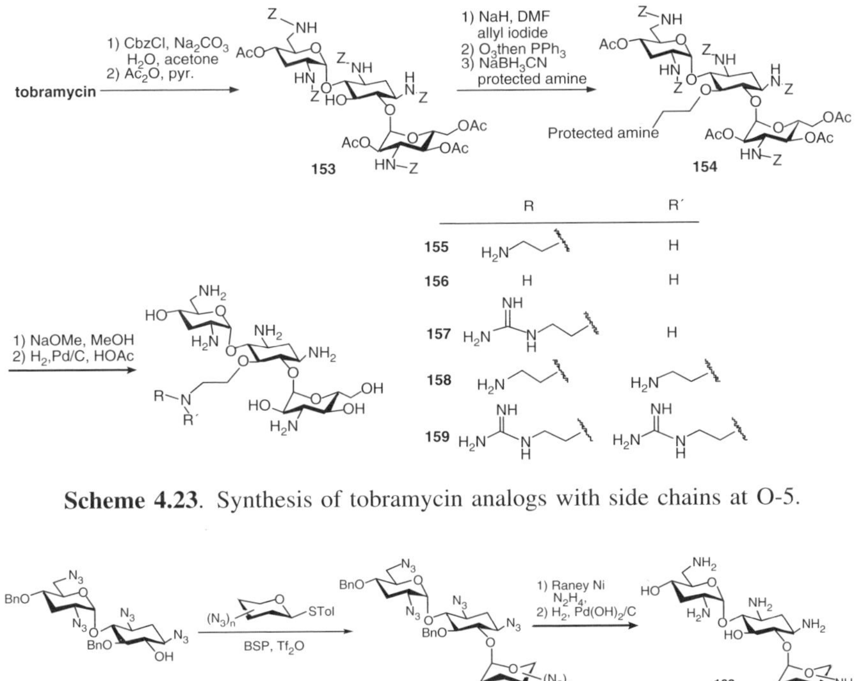

Scheme 4.23. Synthesis of tobramycin analogs with side chains at O-5.

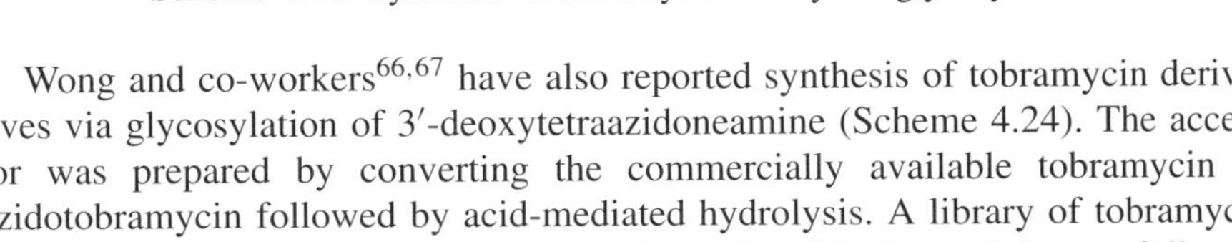

Scheme 4.24. Synthesis of tobramycin library via glycosylation.

Wong and co-workers[66,67] have also reported synthesis of tobramycin derivatives via glycosylation of 3′-deoxytetraazidoneamine (Scheme 4.24). The acceptor was prepared by converting the commercially available tobramycin to azidotobramycin followed by acid-mediated hydrolysis. A library of tobramycin analogs was synthesized from glycosylation using thioglycosyl donors followed by global deprotection. However, the antibacterial activities of these compounds were not reported in this article.

TABLE 4.15. MIC of Kanamycin with Different Side Chains at N-1

Compounds	*S. aureus* (ATCC 13709)	*E. coli* (ATCC 25922)
Kanamycin A	1.2–2.5	2.5–5
Kanamycin B	0.3–0.6	1.2–2.5
Tobramycin	0.3–0.6	0.6–1.2
Amikacin	1.2–2.5	1.2–2.5
Paromomycin	1.2–2.5	2.5–5
Neamine	>10	>10
155	>10	2.5–5
156	>10	5–10
157	>10	2.5–5
158	5–10	1–2
159	2.5–5	0.6–1.2

Unit, μM.

4.7. STUDIES OF OTHER AMINOGLYCOSIDES

4.7.1. Dimer of Neamine, Neomycin, and Kanamycin

By employing a 27-nucleotide RNA that represents the binding site of aminoglycoside toward the 16S rRNA, it has been found that neamine binds to such an rRNA sequence in a 2:1 ratio. Several neamine dimers were then constructed to investigate their antibacterial activity and their capability to resist or inhibit the action of AME (Scheme 4.25).[68–70] These neamine dimers displayed modest to excellent antibacterial activity (Table 4.16). In addition, these dimers have also been noted for their inhibitory effect against AAC(6′)-APH(2″).

Upon further screening and modification, two neamine/nebramine dimers were synthesized and assayed (Figure 4.13).[69,70] A much improved activity was observed even against various clinically isolated strains of *P. aeruginosa* (Table 4.17).

Dimers of kanamycin and neomycin have also been synthesized (Figure 4.14).[71,72] The 5″-OH and 6′-OH are the only primary hydroxy groups on neomycin and kanamycin, respectively, and often served as the point of modification. The studies in these articles, however, focus on the binding affinities of these kanamycin and neomycin dimers toward various RNA sequences of biological interest.

4.7.2. Aminoglycoside-Based Inhibitor or Inactivator Against Aminoglycoside-Modifying Enzymes

Design of aminoglycoside to function as the inhibitor (or inactivator) of AME has been reported less frequently than those involved in the modification of aminoglycosides to be used directly as antibiotic. However, these studies may provide

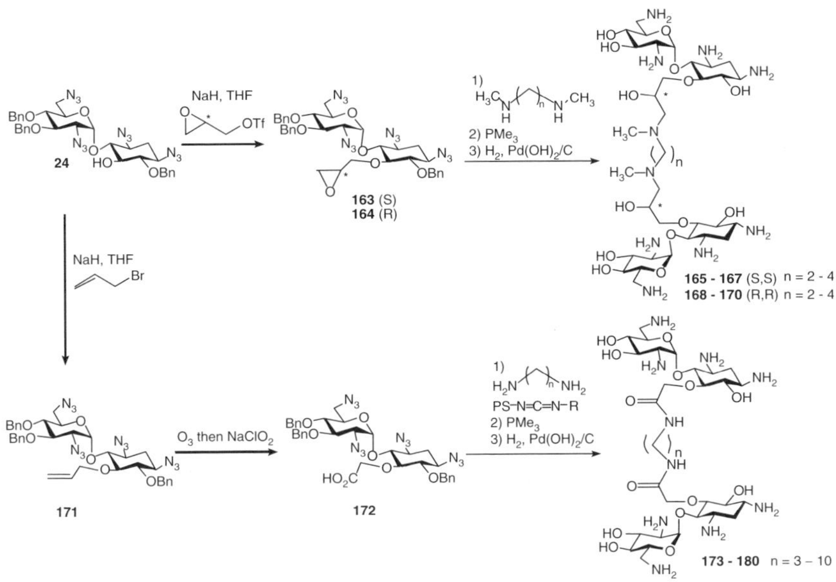

Scheme 4.25. Synthesis of neamine dimers.

TABLE 4.16. MIC of Neamine Dimers

Compounds	*E. coli* (ATCC 25922)
Neamine	50–100
Neomycin	3.1
Gentamicin	1.6
Ribostamycin	12.5
Kanamycin	12.5
Paromomycin	6.25
Tobramycin	3.1
165	25–50
166	12.5–25
167	100
168	100
169	12.5
170	6.25
173	31
175	125

Unit, μM.

valuable insights as well as new options for therapeutic designs. For example, neamine has been modified into affinity labeling molecules for AME by attaching the α-bromoacetyl group regioselectively at the amino group (Scheme 4.26).[73,74] The binding mode of these neamine derivatives, **186–189**, was not impaired by the incorporation of the α-bromoacetyl group, thus enabling these compounds to inactivate the AME of interest, (APH(3′)-IIa).

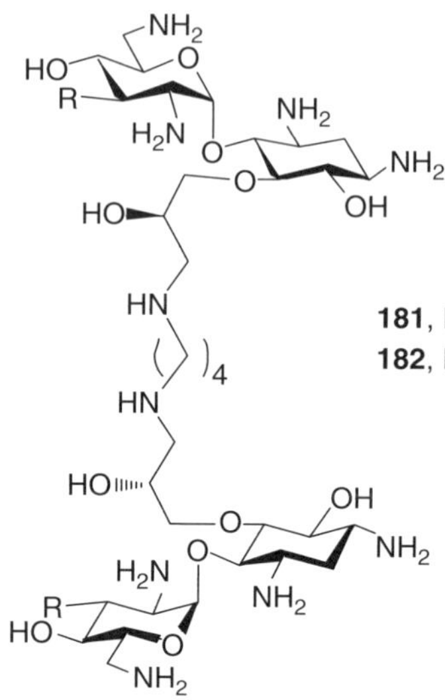

181, R = OH
182, R = H

Figure 4.13. Synthesized neamine dimers.

TABLE 4.17. MIC of Neamine Dimers

Compounds	Resistance	Tob.	**167**	**181**	**182**
S. aureus (ATCC 33591)	MRSA	>64	16	16	—
P. aeruginosa (strain 609)	Resistant	128	8	4	—
P. aeruginosa (strain 663)	Resistant	128	8	8	—
P. aeruginosa (PAE_NUH05)	Resistant	100	6.3	3.1	—
P. aeruginosa (PAE_NUH06)	Resistant	100	12.5	6.3	—
P. aeruginosa (27 EN)	Resistant	4	—	—	1
P. aeruginosa (98 EN)	Resistant	16	—	—	0.5
P. aeruginosa (89 EN)	Resistant	64	—	—	4

Unit, μg/mL; Tob, tobramycin.

Another interesting design against APH(3′) by Mobashery and co-workers[75] is 3′-ketokanamycin, **190**. The keto group is known to exist in equilibrium with its ketal form (Scheme 4.27). After phosphorylation catalyzed by APH(3′), the phosphorylated ketal can be transformed into a keto form by eliminating a dibasic phosphate, and then it can further re-regenerate the ketal. Therefore, the design of ketokanamycin is immune to the modification of APH(3′). In addition, during the catalysis cycle, a molecule of ATP is consumed. As a result, compound **190** may manifest multiple antibacterial modes of actions, such as binding toward the rRNA, inhibiting the APH(3′), and depleting the valuable ATP. The increased activity of **190** against resistant bacteria verifies the merit of this design (Table 4.18).

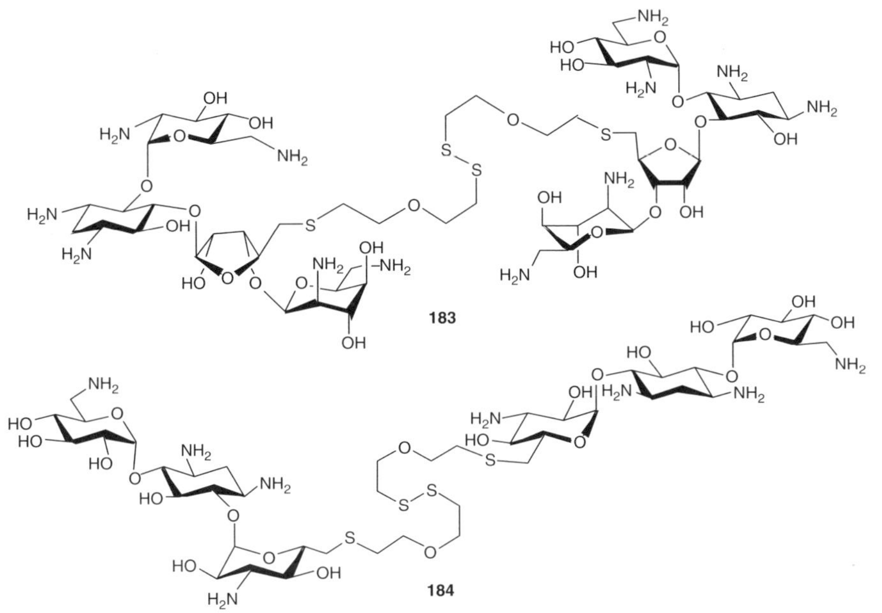

Figure 4.14. Synthesized kanamycin and neomycin dimers.

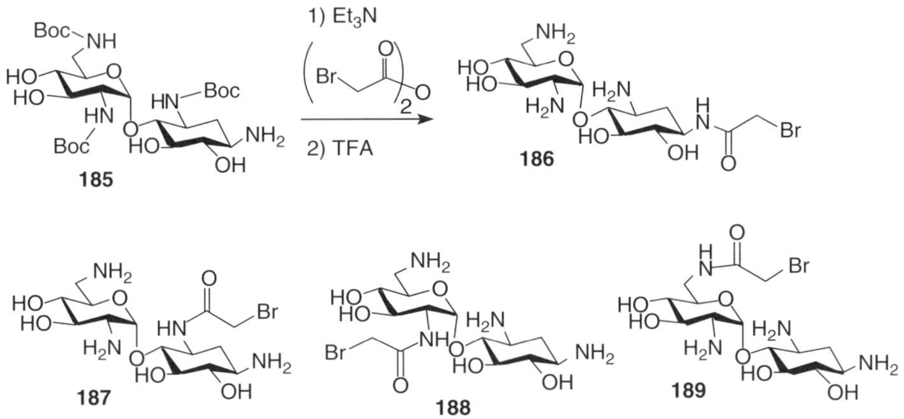

Scheme 4.26. Synthesis and structures of neamine-based inactivators.

Mobashery and co-workers[76] also reported the synthesis and inhibitory application of 2′-NO_2 derivatives of neamine and kanamycin (Scheme 4.28). Using the metal-chelating method, the 2′-NH_2 was selectively unmasked and then oxidized into 2′-NO_2, which will increase the acidity of 2′-H. Upon phosphorylation at the 3′-OH, elimination of phosphate will lead to the formation of a nitroalkene intermediate, **198**, that can function as a Michael acceptor and

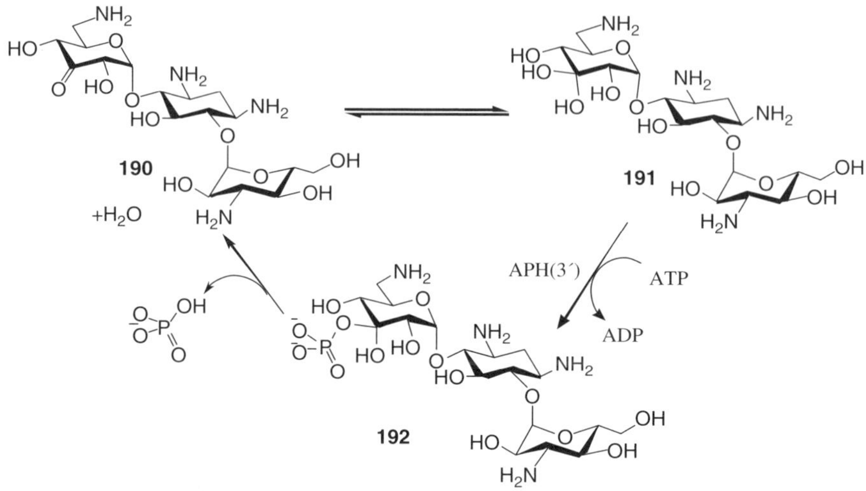

Scheme 4.27. Design and mode of action of ketokanamycin.

TABLE 4.18. MIC of 190

Compounds	Kan A	190
E. coli JM83	8	250
E. coli (APH(3′))	4000	1000

Unit, μg/mL

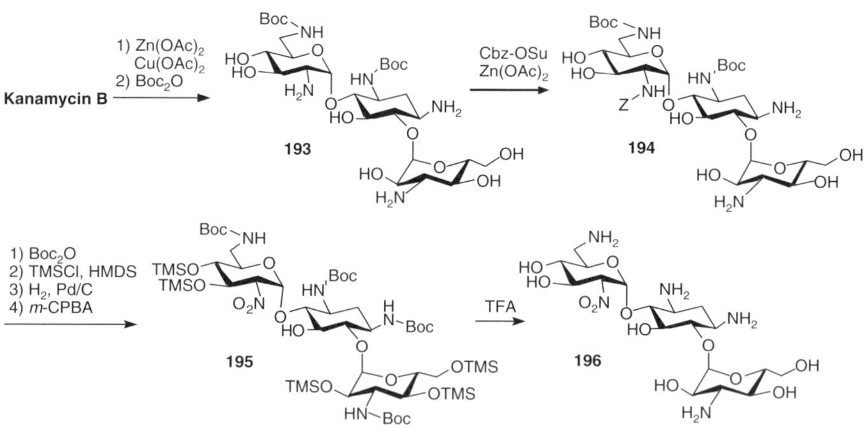

Scheme 4.28. Synthesis of 2′-nitrokanamycin.

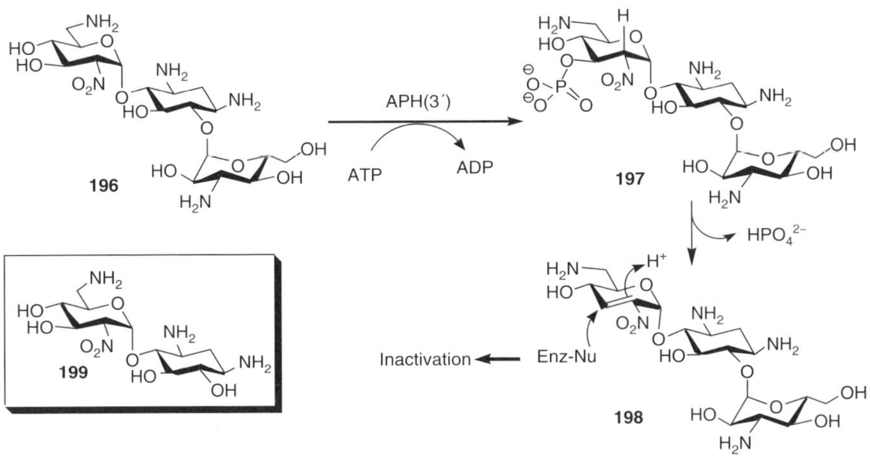

Scheme 4.29. Mode of action of 2′-nitrokanamycin.

inactivate the APH(3′) by trapping the nucleophilic residue of AME in the aminoglycoside binding site (Scheme 4.29). Both 2′-nitrokanamycin, **196**, and 2′-nitroneamine, **199**, inactivated APH(3′). However, it was noted that the kinetic parameters of **196** cannot be studied as they were for **199** ($K_I = 11.7 \pm 5.5$ mM and $K_{cat} = 0.06 \pm 0.02$ min^{-1}).

Inhibitor design that mimics an ATP-aminoglycoside complex has also been studied (Figure 4.15).[77] Such a tethered bisubstrate design contains a neamine core with the 3′-OH linked to adenosine via a nonhydrolyzable $(CH_2)_n$ tether in place of the triphosphate group. The synthesis began from a Boc-protected neamine (Scheme 4.30). Although the synthesis is not concise, several interesting transformations were employed, for example, the selective protection of the 3′-OH using TBS and the benzylation accompanied with the cyclic carbamate formation (from **204** to **205**).

4.7.3. Guanidinoaminoglycoside

In light of their capability of binding to RNA molecules, aminoglycosides have also been explored for their activity against RNA viruses.[78] Besides employing

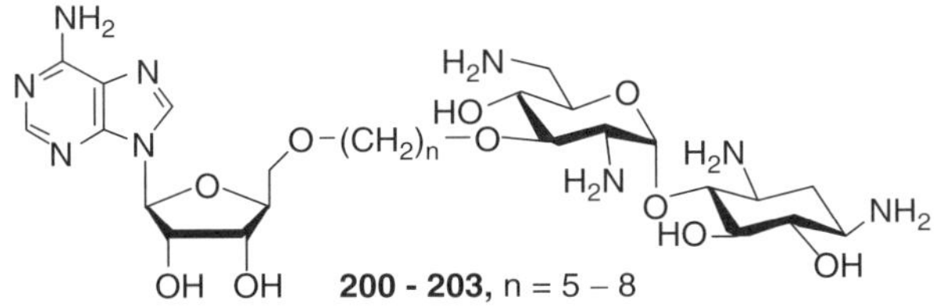

Figure 4.15. Structures of tethered bisubstrate derivatives.

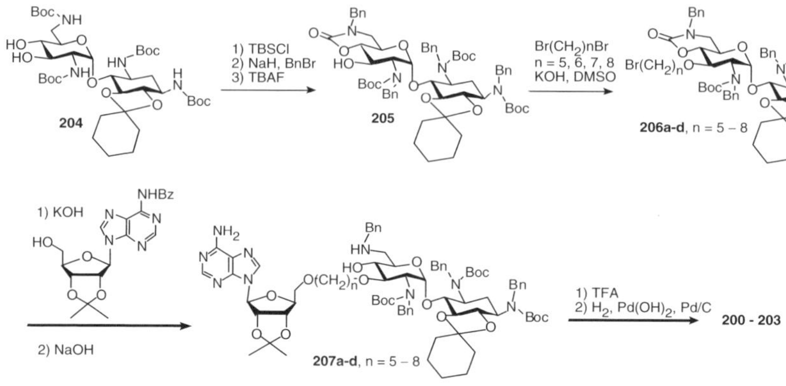

Scheme 4.30. Synthesis of tethered bisubstrates.

aminoglycoside directly for interacting with RNA molecules of viral origin, converting the amino groups into guanidine groups is the other popular design. It is reasoned that guanidinoaminoglycosides will have improved or altered interactions with RNA molecules of interest since the guanidine group has more H-bonding probability than does amino group. The transformation can be carried out in a two-step reaction using N, N'-di-Boc-N''-triflylguanidine, **208** (Scheme 4.31).[79] Kanamycin, neomycin, and pyranmycin have all been transformed into the corresponding guanidinoaminoglycosides (Figure 4.16).[80–82] However, these types of molecules are, in general, inactive against bacteria.

4.7.4. Heterocycle-Substituted Aminoglycosides

Design of using heterocyclic rings as a substitution for ring I has been explored (Figure 4.17 and Scheme 4.32).[83,84] These compounds were studied primarily for their binding affinity toward the fragment of the targeted 16S rRNA. However, compound **231** was found to be active against *E. coli* (ATCC25922) (MIC = 3 μM).

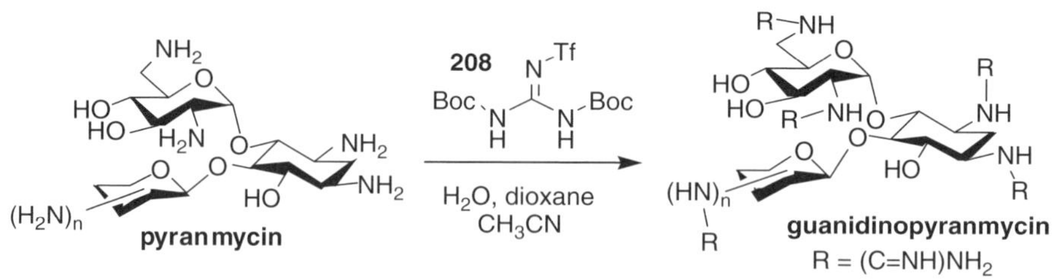

Scheme 4.31. Synthesis of guanidinoaminoglycosides.

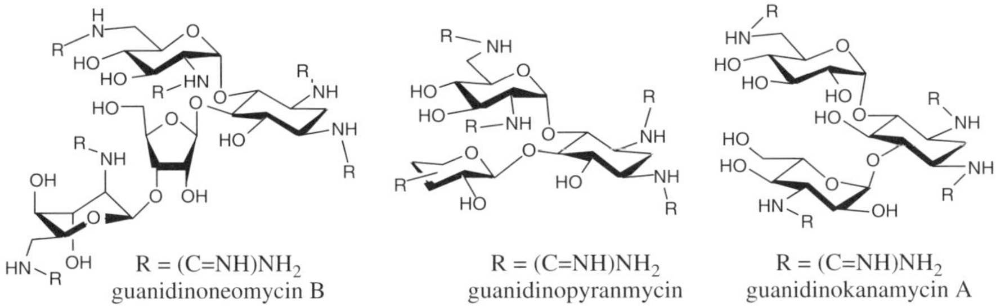

Figure 4.16. Structures of guanidinoaminoglycosides.

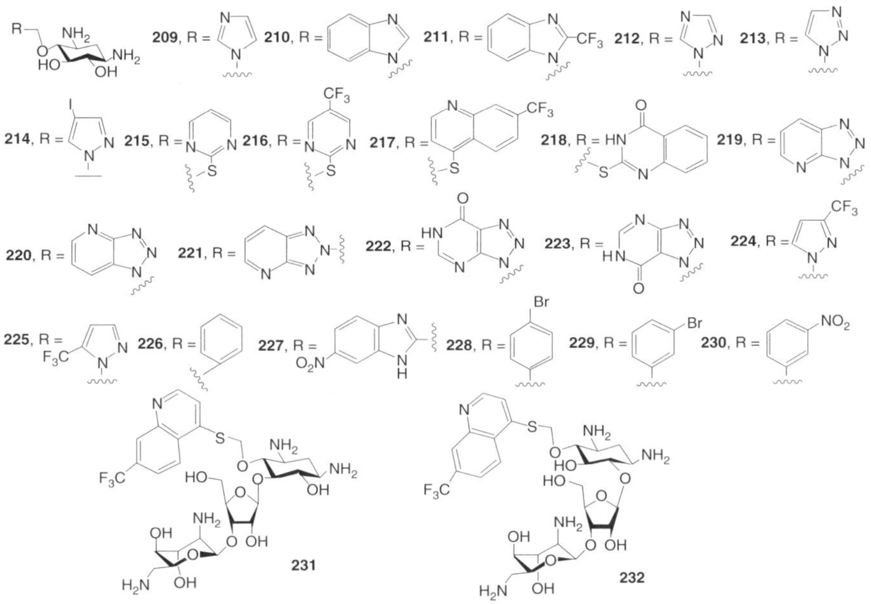

Figure 4.17. Structures of heterocycle-substituted aminoglycosides.

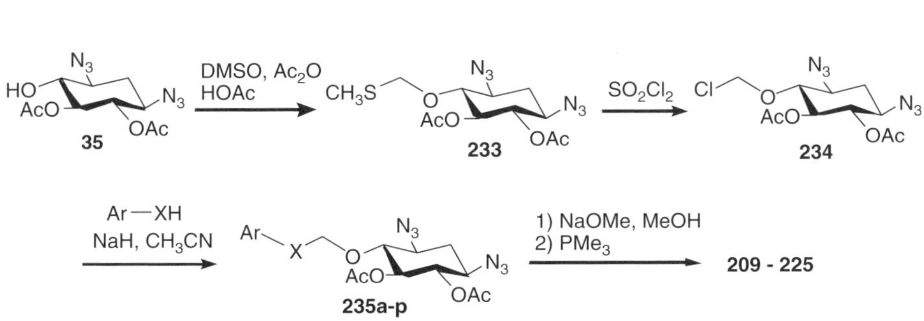

Scheme 4.32. Synthesis of heterocycle-substituted aminoglycosides.

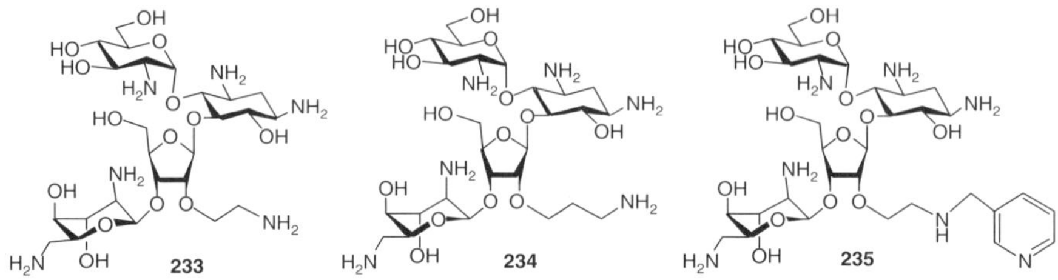

Figure 4.18. Structures of paromomycin with O-2″ side chains.

TABLE 4.19. MIC of Paromomycin with O-2″ Side Chains

Compounds	Par	233	234	235
E. coli (ATCC 25922)	5	6	25	5
S. aureus (ATCC 13709)	3	2	3	1
Streptococcus pyogenes (ATCC 49399)	3	—	—	6
Proteus vulgaris (ATCC 8427)	1	—	—	3
Klebsiella pneumoniae (ATCC 13883)	1	—	—	3

Unit, μM; Par, paromomycin.

Attaching a heterocyclic side chain at the O-2″ of paromomycin has also been reported (Figure 4.18).[85] These compounds have similar activity as compared to paromomycin (Table 4.19).

4.8. CONCLUSION

Despite their known high cytotoxicity, aminoglycosides remain a valuable resource against formidable pathogens, and their importance has been corroborated by the re-emerging interest documented in the literature. Meanwhile, structural knowledge from the studies of aminoglycoside resistance and mode of action has yielded new concepts and alternative designs against aminoglycoside resistant microorganisms. While generating these novel designs using biosynthetic approach and fermentation may not be feasible, the chemical syntheses of new aminoglycosides are often too complicated to be a practical option. The semisynthetic approach should still be the optimal approach for providing new drugs. Nevertheless, the merit of synthesizing real compounds for studying their antibacterial activity is irreplaceable. By combining the advancements in the studies of aminoglycoside resistance, structural information of the targeted rRNA, bacterial uptake of aminoglycoside, improved administration procedure, and advancements in these areas combined with synthetic achievements, we believe that aminoglycoside antibiotics will continue to be a versatile therapeutic for our community.

REFERENCES

1. Umezawa, S.; Tsuchiya, T. In *Aminoglycoside Antibiotics*; Umezawa, H.; Hooper, I. R., Ed.; New York: Springer-Verlag, 1982, pp. 37–110.
2. Hanessian, S.; Haskell, T. H. In *Antibiotics Containing Sugars*; Pigmann, H.; Horton, D., Eds.; New York/London: Academic Press, 1970, pp. 139–211.
3. Haddad, J; Kotra, L. P.; Mobashery, S. In *Glycochemistry Principles, Synthesis, and Applications;* Wang, P.G.; Bertozzi, C. R., Eds.; New York/Basel: Marcel Dekker, 2001, pp. 307–424.
4. Mingeot-Leclercq, M.-P.; Glupczynski, Y.; Tulkens, P. M. *Antimicrob. Agents Chemother*. **1997**, *43*, 727–737.
5. Kotra, L. P.; Haddad, J.; Mobashery, S. *Antimicrob. Agents Chemother*. **2000**, *44*, 3249–3256.
6. Wright, G. D. *Curr. Opin. Microbiol*. **1999**, *2*, 499–503.
7. Fourmy, D.; Recht, M. I.; Blanchard, S. C.; Puglisi, J. D. *Science* **1996**, *274*, 1367–1371.
8. Fourmy, D.; Recht, M. I.; Puglisi, J. D. *J. Mol. Biol*. **1998**, *277*, 347–362.
9. Vicens, Q.; Westhof, E. *Structure* **2001**, *9*, 647–668.
10. Fong, D. H.; Berghuis, A. M. *EMBO J*. **2002**, *21*, 2323–2331.
11. Suami, T.; Nashiyama, S.; Ishikawa, Y.; Katsura, S. *Carbohydr. Res*. **1976**, *52*, 187–196.
12. Georgiadis, M. P.; Constantinou-Kokolou, V. *J. Carbohydr. Chem*. **1991**, *10*, 739–748.
13. Kumar, V.; Remers, W. A. *J. Org. Chem*. **1978**, *43*, 3327–3331.
14. Tsuchiya, T.; Takagi, Y.; Umezawa, S. *Tetrahedron Lett*. **1979**, *51*, 4951–4954.
15. Nagabhushan, T. L.; Cooper, A. B.; Turner, W. N.; Tsai, H.; McCombie, S.; Mallams, A. K.; Rane, D.; Wright, J. J.; Reichert, P.; Boxler, D. L.; Weinstein, J. *J. Am. Chem. Soc*. **1978**, *100*, 5253–5254.
16. Roestamadji, J.; Grapsas, I.; Mobashery, S. *J. Am. Chem. Soc*. **1995**, *117*, 11060–11069.
17. Haddad, J.; Kotra, L. P.; Llano-Sotelo, B.; Kim, C.; Azucena, E. F., Jr.; Liu, M.; Vakulenko, S. B.; Chow, C. S.; Mobashery, S. *J. Am. Chem. Soc*. **2002**, *124*, 3229–3237.
18. Kumar, V.; Remers, W. A. *J. Org. Chem*. **1981**, *46*, 4298–4300.
19. Chou, C.-H.; Wu, C.-S.; Chen, C.-H.; Lu, L.-D.; Kulkarni, S. S.; Wong, C.-H.; Hung, S.-C. *Org. Lett*. **2004**, *6*, 585–588.
20. Greenberg, W. A,; Priestley, E. S.; Sears, P. S.; Alper, P. B.; Rosenbohm, C.; Hendrix, M.; Hung, S.-C.; Wong, C.-H. *J. Am. Chem. Soc*. **1999**, *121*, 6527–6541.
21. Li, J.; Wang, J.; Czyryca, P. G.; Chang, H.; Orsak, T. W.; Evanson, R.; Chang, C.-W. T. *Org. Lett*. **2004**, *6*, 1381–1384.
22. Elchert, B.; Li, J.; Wang, J.; Hui, Y.; Rai, R.; Ptak, R.; Ward, P.; Takemoto, J. Y.; Bensaci, M.; Chang, C.-W. T. *J. Org. Chem*. **2004**, *69*, 1513–1523.
23. Venot, A.; Swayze, E. E.; Griffey, R. H.; Boons, G.-J. *ChemBioChem* **2004**, *5*, 1228–1236.
24. Kumar, V.; Remers, W. A. *J. Med. Chem*. **1979**, *22*, 432–436.

25. Seeberger, P. H.; Baumann, M.; Zhang, G.; Kanemitsu, T.; Swayze, E. E.; Hofstadler, S. A.; Griffey, R. H. *Synlett*. **2003**, 1323–1326.
26. Fridman, M.; Belakhov, V.; Yaron, S.; Baasov, T. *Org. Lett*. **2003**, *5*, 3575–3578.
27. Fridman, M.; Belakhov, V.; Lee, L. V.; Liang, F.-S.; Wong, C.-H.; Baasov, T. *Angew. Chem. Int. Ed*. **2005**, *44*, 447–452.
28. Asensio, J. L.; Hidalgo, A.; Bastida, A.; Torrado, M.; Corzana, F.; Chiara, J. L.; Garcia-Junceda, E.; Canada, J.; Jimenez-Barbero, J. *J. Am. Chem. Soc*. **2005**, *127*, 8278–8279.
29. Blount, K. F.; Zhao, F.; Hermann, T.; Tor, Y. *J. Am. Chem. Soc*. **2005**, *127*, 9818–9829.
30. Pedersen, L. C.; Benning, M. M.; Holden, H. M. *Biochemistry* **1995**, *34*, 13305–13311.
31. Alper, P. B.; Hendrix, M.; Sears, P.; Wong, C.-H. *J. Am. Chem. Soc*. **1998**, *120*, 1965–1978.
32. Wang, J.; Li, J.; Tuttle, D.; Takemoto, J.; Chang, C.-W. T. *Org. Lett*. **2002**, *4*, 3997–4000.
33. Chang, C.-W. T.; Hui, Y.; Elchert, B.; Wang, J.; Li, J.; Rai, R. *Org. Lett*. **2002**, *4*, 4603–4606.
34. Wang, J.; Li, L.; Czyryca, P. G.; Chang, H.; Kao, J.; Chang, C.-W. T. *Bioorg. Med. Chem. Lett*. **2004**, *14*, 4389–4393.
35. Bochkov, A. F.; Zaikov, G. E. *Chemistry of the O-Glycosidic Bond: Formation and Cleavage*; Elmsford, NY: Pergamon Press, 1979.
36. Shallenberger, R. S.; Birch G., G. *Sugar Chemistry*; Westport, Conn.: Avi Pub. Co., 1975.
37. Shaw, K. J.; Rather, P. N.; Hare, R. S.; Miller, G. H. *Microbiol. Rev*. **1993**, *57*, 138–163.
38. Kondo, S. *Jpn. J. Antibiot*. **1994**, *47*, 561–574.
39. Umezawa, H.; Miyasaka, T.; Iwasawa, H.; Ikeda, D.; Kondo, S. *J. Antibiot*. **1981**, *34*, 1635–1640.
40. Jikihara, T.; Tsuchiya, T.; Umezawa, S.; Umezawa, H. *Bull. Chem. Soc. Jpn*. **1973**, *46*, 3507–3510.
41. Tsuchiya, T.; Takahashi, Y.; Endo, M.; Umezawa, S.; Umezawa, H. *J. Carbohydr. Chem*. **1985**, *4*, 587–611.
42. Umezawa, S.; Umezawa, H.; Okazaki, Y.; Tsuchiya, T. *Bull. Chem. Soc. Jpn*. **1972**, *45*, 3624–3628.
43. Umezawa, H.; Umezawa, S.; Tsuchiya, T.; Okazaki, H. *J. Antibiot*. **1971**, *24*, 485–487.
44. Umezawa, S.; Okazaki, Y.; Tsuchiya, T. *Bull. Chem. Soc. Jpn*. **1972**, *45*, 3619–3624.
45. Canas-Rodriguez, A.; Martinez-Tobed, A. *Carbohydr. Res*. **1979**, *68*, 43–53.
46. Matsuno, T. Yoneta, T.; Fukatsu, S. *Carbohydr. Res*. **1982**, *109*, 271–275.
47. Woo, P. W. K.; Haskell, T. H. *J. Antibiot*. **1982**, *35*, 692–702.
48. Koch, K. F.; Rhoades, J. A. *Antimicrob. Agents Chemother*. **1970**, 309–313.
49. Koch, K. F.; Rhoades, J. A.; Hagaman, E. W.; Wenkert, E. *J. Am. Chem. Soc*. **1974**, *96*, 3300–3305.
50. Kondo, S.; Iinuma, K.; Yamamoto, H.; Maeda, K.; Umezawa, H. *J. Antibiot*. **1973**, *26*, 412–415.

51. Tipson, R. S.; Cohen, A. *Carbohydr. Res.* **1965**, *1*, 338–340.
52. Suami, T.; Nishiyama, S.; Ishikawa, Y.; Katsura, S. *Carbohydr. Res.* **1977**, *53*, 239–246.
53. Rai, R.; Chang, H.; Chang, C.-W. T. *J. Carbohydr. Chem.* **2005**, *24*, 131–143.
54. Cappelletti, L. M.; Spagnoli, R. *J. Antibiot.* **1983**, *36*, 328–330.
55. Torii, T.; Tsuchiya, T.; Umezawa, S. *J. Antibiot.* **1982**, *35*, 58–61.
56. Takagi, Y.; Komuro, C.; Tsuchiya, T.; Umezawa, S.; Hamada, M.; Umezawa, H. *J. Antibiot.* **1981**, *34*, 1–4.
57. Iwasawa, H.; Ikeda, D.; Kondo, S.; Umezawa, H. *J. Antibiot.* **1982**, *35*, 1715–1718.
58. Shitara, T.; Umemura, E.; Tsuchiya, T.; Matsuno, T. *Carbohydr. Res.* **1995**, *276*, 75–89.
59. Sharma, M. N.; Kumar, V.; Remers, W. A. *J. Antibiot.* **1982**, *35*, 905–910.
60. Akita, E.; Horiuchi, Y.; Miyazawa, T.; Umezawa, H. *J. Antibiot.* **1983**, *36*, 745–748.
61. Torii, T.; Tsuchiya, T.; Watanabe, I.; Umezawa, S. *Bull. Chem. Soc. Jpn.* **1982**, *55*, 1178–1182.
62. Hanessian, S.; Kornienko, A.; Swayze, E. E. *Tetrahedron* **2003**, *59*, 995–1007.
63. P. T. Nyffeler, C.-H. Liang, K. M. Koeller, C.-H. Wong, *J. Am. Chem. Soc.* **2002**, *124*, 10773–10778.
64. Li, L.; Chen, H.-N.; Chang, H.; Wang, J.; Chang, C.-W. T. *Org. Lett.* **2005**, *7*, 3061–3064.
65. Hanessian, S.; Tremblay, M.; Swayze, E. E. *Tetrahedron* **2003**, *59*, 983–993.
66. Liang, F.-S.; Wang, S.-K.; Nakatani, T.; Wong, C.-H. *Angew. Chem. Int. Ed.* **2004**, *43*, 6496–6500.
67. Yao, S.; Sgarbi, P. W. M.; Marby, A. K. A.; Rabuka, D.; O'Hare, S. M.; Cheng, M. L.; Bairi, M.; Hu, C.; Hwang, S.-B. Hwang, C.-K.; Ichikawa, Y.; Sears, P.; Sucheck, S. J. *Bioorg. Med. Chem. Lett.* **2004**, *14*, 3733–3738.
68. Sucheck, S. J.; Wong, A. L.; Koeller, K. M.; Boehr, D. D.; Draker, K.-A.; Sear, P.; Wright, G. D.; Wong, C.-H. *J. Am. Chem. Soc.* **2000**, *122*, 5230–5231.
69. Agnelli, F.; Sucheck, S. J.; Marby, K. A.; Rabuka, D.; Yao, S.-L.; Sears, P. S.; Liang, F.-S.; Wong, C.-H. *Angew. Chem. Int. Ed.* **2004**, *43*, 1562–1566.
70. Liang, C.-H.; Romero, A.; Rabuka, D.; Sgarbi, P. W. M.; Marby, K. A.; Duffield, J.; Yao, S.; Cheng, M. L.; Ichikawa, Y.; Sears, P.; Hu, C.; Hwang, S.-B. Shue, Y.K.; Sucheck, S. J. *Bioorg. Med. Chem. Lett.* **2005**, *15*, 2123–2128.
71. Michael, K.; Wang, H.; Tor, Y. *Bioorg. Med. Chem.* **1999**, *7*, 1361–1371.
72. Luedtke, N. W.; Liu, Q.; Tor, Y. *Biochemistry* **2003**, *42*, 11391–11403.
73. Roestamadji, J.; Mobashery, S. *Bioorg. Med. Chem. Lett.* **1998**, *8*, 3483–3488.
74. Yang, Y.; Roestamadji, J.; Mobashery, S.; Orlando, R. *Bioorg. Med. Chem. Lett.* **1998**, *8*, 3489–3494.
75. Haddad, J.; Vakulenko, S.; Mobashery, S. *J. Am. Chem. Soc.* **1999**, *121*, 11922–11923.
76. Roestamadji, J.; Grapsas, I.; Mobashery, S. *J. Am. Chem. Soc.* **1995**, *117*, 80–84.
77. Liu, M.; Haddad, J.; Azucena, E.; Kotra, L. P.; Kirzhner, M.; Mobashery, S. *J. Org. Chem.* **2000**, *65*, 7422–7431.
78. Tok, J. B.-H.; Bi, L. *Curr. Top. Med. Chem.* **2003**, *3*, 1001–1019.
79. Feichtinger, K.; Zapf, C.; Sings, H. L.; Goodman, M. *J. Org. Chem.* **1998**, *63*, 3804–3805.

80. Baker, T. J.; Luedtke, N. W.; Tor, Y.; Goodman, M. *J. Org. Chem.* **2000**, *65*, 9054–9058.

81. Luedtke, N. W.; Baker, T. J.; Goodman, M.; Tor, Y. *J. Am. Chem. Soc.* **2000**, *122*, 12035–12036.

82. Hui, Y.; Ptake, P.; Paulman, R.; Pallansch, M.; Chang, C.-W. T. *Tetrahedron Lett.* **2002,** 43, 9255–9257.

83. Ding, Y.; Hofstadler, S. A.; Swayze, E. E.; Griffey, R. H. *Org. Lett.* **2001**, *3*, 1621–1623.

84. Ding, Y.; Hofstadler, S. A.; Swayze, E. E.; Risen, L.; Griffey, R. H. *Angew. Chem. Int. Ed.* **2003**, *42*, 3409–3412.

85. Francois, B.; Szychowski, J.; Adhikari, S. S.; Pachamuthu, K.; Swayze, E. E.; Griffey, R. H.; Migawa, M. T.; Westhof, E.; Hanessian, S. *Angew. Chem. Int. Ed.* **2004**, *43*, 6735–6738.

5

NMR STRUCTURAL STUDIES OF AMINOGLYCOSIDE: RNA INTERACTION

R. ANDREW MARSHALL
Department of Chemistry, Stanford University, Stanford, CA 94305

JOSEPH D. PUGLISI
Department of Structural Biology, Stanford University, Stanford, CA 94305

Aminoglycoside Antibiotics: From Chemical Biology to Drug Discovery, Second Edition, by Dev P. Arya

5.1. INTRODUCTION

The aminoglycosides are a family of diverse antimicrobial compounds that bind the 16S RNA of the bacterial small ribosomal subunit and inhibit protein synthesis.[1–3] Aminoglycosides can be categorized structurally into two groups based upon the identity of the highly conserved aminocyclitol ring around which they are built: streptamine or 2-deoxystreptamine (2-DOS). Aminoglycosides that contain 2-DOS are well-characterized by biochemical and biophysical techniques and are the focus of this review (Figure 5.1). Chemically, the aminoglycosides are cationic oligosaccharides composed of between two and five amino sugars. At physiological pH, the amino groups are nearly all protonated, giving the aminoglycosides a net positive charge.[4,5] The amino sugar moieties are distributed about the 2-deoxystreptamine ring in two major substitution patterns: 4,5-disubstituted, which include neomycin and paromomycin, and 4,6-disubstituted, which include gentamicins and kanamycins. Members of the 4,6-disubsituted class are most used clinically. Aminoglycosides that contain a 2-deoxystreptamine ring target the same region of rRNA and interfere with the ribosomal processes involved in

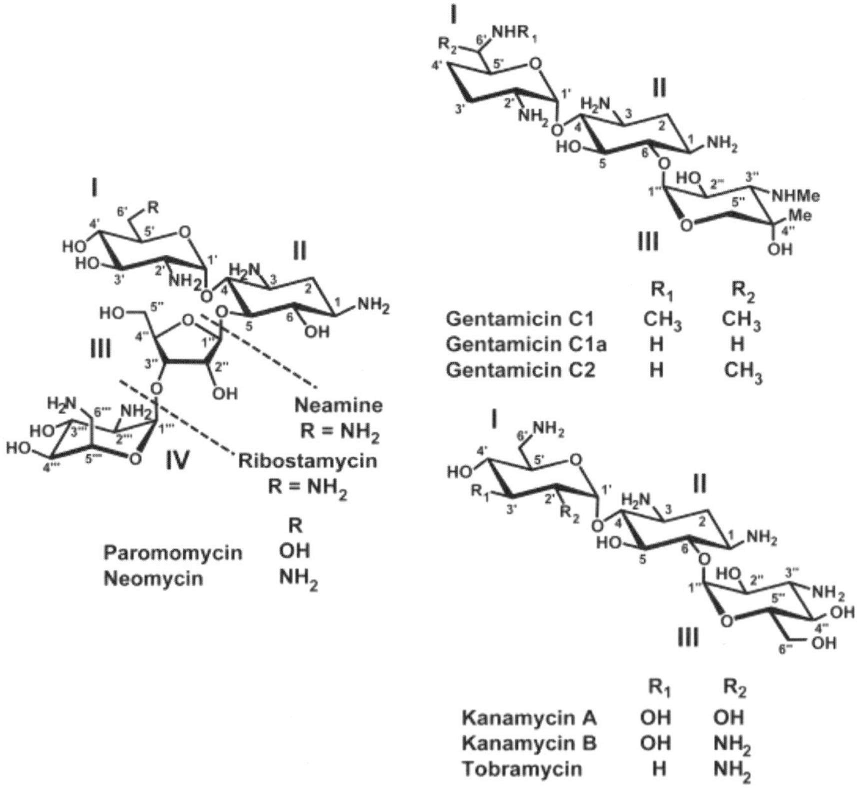

Figure 5.1. Representative aminoglycoside antibiotics with a 2-deoxystreptamine ring.

decoding and processivity. In addition to the ribosome, aminoglycosides have been shown to bind to numerous RNAs and modulate their structure and function.[6–10]

The 70S ribosome is a 2.5-MDa complex that contains three functional RNA components, 5S, 16S, and 23S rRNAs, and over 50 proteins. It comprises two subunits, the 30S and 50S, which work in concert with aminoacylated tRNAs and a number of protein factors to read the genetic code on the mRNA and synthesize proteins.[11] The 30S ribosomal subunit, which incorporates 16S rRNA and 21 proteins, is the site of codon–anticodon interaction as well as the target of aminoglycosides. Biochemical techniques locate the binding sites of the peptidyl (P site) and aminoacyl (A site) tRNAs and show a number of contacts between mRNA–tRNA–rRNA.[12,13] The region of the 16S rRNA that comprises the binding sites of the A- and P-site tRNAs is the decoding site and consists of nucleotides from Helix 44 and the 530 loop that are highly or universally conserved.[14–17] Mutations in the decoding site are deleterious to ribosome function, supporting the fundamental role it plays in translation. Structural and biochemical methods demonstrate that the nucleic acid component of the decoding site makes minimal contacts with ribosomal proteins, highlighting the catalytic activities of rRNA in the ribosome.[18,19]

The decoding site is the target of aminoglycoside antibiotics. Single-nucleotide mutations and base modifications in the decoding site confer a high degree of resistance to aminoglycosides.[20–23] Biochemical and biophysical techniques locate the site of aminoglycoside binding to an adenine-rich internal loop at the 3′ end of Helix 44.[3,24] The decoding site of the 30S subunit is a domain comprised primarily of rRNA that is critical in the binding of tRNA and the function of the ribosome. Aminoglycosides bind to this functional piece of rRNA and disrupt the function of the ribosome.

NMR is a powerful and versatile tool for structural studies of biological RNAs and complexes they form with other nucleic acids, proteins, and small molecules. The goal of these studies is to determine the role that structure and dynamics play in biological function. NMR has the capacity to determine high-resolution structures, as well as to map RNA:ligand interfaces at low resolution. Most structures of RNA and RNA–ligand complexes are under 20 KDa in size[25–27]; however, recent advances allow for determination of solution structures of complexes up to 40 kDa.[28–32] NMR can also probe dynamic motions in RNA on micro- to millisecond time scales.[33] A number of biologically relevant internal motions such as stem repositioning,[34] base-pair breathing,[35] sugar-pucker interconversion,[36] and base flipping are observable by these techniques.[37] The development of 800- and 900-MHz instruments and cryoprobe technologies continue to push the limits of signal to noise and spectral resolution, allowing for the study of larger systems. In addition, recent advances in isotopic-labeling strategies, pulse sequence methodology, and the acquisition and analysis of residual dipolar coupling data for the determination of global geometries of RNA relieve limitations in RNA structural studies.[38–42]

This chapter highlights the impact of NMR in the study of aminoglycoside antibiotics. The elucidation of the molecular framework behind the biological function of aminoglycosides garnered from NMR studies of aminoglycoside–rRNA interaction serves as the primary focus. The structural information is viewed in the context of biochemical and biophysical data as well as the high-resolution X-ray crystallography structures of aminoglycoside complexes with rRNA. Since aminoglycosides are an archetype for small-molecule recognition of RNA, the binding of aminoglycosides to other biological RNAs and nonbiological RNAs studied by NMR is also addressed. Finally, the application of NMR in the design of new ribosome-targeted antibiotics is discussed.

5.2. AN OLIGONUCLEOTIDE MODEL OF THE RIBOSOMAL A-SITE

Like many biological RNAs, the size of the 16S rRNA precludes it from high-resolution NMR study; this obstacle is overcome via the design of a model RNA oligonucleotide. Genetic and biochemical methods pinpoint aminoglycoside binding to helix 44 of the 30S subunit in a region comprised almost entirely of rRNA. These properties make the decoding site a good candidate for the development of oligonucleotide mimetics. In addition, the high thermodynamic stability of small RNAs allows for the creation of small model domains of RNA secondary and tertiary structures.[43]

Purohit and Stern[44] demonstrate the feasibility of the reduction strategy with the design of a 44-nucleotide RNA oligonucleotide corresponding to the 3′ and 5′ ends of helix 44. This oligonucleotide folds and specifically binds aminoglycosides outside of the context of the 30S subunit. Truncation of the Purohit and Stern oligonucleotide to nucleotides 1403 to 1412 and 1488 to 1497 proves optimal for NMR study. The shortened construct consists of two short helical regions separated by an asymmetric internal loop containing A1408, A1492, and A1493 (Figure 5.2). Extension of the upper helical stem by two additional G–C base pairs, along with capping of the lower helical stem with a stable UUCG tetraloop of known 3D structure, gives the final construct.[45] Importantly, the nucleotides in the decoding site construct display similar patterns of reactivity toward dimethyl sulfate (DMS) modification as they do in the 30S subunit. In the absence of aminoglycosides, the N1 positions of A1408, A1492, and A1493 as well as the N7 positions of G1491 and G1494 are reactive toward DMS.[3,46] The addition of aminoglycosides to the modification reaction results in a "footprint" region of decreased reactivity for A1408, G1494, and G1491 similar to that observed with the 30S.[3,46,47] Quantitative chemical footprinting determines the K_d for paromomycin to be in the range of 200 nM, which is consistent with the binding of paromomycin to the 30S subunit.[46] These results validate the decoding site oligonucleotide as a genuine model of the ribosomal decoding site for use in both biochemical and structural studies.

The structure of the A-site RNA oligonucleotide reveals the nature of the decoding site in solution (Figure 5.3).[48] Globally, the upper and lower stems of

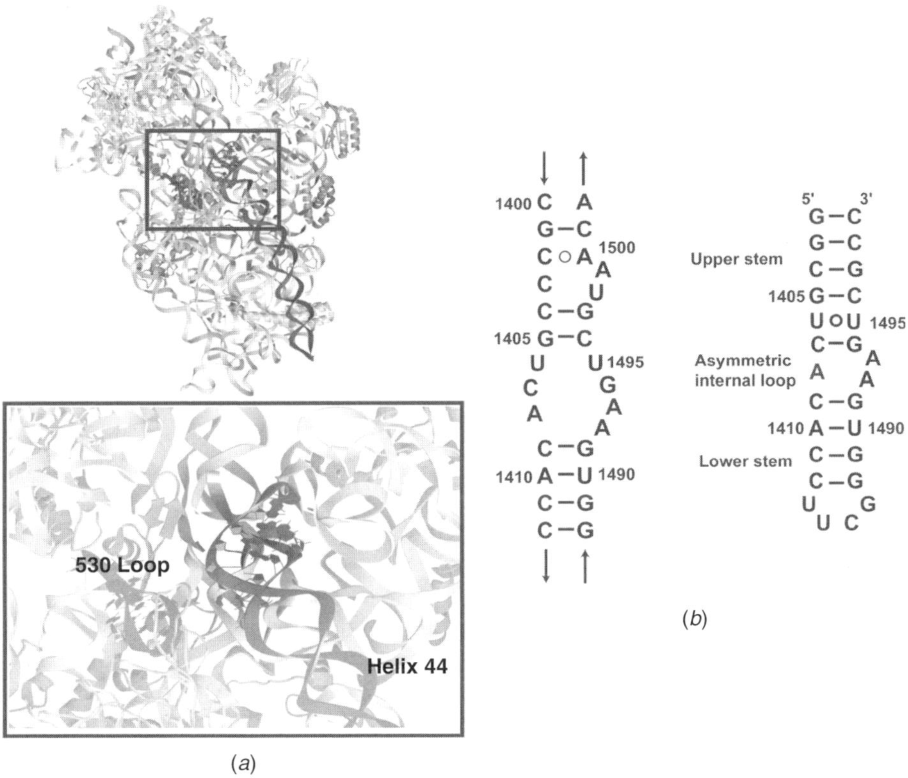

Figure 5.2. Design of a model oligonucleotide for the decoding site. (A) A rendering of the *E. coli* 30S subunit (PDB-2AVY). Helix 44 and the 530 loop of the decoding site are labeled. The zoomed image of the decoding site from the boxed region shows explicitly the highly conserved nucleotides in this site. The sequence of nucleotides shown in helix 44 is identified in part B. (B) Secondary structure of *E. coli* 16S rRNA from the region of helix 44 within the decoding site and A-site oligonucleotide. See color plates.

the construct form two continuous A-form helices of six and four base pairs, respectively. An asymmetric internal loop consisting of adenines 1408, 1492, and 1493 separates the two helical segments. Base pairs between C1407–G1494 and U1406–U1495 form the ceiling of the asymmetric loop, while the base pair between C1409 and G1491 comprises the floor of the loop.

Most interestingly, the NMR data demonstrate the asymmetric intern loop to be loosely structured and dynamic in nature. Residues A1408 and A1493 stack between the upper and lower stems and interact through a single hydrogen bond between N1 of A1408 and N6 of A1493. Unpaired A1492 stacks between A1493 and G1491 of the lower stem. Nuclear Overhauser effect (NOE) cross peaks between the H2 protons of A1408 and A1493, in combination with characteristic of base stacking NOEs, are consistent with the orientation of A1408 and A1493.[48] In addition, internucleotide NOE connectivities suggest that both A1492

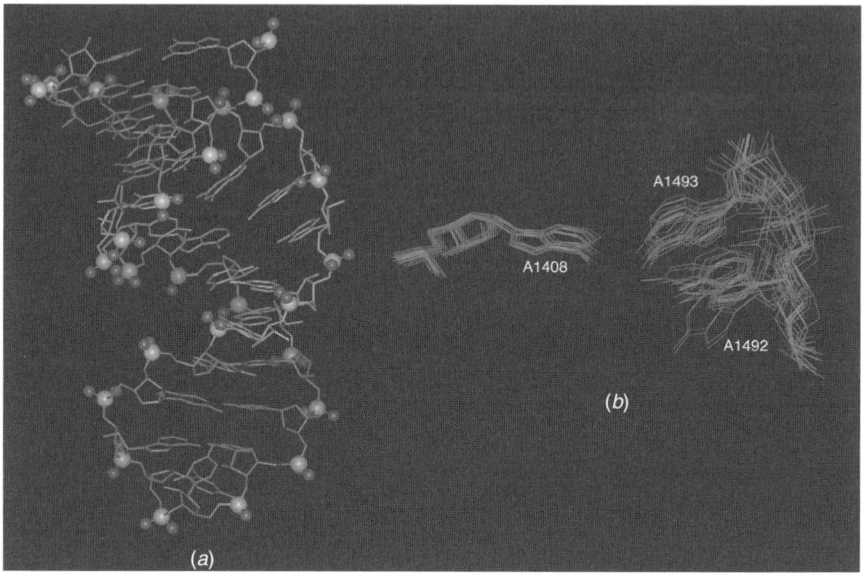

Figure 5.3. NMR structure of the A-site oligonucleotide. (A) Single representative structure of free A-site oligonucleotide.[48] All heavy atoms are displayed; the core of the antibiotic binding site (nucleotides U1406 to A1410, and U1490 to U1495) is lightly shaded, while the residues outside the core are more darkly shaded. Phosphate and phosphoryl oxygen atoms are shown as spheres. (B) A1408, A1492, and A1493 from the heavy atom superposition (residues U1406–A1410; U1490–U1495) of 20 representative structures of the A-site oligo viewed from the major groove.[48] The RMSD of the heavy atoms for A1492 and A1493 is 1.7 Å. See color plates.

and A1493 stack inside the helix. Unusual upfield-shifted C1′ resonances and H1′–H2′ cross peaks in COSY and TOCSY spectra for ribose sugars of A1492 and A1493 indicate a dynamic equilibrium between C2′-*endo* and C3′-*endo* sugar conformations for these residues. Probing the conformation of A1408, A1492, and A1493 with ^{15}N-correlated NOESY experiments for exchange-broadened protons reveals that the amino protons of A1408, A1492, and A1493 are in fast exchange with the solvent due to the conformational dynamics of the internal loop.[49] Furthermore, the dynamic nature of A1492 and A1493 manifests itself in the ensemble of NMR structures (Figure 5.3). The RMSD values for A1492 and A1493 exceed those of neighboring nucleotides by greater than 0.5 Å.[48] It is likely that the disorder of these residues is indicative of conformational flexibility in the asymmetric loop that is necessary for function.

X-ray crystallography gives further evidence of dynamic nature of A1492 and A1493. In the 3-Å resolution structure of the 30S subunit from *T. thermophilus*, Ramakrishnan and co-workers[18] are unable to fit the electron density assigned to A1492 and A1493 to a single conformation, implying residual dynamics in the crystal form. Likewise, Hermann and co-workers[50] capture two conformations of

A1492 and A1493 in the structure of an oligonucleotide mimic of the decoding site. In this structure, A1492 and A1493 adopt a 50% mixture of conformations in which both nucleotides adopt an extrahelical position, or A1492 stacks within the RNA helix and A1493 adopts an extrahelical conformation. The extrahelical conformation of A1493 is stabilized by intermolecular crystal packing contacts in both forms, while A1492 makes no crystal packing interactions in either form.

Fluorescence studies of 2-aminopurine (2-AP)-substituted oligonucleotides offer a complementary approach to characterize the conformational dynamics of A1492 and A1493. These experiments confirm that A1492 and A1493 are in conformational equilibrium between intra- and extrahelical states. The fluorescence signal of 2-AP is highly sensitive to local chemical environment. When introduced into nucleic acids, 2-AP fluorescence, which is quenched within a helical segment, reports on the degree of base stacking.[51,52] Using 2-AP fluorescence, Hermann and co-workers have verified the conformations of A1492 and A1493 in the oligonucleotide mimic characterized by X-ray crystallography.[50] Observing 2-AP fluorescence in a combination of steady state and time resolved modes, Pilch and co-workers[53] show that A1492 is in dynamic equilibrium between stacked and de-stacked forms. These studies reveal that A1492 is in conformational equilibrium between intra- and extrahelical states, with the intrahelical state being the favored form.

Biochemical and NMR studies characterize the A-site oligonucleotides as a valid model of the decoding site of the 30S subunit. It consists of two rigid helical domains connected through a loosely structured internal loop. In conjunction with other biophysical techniques, NMR demonstrates that A1492 and A1493 are the most dynamic residues in the internal loop. The conformational equilibrium of these residues is crucial to the function of the ribosome during decoding.

5.3. STRUCTURAL ASPECTS OF AMINOGLYCOSIDE BINDING TO A-SITE OLIGONUCLEOTIDE

NMR allows for characterization of ligand–macromolecule complexes at both low and high resolution. Application of high-resolution structure and low-resolution chemical shift mapping techniques to the complexes of aminoglycosides and the A-site oligo produces a wealth of information concerning the mechanism of aminoglycoside binding and action. As discussed above, confirmation of the validity of short oligonucleotide models via biochemical techniques is an essential first step in the investigation of aminoglycoside–RNA complex formation.

5.3.1. Low-Resolution Study of Aminoglycoside Binding

The well-resolved imino proton resonances from base-paired uracil and guanine resonances serve as powerful probes to characterize RNA–ligand interaction by NMR. The chemical shift of imino proton resonances is responsive to perturbations in the local chemical environment that result from RNA conformational changes and proximity to ligands. The shift of existing resonances or appearance of new resonances signals ligand binding, allowing for determination of the nucleotides in the vicinity of the RNA:ligand interaction. Titration of the A-site oligonucleotide with paromomycin to 1:1 stoichiometry leads to shifts of imino proton resonances from base pairs in the region about the asymmetric internal loop (Figure 5.4).[46] At substoichiometric amounts of paromomycin, the new imino proton resonances appear, representing paromomycin-bound RNA. This titration behavior is consistent with the dissociation constant determined for paromomycin and A-site oligo. The most significant shifts occur for U1490 and G1491, which border the asymmetric loop and shift downfield upon complex formation. Analysis of the nonexchangeable base and ribose proton NMR spectra

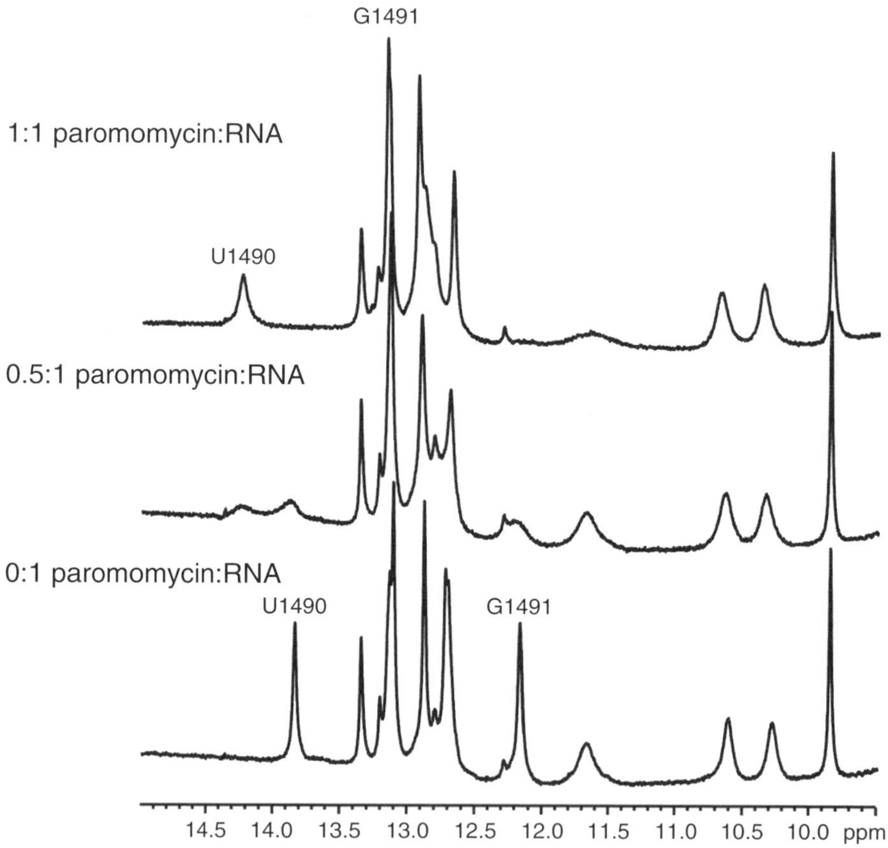

Figure 5.4. Titration of A-site oligonucleotide with paromomycin. Titration of the A-site oligonucleotide with paromomycin yields a specific 1:1 complex. Imino proton resonances for U1490 and G1491, which shift significantly upon binding of paromomycin, are labeled. The intermediate point in the titration demonstrates the appearance of new peaks from the bound form of the RNA.

confirm these chemical shift changes.[46] Streptomycin and spermine do not cause large imino proton shifts, demonstrating the specificity of the A-site oligo for 2-DOS containing aminoglycosides.

Observation of the titration behavior of RNA imino proton resonances allows for comparative studies of the binding of aminoglycosides to rRNA and shows 2-DOS containing aminoglycosides to bind the A-site oligonucleotide in a similar fashion. Neomycin, ribostamycin, neamine, and gentamicin C1a demonstrate the structural diversity of the aminoglycoside antibiotics that contain a 2-deoxystreptamine ring. Each of the aforementioned aminoglycosides contains the highly conserved ring I and II motif that is present in all therapeutically active aminoglycosides. Despite the unique structural properties of each antibiotic, titration of the free A-site oligonucleotide with each aminoglycoside results in the downfield shift of the imino protons of G1491 and U1490 as in the paromomycin titration.[46,54,55] In addition, the chemical shift changes of nonexchangeable proton resonances for the RNA and the antibiotic rings I and II are similar in each complex. This indicates that in each complex the chemical environment at the drug–RNA interface is nearly identical and that the chemical diversity of aminoglycosides does not dramatically affect rRNA binding.

5.3.2. High-Resolution Study of Aminoglycoside Binding

High-resolution NMR studies of paromomycin and gentamicin C1a in complex with the A-site oligonucleotide give insight into the general chemical principles that govern aminoglycoside binding. Paromomycin and gentamicin C1a represent the two major subclasses of aminoglycosides that have a 2-DOS ring: the 4′,5′- and 4′,6′-disubstituted.

The solution structures of 1:1 complex between paromomycin and gentamicin C1a and the A-site oligonucleotide have been solved at high resolution using heteronuclear NMR techniques.[55,56] These approaches require the preparation of uniformly labeled ^{15}N- and ^{13}C-RNA oligonucleotides via *in vitro* transcription with labeled nucleoside triphosphates. The use of RNAs labeled with NMR-active nuclei allows for the application of sophisticated heteronuclear NMR methods.[38] These methods facilitate the assignment of NMR resonances and the acquisition of structural restraints for detailed structure determinations.

The high-resolution structures of paromomycin and gentamicin C1a with the A-site oligonucleotide demonstrate the similar binding mode of both antibiotics and the stabilization that aminoglycoside binding brings to the asymmetric internal loop. In both structures, the antibiotics bind to the major groove of the oligonucleotide at the asymmetric internal loop.[55,56] The superposition of nucleotides, G1405 to A1410 and U1490 to C1496, in each structure shows that rings I and II bind in nearly an identical manner and induce a change in the conformation of the asymmetric internal loop (Figure 5.5). The general characteristics of the upper and lower stems remain unchanged in the free and drug-bound forms of the RNA. In the asymmetric internal loop, A1408 and A1493 stack between the upper and lower stems and base pair. In the complex with paromomycin, the internuclear distances between A1408(N6)–A1493(N7) and A1408(N1)–A1493(N6)

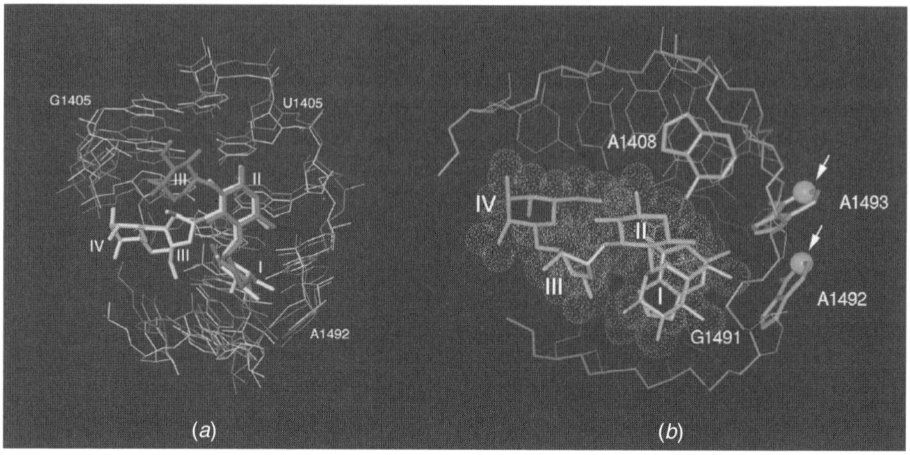

Figure 5.5. NMR structures of paromomycin and gentamicin C1a in complex with the A-site oligonucleotide. (A) Best-fit superposition of the paromomycin-RNA and gentamicin C1a-RNA complexes, viewed from the major groove side of the RNA.[57] The heavy atoms of the core (nucleotides U1406 to A1410, and U1490 to U1495) of the RNA are superimposed. Only the core is represented. The aminosugar rings of each antibiotic and several nucleotides are labeled for clarity. (B) View of paromomycin bound to the A-site oligonucleotide viewing from the upper stem looking down the helical axis. The phosphoidester backbone is highlighted.[56] Paromomycin binding displaces A1492 and A1493 toward the minor groove. The N1 positions of A1492 and A1493 are shown as spheres and, in the context of the 30S subunit, point toward the location tRNA–mRNA complex formation in the A-site. See color plates.

are in agreement with the formation of two hydrogen bonds. In the gentamicin complex, the pattern of hydrogen-bonding in the noncanonical A–A base pair is consistent with two possible A1408–A1493 base-pairing schemes in which the exocyclic N6 is hydrogen-bonded to either N1 or N3 of A1408. This agrees with biochemical studies showing that paromomycin binding protects the N1 position of A1408 from chemical modification with dimethyl sulfate.[3,46] In both structures, A1492 is de-stacked from the helix and directed toward the minor groove, opening the major groove for antibiotic binding. The reduction of the RMSD values for A1492 and A1493 in the antibiotic bound structures suggests that drug binding induces a more structured conformation of A1492 and A1493.[55,56] Residual dipolar coupling measurements performed on the free oligonucleotide and the 1:1 complexes of the A-site oligonucleotide with paromomycin and gentamicin C1a confirm the conformational change in A1492 and A1493.[57] The base pair between A1408 and A1493 and the extrahelical conformation of A1492 results in a distortion of the RNA backbone and widened major groove, which forms a binding pocket for ring I and II of aminoglycosides.

NMR structures also reveal that rings I and II of both antibiotics bind in a similar orientation with respect to the RNA, direct the specific interaction with the decoding site, and stabilize the conformation of A1492 and A1493. The N1 and N3 amino groups of ring II recognize the pattern of hydrogen bond donors presented in the major groove by G1494 and U1495. Mutagenesis and chemical modification of these residues drastically reduce the affinity of aminoglycosides for the A-site.[58] Ring I occupies the space vacated by A1492 and stacks over G1491. The hydroxyl and amino chemical groups on ring I contact the phosphate backbone in the vicinity of A1492 and A1493 and stabilize the conformation of the asymmetric internal loop. Substitution of the nonbridging, major groove phosphoryl oxygens at positions 1492 and 1493 with sulfur reduces the affinity of the A-site oligonucleotide for paromomycin, validating the importance of these interactions.[59] In the structure with paromomycin, the 2′-amino group of ring I forms an intramolecular hydrogen bond with the endocyclic oxygen of ring III, which orients the four rings of paromomycin into a L-shape in which rings II, III, and IV are positioned at a right angle to ring I. Ring I of gentamicin C1a lacks the 3′-, 4′-, and 6′-hydroxyl moieties of paromomycin and compensates for this loss through favorable electrostatic interactions with the phosphate backbone through the 2′- and 6′-amino groups and hydrophobic interactions with G1491.[55]

While rings I and II direct specific binding and stabilize A1492 and A1493, the additional rings make additional contacts with RNA and increase the affinity of antibiotic binding. Ring III and IV of paromomycin are directed toward the lower stem of the RNA. In contrast to the well-defined conformations of rings I and II, rings III and IV are more disordered, indicating a higher degree of conformational dynamics in these residues. While these residues do not appear to make specific interactions with RNA in the paromomycin structure, the position of ring IV suggests that it makes favorable contacts with the phosphate backbone. Pilch and co-workers[5,60] observe that protonation of the 2‴-amino group of ring IV is linked to drug binding. Further demonstrating the importance of ring IV in the affinity of paromomycin for the A-site, thermodynamic studies show that ribostamycin, which lacks ring IV, has a significantly reduced affinity for the A-site.[54,61] The third ring of gentamicin is well-defined in the ensemble of structures and forms specific interactions with nucleotides of the upper stem, unlike rings III and IV of paromomycin. The ether and hydroxyl groups at the 3″ and 2″ position of ring III form hydrogen bonds with G1405. Changing the G1405–C1496 base pair to a C1405–G1496 greatly reduces the affinity of gentamicin C1a for the A-site oligonucleotide, while the same mutation has no effect on the binding of paromomycin.[55] The additional rings in both the 4′,5′- and 4′,6′-disubstituted aminoglycosides enhance the binding affinity of these classes of antibiotics with respect to neamine (rings I and II), which binds the A-site oligonucleotide with lower affinity than both paromomycin and gentamicin C1a.[54,61]

High-resolution structures of paromomycin and gentamicin C1a show that the antibiotics bind in a similar fashion to the asymmetric loop of the A-site oligonucleotide. Rings I and II set the register for binding by recognizing the G1494–U1495 step and stabilize a de-stacked conformation of A1492 and A1493.

Additional rings on paromomycin and gentamicin C1a point to a molecular explanation of why 4′,6′-disubstituted aminoglycosides are more specific for the A-site binding pocket and enhance the affinity of aminoglycosides for the decoding site.

5.4. STRUCTURAL INSIGHTS INTO AMINOGLYCOSIDE RESISTANCE

The insights into aminoglycoside binding to the A-site oligonucleotide gained through NMR studies explain the structural origins of bacterial resistance. Bacteria achieve resistance to aminoglycosides by actively transporting aminoglycosides out of the cell, modifying the rRNA at the decoding site, and covalently modifying the antibiotics.[62] The latter two mechanisms can be understood in the context of the structures of paromomycin and gentamicin in complex with the A-site oligonucleotide (Figure 5.6).

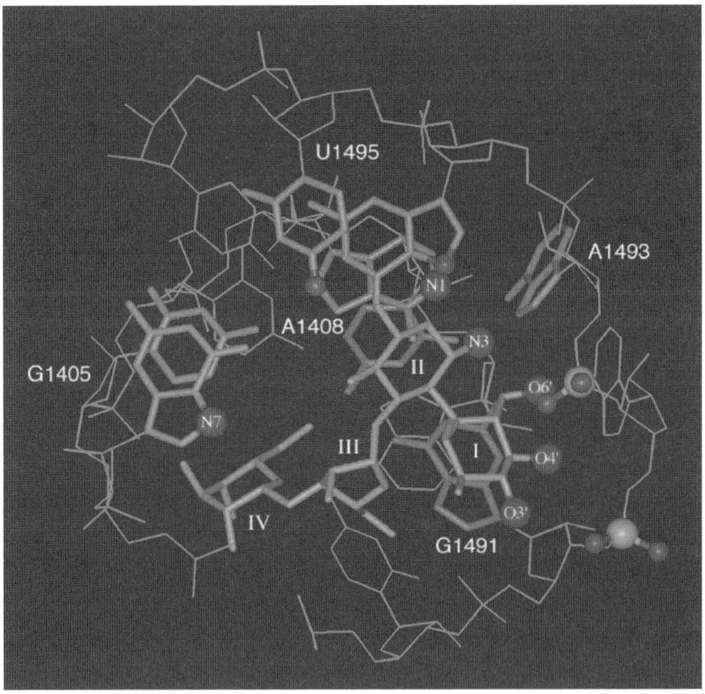

Figure 5.6. Structural map of modifications that lead to aminoglycoside resistance. The RNA is darkly shaded and paromomycin is lightly shaded. G1405, U1406•U1495, A1408•A1493, C1409•G1491, and G1494 are highlighted as bold lines, and the rRNA functional groups that make specific contacts with aminoglycosides are explicitly labeled.[56] Aminoglycoside modification enzymes and RNA methylases covalently transform atoms shown as spheres on both the aminoglycoside and the RNA. Disruption of the G1491–C1409 base-pair (labeled) leads to resistance as well.

Mutations and covalent modification of nucleotides in the decoding region decrease the affinity of aminoglycosides for the rRNA target by disrupting key drug–RNA interactions and result in antibiotic resistance. The most common resistance mutations include (a) a mispair at position 1409–1491 of the lower stem[20,21] and (b) the point mutations U1495C and A1408G.[22,23,63] The mutation of the C1409–G1491 base pair disrupts the binding pocket of ring I. Mutation of A1408 to guanine negates the hydrogen-bonding interactions of A1408 and 1493, while also introducing the exocyclic N2 amino group of guanine into the major groove. Covalent modification of nucleotides in the decoding site is the resistance mechanism of choice for bacteria that produce aminoglycosides. Methylation of the N1 position of A1408, like the A1408G mutation, disrupts the noncanonical A–A base pair, resulting in a potential deformation of the aminoglycoside binding pocket. Methylation of the N7 position of G1405 is common in bacteria that produce 4′,5′-disubstituted aminoglycosides, and the structure of gentamicin C1a illustrates that methylation of this position hampers specific interactions between ring III and G1405.

Enzymatic modification of the aminoglycosides disrupts the binding of aminoglycosides by targeting chemical groups on the antibiotics that make intimate contact with the target.[64] The enzymes that modify aminoglycosides are plasmid encoded and readily transferred between bacterial strains, making this resistance mechanism prevalent in aminoglycoside resistant bacteria. The modification of aminoglycosides results in a dramatic decrease in the affinity of the antibiotic for rRNA.[65] The primary targets of the aminoglycoside-modifying enzymes are the functional groups on the highly conserved rings I and II, which are essential for specific binding and function of all classes of aminoglycosides. Enzymatic modifications generally disrupt favorable electrostatic and hydrogen-bonding interactions between aminoglycosides and rRNA, while at the same time introducing steric and electrostatic impairments to binding. Acetylation of the amino groups on ring II disrupts the recognition motif of hydrogen bond donors in the major groove as well as favorable electrostatic interactions with the phosphate backbone. Amino groups on ring I are also the target of acetylation and negate electrostatic interactions between these residues and the phosphate residues of A1492 and A1493. The hydroxyl groups on ring I, which make important hydrogen bond contacts with the phosphate backbone, are the target of adenylation and phosphorylation.[65] Adenylation and phosphorylation introduce steric and electrostatic penalties to the binding of aminoglycosides. Overall, aminoglycoside modifying enzymes disable the key electrostatic and hydrogen-bonding interactions between rings I and II and rRNA, creating energetic penalties to aminoglycoside binding.

5.5. STRUCTURAL ORIGINS OF AMINOGLYCOSIDE-INDUCED MISCODING

The NMR structures of aminoglycosides in complex with the A-site oligo also elucidate the structural origins of antibiotic-induced miscoding. To ensure the fidelity of translation, the decoding region of the 30S subunit plays an active role in monitoring the binding and Watson–Crick base-pairing interaction between the tRNA anticodon and the mRNA codon at the A-site. The binding of aminoglycosides decreases the rate of aminoacyl-tRNA dissociation from the ribosomal A-site, slowing translation and increasing the error rate of amino acid incorporation.[66] This suggests that aminoglycosides favor a conformation of rRNA that displays indiscriminant high-affinity interaction with the mRNA–tRNA codon–anticodon complex. The NMR structures of aminoglycosides bound to the A-site oligonucleotide demonstrate that the binding of these drugs stabilizes a conformational state of the normally dynamic asymmetric internal loop. In this state, A1492 and A1493 adopt an extrahelical position directed toward the minor groove of the A-site oligonucleotide. Biochemical experiments and X-ray crystallography show that upon the formation of a cognate mRNA–tRNA interaction, A1492 and A1493 de-stack into a position to make contacts with the codon–anticodon complex.[67,68] It follows that aminoglycoside binding induces a conformation of A1492 and A1493 that mimics the decoding site during the selection of the cognate codon–anticodon complex. This incorrectly signals the formation of a cognate codon–anticodon complex, thereby increasing the probability of tRNA misincorporation and miscoding.

5.6. PROKARYOTIC SPECIFICITY OF AMINOGLYCOSIDES

Despite the high degree of nucleotide conservation between the prokaryotic and eukaryotic decoding regions, aminoglycosides display a greater propensity for interfering with translation in prokaryotes.[69] A major difference in the sequences of the decoding regions of the prokaryotic and eukaryotic organisms is an A1408 to G1408 change in the asymmetric internal loop. The mutation of A1408G in *E. coli* ribosomes leads to a high level of resistance to many aminoglycosides by rendering the mutant ribosomes less susceptible to aminoglycoside-induced inhibition of translation.[70] To investigate the structural aspects of aminoglycoside specificity, the A1408G mutation was introduced into the A-site oligonucleotide, and both free and paromomycin-bound structures were solved by heteronuclear NMR.

The structure of the free A1408G mutant oligonucleotide shows significant differences in the conformation of the asymmetric internal loop.[71] A1492 and A1493 adopt similar orientations with respect to each other, yet both are shifted toward C1409 of the lower stem. G1408 is also shifted toward the major groove with respect to A1408 in the wild-type construct. The shift helps to maintain the 1408–1493 base-pairing interaction that occurs between the 2-amino group of G1408 and the N1 position of A1493 in the G1408 oligonucleotide. The structural differences are consistent with chemical shift differences observed in the nucleotides surrounding the asymmetric internal loop, which result from the different chemical environments seen in each structure. Overall, the different geometries of the nucleotides in the G1408 oligonucleotide result in a major groove that is shallower in comparison to the wild-type structure.

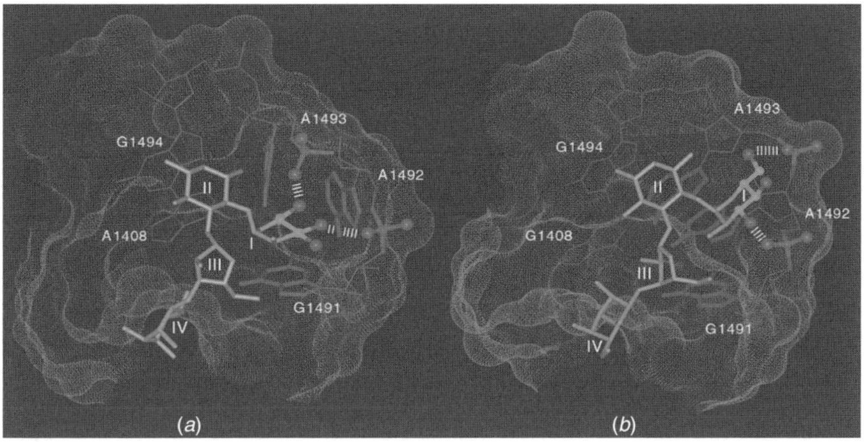

Figure 5.7. Structural origins of aminoglycoside specificity. A representation of the RNA surface of the prokaryotic (red) and eukaryotic (blue) A-site construct in complex with paromomycin highlights the difference in the position of ring I.[72] Dashed lines highlight potential hydrogen bond contacts between ring I and the A-site. In the complex with the prokaryotic A-site construct, ring I engages the binding pocket vacated by A1492, while in the eukaryotic construct, ring I interacts with the edge of the major groove. See color plates.

Paromomycin binds the G1408 oligonucleotide, yet the structure reveals that paromomycin binding does not induce the same conformational change in the eukaryotic A-site as it does in the prokaryotic construct (Figure 5.7).[72] Chemical footprinting and isothermal titration calorimetry demonstrate a decreased affinity of paromomycin for the eukaryotic versus the prokaryotic A-site oligonucleotide.[70,73] In the structure of the paromomycin:eukaryotic A-site complex, rings I and II contact RNA nucleotides G1491 to U1495, with ring II guiding the sequence-specific recognition of the G1494 and U1495. However, ring I does not penetrate deeply in the major groove and displace A1492 as it does in the prokaryotic construct. It is likely that geometry imposed by the G1408–A1493 base pair disrupts the binding pocket of ring I in the eukaryotic construct by introducing unfavorable steric interactions in the major groove. Residual dipolar coupling and 2-AP fluorescence studies verify the lack of conformation change upon paromomycin binding in the eukaryotic construct.[58,73,74] Therefore, it is likely that a combination of reduced affinity and ineffective binding of paromomycin account for the specific action of aminoglycosides on the prokaryotic ribosomes.

5.7. COMPARISON OF STRUCTURES SOLVED BY NMR AND X-RAY CRYSTALLOGRAPHY

NMR spectroscopy and X-ray crystallography are the two most common biophysical methods used in the determination of high-resolution structures of macromolecules. Each method has inherent strengths and weakness due to the nature of the physical phenomena that govern each technique. Solution-phase NMR is particularly adept at characterizing dynamic structural processes that have biological significance, yet NMR also has a size limitation that significantly hinders study of biological molecules larger that 50 kDa. In addition, the precision of NMR structures is often lower than those of high-quality crystal structures, because the number of conformational restraints defined by NMR is limited. In contrast, crystallography is relatively unhindered by the size of the biological molecule of interest, but is limited by the inability to detect structurally important dynamics and the complications brought about by crystal packing. In cases where both NMR and crystal structures exist for a biological molecule, it is important to take into account the strengths and weaknesses of each technique in the analysis of the structure.

Due to the clinical importance of aminoglycoside antibiotics, the complexes of these drugs with rRNA have been popular targets for structural characterization. In addition to the NMR structures presented in this chapter, a number of high-resolution X-ray crystal structures of aminoglycosides in complex with rRNA have been solved.[75–82] In the context of this review, it is important to draw comparisons of three of these structures and the NMR structures of the complexes between paromomycin and gentamicin C1a. The first of these the high-resolution structures is of paromomycin in complex with the 30S subunit.[75] The second and third structures are of paromomycin and genticin (G418) in complex with an RNA oligonucleotide mimetic of the A-site designed expressly for X-ray crystallography.[71,79]

The structure of paromomycin with the 30S subunit shows great similarity to the NMR structure, yet there are key differences in the two structures that have biological implications. In the crystal structure, paromomycin binds at the asymmetric internal loop and makes nearly identical interactions with the rRNA component of the decoding site as is seen in the A-site oligonucleotide. Superposition of the lower-stem nucleotides from both structures illustrates the similarity of both structures, yet also points to the important differences (Figure 5.8). The positions of ring IV, upper-stem nucleotides U1406 and G1405, and A1492 and A1493 are three such differences that highlight the strengths and weaknesses of the two techniques. Ring IV in the solution structure is disordered and may represent residual dynamics, while in the crystal structure the drug adopts a single conformation that is the most thermodynamically stable under experimental conditions. The difference between the position of U1406 and G1405 in the NMR and crystal structure is due to a global bend in the RNA, which is difficult to determine in NMR due to the local nature of the restraints used in NMR structure calculations. However, the introduction of residual dipolar coupling restraints into the structure calculation protocol has improved this shortcoming.[83] The most dramatic difference between the two structures is the position of A1492 and A1493, which adopt an exaggerated extrahelical conformation in the crystal structure. The positions of A1492 and A1493 in the X-ray crystal structure are inconsistent with intra- and intermolecular NOEs observed in solution NMR experiments. However, in dynamic regions of RNA such as the A-site, it is difficult for NMR

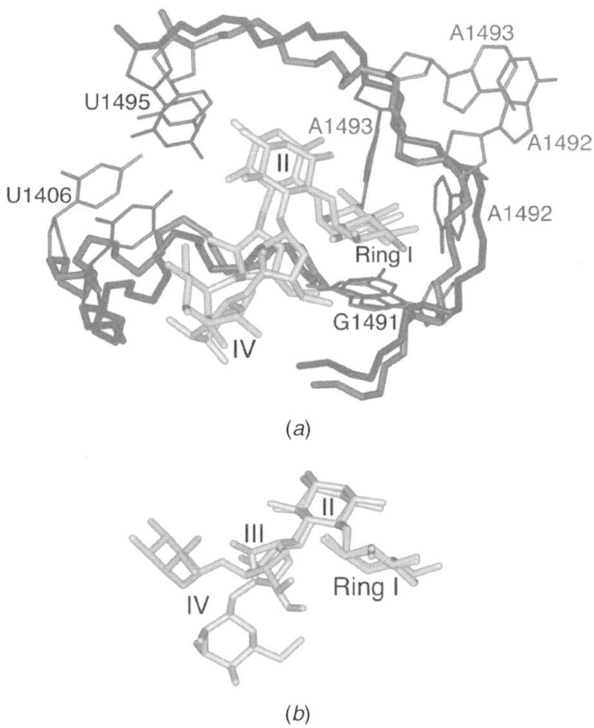

Figure 5.8. Superposition of NMR and X-ray structures of the paromomycin–rRNA complex. Superposition of lower and upper stem residues C1409–C1411 and G1488–G1491 compares the position of internal loop residues, U1406–A1408 and A1492–U1495.[83] In each structure, paromomycin is lightly shaded while the rRNA is darkly shaded. Important residues are explicitly labeled for clarity. (A) Superposition clearly demonstrates the major differences in the positioning of U1406 and A1492/1493. (B) Omitting rRNA for clarity, the difference in ring IV conformation is apparent.

to completely validate or rule out the conformation observed in the crystal structure. The exaggerated extrahelical conformation of A1492 and A1493 in the crystal structure would be difficult to observe using NMR techniques because few, if any, unique NOEs would be observed in this conformation, and the analysis of RDCs for A1492 and A1493 is complicated by the dynamic nature of the asymmetric internal loop.

While the positions of A1492 and A1493 are drastically different in both structures, it is likely that both represent valid and biologically relevant conformations of the decoding site. NMR data taken on the A-site oligonucleotide in solution at temperatures closer to those experienced by active ribosomes illustrate that the residues of the asymmetric internal loop undergo dynamic motions with microseconds to milliseconds time scales. NMR shows that the predominant conformation of A1492 and A1493 in the paromomycin–A-site oligonucleotide complex is intermediate between that demonstrated by the NMR structure of the free A-site oligonucleotide and that of the X-ray crystal structure. The conformation of A1492 and A1493 seen in the X-ray crystal structure may likely represent a thermodynamically stable endpoint of a dynamic process that is captured in the crystal form. Overall, NMR and X-ray crystal structures of paromomycin in complex with rRNA complement each other, helping to elucidate the mechanism of aminoglycoside-induced inhibition of translation.

In addition to the structure of paromomycin with the 30S subunit, X-ray crystal structures of paromomycin and genticin (G418) in complex with an oligonucleotide model of the decoding site have been solved at high resolution.[77,79] These structures display a largely similar interaction between each antibiotic and rRNA when compared to the paromomycin and gentamicin C1a NMR structures, respectively. The high-resolution structures also give insight into the potential role of bound water molecules in the interaction between aminoglycosides and rRNA.

As in the structure of paromomycin in complex with the 30S subunit, the position of A1492 and A1493 is the major difference between the X-ray crystal and NMR structures of aminoglycoside–rRNA complexes. A1492 and A1493 adopt an exaggerated extrahelical conformation in both antibiotic–oligonucleotide structures that is similar to that seen in these 30S structure. However, unlike the structure of the 30S subunit, A1492 and A1493 engage in crystal packing interactions with oligonucleotides at adjacent positions in the crystal lattice, which complicates the evaluation of the conformation of these residues.[77,79] In these structures, crystal packing may have an influence on the conformation of A1492 and A1493 that stabilizes an exaggerated conformation. Fluorescent studies of aminoglycosides complexed to 2-AP substituted oligonucleotides are consistent with the extrahelical conformation seen in the X-ray structures of these complexes, however, these studies are unable to report on the extent to which A1492 and A1493 are de-stacked.[50] Since A1492 and A1493 adopt similar conformations in the oligonucleotide and 30S structures, it is likely that their position is valid.

5.8. STRUCTURES OF AMINOGLYCOSIDES WITH OTHER RNAs

The rich chemical functionality and the flexible sugar scaffolds that aminoglycosides use in the specific recognition the rRNA component of the decoding site also helps them to bind a number of biological and non-biologically derived RNAs. Aminoglycosides bind to a number of different biological RNAs, including group I introns,[6] HIV RRE RNA,[8,84,85] and HIV TAR RNA,[86] as well as artificially selected RNAs, known as aptamers.[87,88] NMR has proven instrumental in the study of aminoglycosides binding to these RNAs helping to demonstrate the versatility of the aminoglycosides as RNA-binding small molecules and to further characterize the chemical principles governing RNA recognition by small molecules.

5.8.1. Low-Resolution Model of Neomycin in Complex with HIV-1 TAR RNA

A low-resolution solution model of the complex between neomycin and the Tat-responsive RNA (TAR) of the human immunodeficiency virus 1 (HIV-1) determined by Rosch and co-workers[89] captures a unique binding mode of neomycin to the minor groove of RNA. In the HIV replication cycle, Tat–TAR interaction plays an integral role in the activation of RNA Polymerase II for transcription of the viral genome, making the Tat–Tar interaction a target of anti-HIV drug design efforts.[90] Neomycin effectively competes with Tat-derived peptides for binding to TAR. Chemical shift mapping locates the site of neomycin binding at the pyrimidine bulge and lower stem in the vicinity of G19-C41. Despite a lack of intermolecular NOE restraints, NMR data obtained from the ribostamycin and neomycin TAR complexes allow for computational docking between neomycin and TAR. The set of representative structures illustrates that neomycin binds the minor groove of TAR in an extended conformation and induces a conformation change in of the phosphate backbone in the vicinity of the pyrimidine bulge. Changes in the CD spectra of the TAR and the simultaneous release of Tat upon titration of the Tat–TAR complex with neomycin suggest that conformational change is the mechanism for the inhibition.[91] In opposition to NMR data, molecular dynamics simulations locate the position of neomycin binding in the major groove of TAR spanning the bulge and making contacts with both the upper and lower stems.[92] Currently, there is no consensus on the binding mode of neomycin to TAR, yet the NMR data are consistent with a weak interaction of neomycin with the minor groove of the RNA.

5.8.2. Structure of Neomycin in Complex with Splicing Regulatory RNA Stem Loop

The structure of neomycin in complex with a regulatory RNA responsible for the regulation of tau exon 10 alternative splicing demonstrates the principles that govern aminoglycoside binding to the major groove of double-stranded RNA.[93] The RNA consists of 18 nucleotides that form a six-base-pair stretch of A-form RNA duplex capped by an unstructured loop of six unpaired nucleotides. Mutations in the regulatory domain decrease the thermodynamic stability of the RNA stem-loop structure and result in misregulation of splicing, which leads to the haphazard production of human tau protein isoforms and pathogenesis.[94] Neomycin binds the regulatory RNA with micromolar affinity and increases the stability of the stem loop.[93]

The structure of the RNA–neomycin complex shows neomycin binding in the narrow major groove of the RNA stem loop, making primarily electrostatic contacts instead of making sequence-dependent contacts in the major groove. Neomycin adopts an L-shaped conformation similar to that of paromomycin in complex with the A-site oligonucleotide. The authors propose that this shape complements the geometry of A-form RNA.[93] Unlike the structures of aminoglycosides with the A-site oligonucleotide in which rings I and II dictate binding, the majority of intermolecular NOEs come from the interaction of ring IV of neomycin with the RNA. This reveals that rings I and II are not a universal guide to high-affinity binding of aminoglycosides to RNA. Additionally, the binding of neomycin to the stem-loop structure results in the burial of 810 Å^2 of drug area compared to the 1070 Å of paromomycin surface area buried in the complex with the A-site oligonucleotide.[93] The burial of additional drug surface area in the decoding site RNA suggests that specific and high-affinity binding of small molecules to RNA is most successful when the molecule makes intimate contact with the target RNA. Overall, this structure highlights aspects of aminoglycosides that make them efficient RNA-binding small molecules and suggests strategies for the design of small molecules that bind RNA.

5.8.3. Structures of Neomycin and Tobramycin in Complex with RNA Aptamers

Aptamers are RNA oligonucleotides created by *in vitro* selection and evolution techniques to bind ligands with high affinity and specificity.[95–97] Although aptamer RNAs are not of biological origin, structures of aptamer–ligand complexes help to elucidate the role of the diverse folding topologies of RNA in molecular recognition. Aminoglycoside–aptamer complexes highlight the structural principles that govern specific and high-affinity aminoglycoside–RNA interactions seen in structures of aminoglycosides bound to biologically derived RNAs. The NMR structures of tobramycin and neomycin B in complex with aptamer RNAs show that high-affinity RNA–aminoglycoside interactions result from the ability of the major groove of RNA to complement the shape and electrostatic properties of aminoglycosides.[98–100]

The two structures of tobramycin–aptamer complexes and the neomycin B-aptamer show striking similarity in the manner in which the RNA recognizes the antibiotic, and they demonstrate the importance of a distorted major groove in aminoglycoside binding. In all three structures, the antibiotic binds the aptamer RNA in the major groove at the junction of a double-helical stem and unpaired hairpin loop. Antibiotic binding results in the formation of both canonical and noncanonical base-pairing arrangements in the major groove, which define the structure of the binding site. In each structure, the base-pairing arrangement distorts the RNA major groove, making it wider, deeper, and more accessible to antibiotics than canonical A-form RNA. Each antibiotic–aptamer complex also contains a nucleotide that adopts an extrahelical conformation, which encapsulates the antibiotics in the RNA binding pocket. The unique folding and base-pairing adopted in the distorted major groove of the aptamer RNA binding pockets maximizes the contact surface area at the RNA–antibiotic interface.

The structures of tobramycin and neomycin B in complex with aptamer RNAs also display the importance of interactions between the amino groups of aminoglycosides and the phosphate backbone of RNA. In the neomycin–aptamer complex, the 6′-amino group on ring I interacts with the phosphate backbone. Paromomycin, in which the 6′ position of ring I is a hydroxyl group, binds the same aptamer with 100-fold lower affinity. Similarly, contacts between the 2′- and 3″-amino groups of tobramycin and the phosphate backbone in the tobramycin:aptamer complex are required for high-affinity binding.[99] Changing either of these amino groups to hydroxyl groups, as in kanamycin A and gentamicin C, reduces binding affinity for the aptamer RNA.[99]

Comparison of the structures of biological RNA–aminoglycoside complexes with the aptamer:aminoglycoside complexes highlights the importance of distorted RNA major groove geometry in the specific and high-affinity interaction with aminoglycosides. Aminoglycosides bind the distorted major groove of the ribosomal A-site and RNA aptamers with high affinity and specificity. Favorable noncovalent interactions between the amino and hydroxyl functional groups of the drugs and the RNA bases, ribose sugars and phosphate backbone are facilitated by the deep and wide major groove. Furthermore, the accessibility of the major groove in these RNAs allows for aminoglycoside recognition of the unique sequence-dependent patterns of hydrogen bond donor and acceptors from which specificity arises. The shallow major groove of the tau exon 10 regulatory RNA binds aminoglycosides with lower affinity than the A-site and aptamer RNAs.

5.9. NMR IN THE DESIGN OF AMINOGLYCOSIDE MIMETICS

The growing problem of antibiotic-resistant bacterial infections creates a need for new antibiotics.[101] Aminoglycosides are one of the critical classes of antibiotics that have seen a reduction in clinical utility due to antibiotic resistance. NMR impinges upon the design of aminoglycoside mimetics for RNA targeted drug therapies in two major ways. First, the characterization of the structural and physical properties that govern aminoglycoside binding, resistance, specificity, and mode of action via NMR and other biophysical techniques has helped to motivate and to guide the development of both RNA- and ribosome-targeted drugs via rational design. Second, adapting NMR methodology traditionally used in protein-targeted drug design for RNA aids in the design of RNA-directed therapeutics. We focus on the development of small molecules targeted at the decoding region of the ribosomal A-site, highlighting a few studies that are representative of work in the field.

The structural studies of define the key elements in aminoglycosides and RNA that are essential for antibiotic binding and function. This knowledge fuels the design of aminoglycoside mimetics that fit into two categories: those that maintain an aminoglycoside-like saccharide scaffold and those that comprise a nonsaccharide scaffold. Aminoglycoside mimetics in the first category take advantage of the well-characterized and validated polysacchride scaffold of the aminoglycosides. Introducing chemical modifications at rationally determined sites on these molecules improves binding affinity and specificity or reduces the potential for modification by aminoglycoside detoxifying enzymes. The second category consists primarily of small molecules chosen from chemical compound libraries via high throughput screening or computational screening protocols and subjected to rounds of lead optimization. NMR plays an essential role in both high-throughput screening of A-site targeted drugs as well as the optimization of lead compounds discovered by screening protocols.

With the goal of optimizing the aminoglycoside scaffold to enhance A-site specificity and reduce modification by detoxifying enzymes, two groups have employed a strategy to restrain the conformational flexibility of aminoglycosides. A constrained paromomycin derivative designed by Tor and co-workers[102] seeks to exploit the different conformations adopted by the neomycin class of antibiotics in complex with the A-site and HIV-1 TAR RNA to increase A-site specificity. Using a similar strategy, a conformationally locked neomycin derivative takes advantage of the different binding modes of aminoglycosides to RNA and resistance-causing modification enzymes.[103] Paromomycin adopts an L-shaped conformation when bound to the A-site, while neomycin adopts a more elongated structure when in complex with HIV TAR RNA.[56,89] Similarly, aminoglycosides do not bind in an L-shape to modifying enzymes, suggesting that limiting the flexibility of aminoglycosides may serve as a strategy for increasing the specificity and reducing antibiotic resistance.[104–106]

The strategy to restrict aminoglycoside flexibility proves to be a promising avenue toward new A-site targeted antibiotics. NMR and X-ray crystal structures of paromomycin in complex with the A-site show that the 2′ amino group of ring II contacts the endocyclic oxygen of ring III and the hydroxyl group at the 5″ position. Covalent linkage of rings I and III would pre-organize paromomycin in an A-site bound conformation that, in theory, would increase its affinity and specificity for the A-site. Unfortunately, covalent attachment of the 2′ and 5″ positions of paromomycin and neomycin yields conformationally resrestricted derivatives that bind to A-site with slightly lower affinities than the unmodified antibiotics.[102,103] Structural studies showing ring III of the restrained neomycin analog making no intermolecular contacts with RNA may explain the decrease in affinity.[107] Unexpectedly, the constrained antibiotics also show only a modest decrease in affinity of the HIV TAR RNA.[102] Since the structural basis of aminoglycoside binding to the HIV TAR RNA is not completely understood, it is difficult to interpret these results in light of current data. However, successful demonstration that constrained aminoglycosides are less susceptible to modification enzymes while maintaining antimicrobial activity toward resistant bacterial strains gives promise that this strategy could alleviate problems with resistance.[103]

The structures of aminoglycosides bound to the A-site reveal the essential role of rings I and II in the specific recognition and antibiotic action of the aminoglycosides. However, rings I and II alone bind the A-site with low affinity and are the most prominent sites of enzymatic modification in aminoglycosides.[58,61,62] This makes the design of new molecules, which maintain the important elements of ring I and II, bind to the A-site with high affinity, as well as lack chemical elements that are susceptible to enzymatic modification, an important goal. Substitution of the saccharide functionality appended to 2-DOS with chemical scaffolds that can recognize RNA and are not susceptible to modification by aminoglycoside-detoxifying enzymes is a popular strategy. Swayze and co-workers[108] have recently designed a set of small molecules that contain the 2-deoxystreptamine moiety and lack carbohydrate functionality. Their approach combines two design strategies: (a) Ring I replacement by an aromatic heterocycle and (b) substitution of the remaining saccharide moieties with open alkylamino chains. Independently, each strategy has proven successful in the design of aminoglycoside mimetics that bind to the A-site with high affinity, inhibit translation *in vitro*, and have antibacterial potency.[109–113] The two most potent translation inhibitors synthesized in this study bind to the A-site with low micromolar affinity and inhibit translation *in vitro* with similar efficiency to paromomycin. The design of noncarbohydrate small molecules that maintain the specific recognition motifs of the aminoglycosides remains a promising avenue for the development of new antibiotics.

In addition to guiding the rational structure-based design process, NMR has been useful in the development of new scaffolds that target the A-site. Fesik and co-workers (Shuker et al., 1996; Yu et al., 2003)[114,115] have adapted the "SAR by NMR" strategy for the design of protein-targeted drugs for RNA. Using the A-site model oligonucleotide as the target, a library of 10,000 compounds was screened using chemical shift perturbations of imino proton resonances to signal binding of a member of the library.[115] The binding affinity of each hit was determined by following the shift of imino resonances as a function of drug concentration. This identified a number of chemical scaffolds that bind the A-site with submicromolar affinity, including the benzimiazole, 2-aminoquinoline, and 2-aminopyridine compounds. These compounds were then subjected to standard medicinal chemistry techniques to modify the scaffolds and determine what functionality leads to increased binding affinity. Chemical shift perturbation of the imino proton resonances was again used to determine the structure–activity relationships among the series of modified compounds.

As an alternative and complementary method to following chemical shift perturbations in NMR-based screening, James and co-workers[116] have used saturation transfer difference (STD) NMR to report on small-molecule-binding RNAs. STD NMR has been used extensively to characterize the binding of small molecules to proteins and relies on transfer of magnetization from macromolecules to bound small molecules.[117] With this methodology, James and co-workers have characterized the binding of a phenothiazine scaffolds to the A-site model oligonucleotide and other RNAs.

Chemical shift perturbations and STD NMR are powerful techniques that can be adapted to high-throughput screening for small-molecule binding to RNA. However, binding to RNA is not always sufficient for drug activity. Aminoglycosides bind to the A-site and induce a conformational change in A1492 and A1493 that has been implicated in the mechanism of drug action.[56,75,77] Therefore, new A-site targeted drugs must not only bind the target specifically and with high affinity, but also demonstrate the ability to displace A1492 and A1493 in a manner equivalent to the aminoglycosides. Advances in the measurement and analysis of residual dipolar couplings (RDCs) have increased the capacity of

NMR to monitor conformational changes. RDCs report on the orientation of bond vectors with respect to the molecular alignment tensor of the macromolecule in a liquid crystalline medium.[118] Changes in the value of RDCs upon small-molecule binding signify that a conformational change in the macromolecule has occurred as a result of drug binding. RDCs have been utilized to report on the conformation change in A1492 and A1493 in the A-site as a function of aminoglycoside binding and represent a third methodology that can be used in high-throughput screening protocols to design new A-site targeted antibiotics.[57]

5.10. PERSPECTIVES

Aminoglycosides remain clinically important antibiotics. NMR provided the initial breakthrough in structural understanding of aminoglycoside action on the ribosome, and it remains a powerful tool for the biophysical characterization of drug–RNA interaction. The combined use of NMR, X-ray crystallography, thermodynamic and functional assays, and computational methods is needed to drive forward the development of new aminoglycosides with improved clinical properties. The rich data described above, combined with the application of new synthetic methods, bode well for the future.

REFERENCES

1. Davies, J.; Gorini, L.; Davis, B. D. *Mol. Pharmacol.* **1965**, *1*, 93–106.
2. Davies, J.; Davis, B. D. *J. Biol. Chem.* **1968**, *243*, 3312–3316.
3. Moazed, D.; Noller, H. F. *Nature* **1987**, *327*, 389–394.
4. Botto, R.; Coxon, B. *J. Am. Chem. Soc.* **1983**, *105*, 1021–1028.
5. Kaul, M.; Barbieri, C. M.; Kerrigan, J. E., et al. *J. Mol. Biol.* **2003**, *326*, 1373–1387.
6. von Ahsen, U.; Davies, J.; Schroeder, R. *Nature* **1991**, *353*, 368–370.
7. Waldsich, C.; Semrad, K.; Schroeder, R. *RNA* **1998**, *4*, 1653–1663.
8. Zapp, M. L.; Stern, S.; Green, M. R. *Cell* **1993**, *74*, 969–978.
9. Mei, H. Y.; Cui, M.; Heldsinger, A., et al. *Biochemistry* **1998**, *37*, 14204–14212.
10. Werstuck, G.; Green, M. R. *Science* **1998**, *282*, 296–298.
11. Green, R.; Noller, H. F. *Annu. Rev. Biochem.* **1997**, *66*, 679–716.
12. Moazed, D.; Noller, H. F. *Cell.* **1986**, *47*, 985–994.
13. Moazed, D.; Noller, H. F. *J Mol Biol.* **1990**, *211*, 135–145.
14. Cunningham, P. R.; Nurse, K.; Bakin, A., et al. *Biochemistry* **1992**, *31*, 12012–12022.
15. Cunningham, P. R.; Nurse, K.; Weitzmann, C. J., et al. *Biochemistry* **1993**, *32*, 7172–7180.
16. Zimmermann, R. A.; Thomas, C. L.; J., W. Structure and function of rRNA in the decoding domain and in the peptidyl transferase center. In *The Ribosome: Structure, Function, and Evolution*; Hill, W. E., Dahlberg, A.; Garrett, R. A.; Moore, P. B.; Schlessinger, D.; Warner, J. R., Eds.; Washington, DC: American Society for Microbiology, 1990, pp. 331–347.

17. Gutell, R. R. *Nucleic Acids Res.* **1994**, *22*, 3502–3507.
18. Wimberly, B. T.; Brodersen, D. E.; Clemons, W. M., et al. *Nature* **2000**, *407*, 327–339.
19. Powers, T.; Noller, H. F. *RNA* **1995**, *1*, 194–209.
20. De Stasio, E. A.; Dahlberg, A. E. *J. Mol. Biol.* **1990**, *212*, 127–133.
21. De Stasio, E. A.; Moazed, D.; Noller, H. F., et al. *EMBO J.* **1989**, *8*, 1213–1216.
22. Alangaden, G. J.; Kreiswirth, B. N.; Aouad, A., et al. *Antimicrob. Agents Chemother.* **1998**, *42*, 1295–1297.
23. Prammananan, T.; Sander, P.; Brown, B. A., et al. *J. Infect. Dis.* **1998**, *177*, 1573–1581.
24. Woodcock, J.; Moazed, D.; Cannon, M., et al. *EMBO J.* **1991**, *10*, 3099–3103.
25. Cromsigt, J.; van Buuren, B.; Schleucher, J., et al. *Methods Enzymol.* **2001**, *338*, 371–399.
26. Varani, G.; Chen, Y.; Leeper, T. C. *Methods Mol. Biol.* **2004**, *278*, 289–312.
27. Wu, H.; Finger, L. D.; Feigon, J. *Methods Enzymol.* **2005**, *394*, 525–545.
28. Lukavsky, P. J.; Kim, I.; Otto, G. A., et al. *Nat. Struct. Biol.* **2003**, *10*, 1033–1038.
29. D'Souza, V.; Dey, A.; Habib, D., et al. *J Mol Biol.* **2004**, *337*, 427–442.
30. D'Souza, V.; Summers, M. F. *Nature* **2004**, *431*, 586–590.
31. Davis, J. H.; Tonelli, M.; Scott, L. G., et al. *J. Mol. Biol.* **2005**, *351*, 371–382.
32. Oberstrass, F. C.; Auweter, S. D.; Erat, M., et al. *Science* **2005**, *309*, 2054–2057.
33. Al-Hashimi, H. M. *ChemBioChem* **2005**, *6*, 1506–1519.
34. Tolman, J.R.; Al-Hashimi, H. M.; Kay, L. E., et al. *J. Am. Chem. Soc.* **2001**, *123*, 1416–1424.
35. Gueron, M.; Leroy, J. L. *Methods Enzymol.* **1995**, *261*, 383–413.
36. Boisbouvier, J.; Bryce, D. L.; O'Neil-Cabello, E., et al. *J. Biomol. NMR* **2004**, *30*, 287–301.
37. Reiter, N. J.; Blad, H.; Abildgaard, F., et al. *Biochemistry* **2004**, *43*, 13739–13747.
38. Furtig, B.; Richter, C.; Wohnert, J., et al. *ChemBioChem* **2003**, *4*, 936–962.
39. Lukavsky, P. J.; Puglisi, J. D. *RNA.* **2004**, *10*, 889–893.
40. Lukavsky, P. J.; Puglisi, J. D. *Methods Enzymol.* **2005**, *394*, 399–416.
41. Latham, M. P.; Brown, D. J.; McCallum, S. A., et al. *ChemBioChem* **2005**, *6*, 1492–1505.
42. Tzakos, A. G.; Grace, C. R.; Lukavsky, P. J., et al. *Annu. Rev. Biophys. Biomol. Struct.* **2006**, *35*, 319–342.
43. Turner, D. H. Conformational changes. In *Nucleic Acids: Structures, Properties and Functions*; Bloomfield, V. A.; Crothers, D. M.; Tinoco, I. Eds.; Herndon, VA: University Science Books, 2000; pp. 259–334.
44. Purohit, P.; Stern, S. *Nature* **1994**, *370*, 659–662.
45. Allain, F. H.; Varani, G. *J. Mol. Biol.* **1995**, *250*, 333–353.
46. Recht, M. I.; Fourmy, D.; Blanchard, S. C., et al. *J. Mol. Biol.* **1996**, *262*, 421–436.
47. Stern, S.; Moazed, D.; Noller, H. F. *Methods Enzymol.* **1988**, *164*, 481–489.
48. Fourmy, D.; Yoshizawa, S.; Puglisi, J. D. *J. Mol. Biol.* **1998**, *277*, 333–345.
49. Mueller, L.; Legault, P.; Pardi, A. *J. Am. Chem. Soc.* **1995**, *117*, 11043–11048.

50. Shandrick, S.; Zhao, Q.; Han, Q., et al. *Angew. Chem. Int. Ed.* **2004**, *43*, 3177–3182.
51. Ward, D. C.; Reich, E.; Stryer, L. *J. Biol. Chem.* **1969**, *244*, 1228–1237.
52. Millar, D. P. *Curr. Opin. Struct. Biol.* **1996**, *6*, 322–326.
53. Kaul, M.; Barbieri, C. M.; Pilch, D. S. *J. Am. Chem. Soc.* **2006**, *128*, 1261–1271.
54. Fourmy, D.; Recht, M. I.; Puglisi, J. D. *J Mol Biol.* **1998**, *277*, 347–362.
55. Yoshizawa, S.; Fourmy, D.; Puglisi, J. D. *EMBO J.* **1998**, *17*, 6437–6448.
56. Fourmy, D.; Recht, M. I.; Blanchard, S. C., et al. *Science.* **1996**, *274*, 1367–1371.
57. Lynch, S. R.; Puglisi, J. D. *J Am Chem Soc.* **2000**, *122*, 7853–7854.
58. Recht, M. I.; Douthwaite, S.; Dahlquist, K. D., et al. *J. Mol. Biol.* **1999**, *286*, 33–43.
59. Blanchard, S. C.; Fourmy, D.; Eason, R. G., et al. *Biochemistry* **1998**, *37*, 7716–7724.
60. Barbieri, C. M.; Pilch, D. S. *Biophys. J.* **2006**, *90*, 1338–1349.
61. Kaul, M.; Pilch, D. S. *Biochemistry.* **2002**, *41*, 7695–7706.
62. Davies, J.; Wright, G. D. *Trends Microbiol.* **1997**, *5*, 234–240.
63. Spangler, E. A.; Blackburn, E. H. *J. Biol. Chem.* **1985**, *260*, 6334–6340.
64. Magnet, J.; Blanchard, J. S. *Chem. Rev.* **2005**, *105*, 477–497.
65. Llano-Sotelo, B.; Azucena, E. F.; Kotra, L. P., et al. *Chem. Biol.* **2002**, *9*, 455–463.
66. Karimi, R.; Ehrenberg, M. *Eur. J. Biochem.* **1994**, *226*, 355–360.
67. Yoshizawa, S.; Fourmy, D.; Puglisi, J. D. *Science* **1999**, *285*, 1722–1725.
68. Ogle, J. M.; Brodersen, D. E.; Clemons, W. M., et al. *Science* **2001**, *292*, 897–902.
69. Wilhelm, J. M.; Jessop, J. J.; Pettitt, S. E. *Biochemistry* **1978**, *17*, 1149–1153.
70. Recht, M. I.; Douthwaite, S.; Puglisi, J. D. *EMBO J.* **1999**, *18*, 3133–3138.
71. Lynch, S. R.; Puglisi, J. D. *J. Mol. Biol.* **2001**, *306*, 1023–1035.
72. Lynch, S. R.; Puglisi, J. D. *J. Mol. Biol.* **2001**, *306*, 1037–1058.
73. Kaul, M.; Barbieri, C. M.; Pilch, D. S. *J. Mol. Biol.* **2005**, *346*, 119–134.
74. Kaul, M.; Barbieri, C. M.; Pilch, D. S. *J. Am. Chem. Soc.* **2004**, *126*, 3447–3453.
75. Carter, A. P.; Clemons, W. M.; Brodersen, D. E., et al. *Nature* **2000**, *407*, 340–348.
76. Brodersen, D. E.; Clemons, W. M.; Carter, A. P., et al. *Cell* **2000**, *103*, 1143–1154.
77. Vicens, Q.; Westhof, E. *Structure* **2001**, *9*, 647–658.
78. Vicens, Q.; Westhof, E. *Chem. Biol.* **2002**, *9*, 747–755.
79. Vicens, Q.; Westhof, E. *J. Mol. Biol.* **2003**, *326*, 1175–1188.
80. Russell, R. J.; Murray, J. B.; Lentzen, G., et al. *J. Am. Chem. Soc.* **2003**, *125*, 3410–3411.
81. Han, Q.; Zhao, Q.; Fish, S., et al. *Angew. Chem. Int. Ed.* **2005**, *44*, 2694–2700.
82. Kondo, J.; Francois, B.; Urzhumtsev, A., et al. *Angew. Chem. Int. Ed.* **2006**, *45*, 3310–3314.
83. Lynch, S. R.; Gonzalez, R. L.; Puglisi, J. D. *Structure.* **2003**, *11*, 43–53.
84. Wang, Y.; Hamasaki, K.; Rando, R. R. *Biochemistry.* **1997**, *36*, 768–779.
85. Hendrix, M.; Priestley, E. S.; Joyce, G. F., et al. *J. Am. Chem. Soc.* **1997**, *119*, 3641–3648.
86. Mei, H.-Y.; Galan, A. A.; Halim, N. S., et al. *Bioorg. Med. Chem. Lett.* **1995**, *5*, 2755–2760.
87. Wang, Y.; Rando, R. R. *Chem. Biol.* **1995**, *2*, 281–290.

88. Patel, D. J. *Curr. Opin. Chem. Biol.* **1997**, *1*, 32–46.
89. Faber, C.; Sticht, H.; Schweimer, K., et al. *J. Biol. Chem.* **2000**, *275*, 20660–20666.
90. Gallego, J.; Varani, G. *Acc. Chem. Res.* **2001**, *34*, 836–843.
91. Wang, S.; Huber, P. W.; Cui, M., et al. *Biochemistry.* **1998**, *37*, 5549–5557.
92. Hermann, T.; Westhof, E. *J. Med. Chem.* **1999**, *42*, 1250–1261.
93. Varani, L.; Spillantini, M. G.; Goedert, M., et al. *Nucleic Acids Res.* **2000**, *28*, 710–719.
94. Varani, L.; Hasegawa, M.; Spillantini, M.G., et al. *Proc. Natl. Acad. Sci. USA.* **1999**, *96*, 8229–8234.
95. Tuerk, C.; Gold, L. *Science* **1990**, *249*, 505–510.
96. Robertson, D. L.; Joyce, G.F. *Nature* **1990**, *344*, 467–468.
97. Ellington, A. D.; Szostak, J.W. *Nature* **1992**, *355*, 850–852.
98. Jiang, L.; Suri, A. K.; Fiala, R., et al. *Chem. & Biol.* **1997**, *4*, 35–50.
99. Jiang, L.; Patel, D. J. *Nat. Struct. Biol.* **1998**, *5*, 769–774.
100. Jiang, L.; Majumdar, A.; Hu, W., et al. *Structure* **1999**, *7*, 817–827.
101. Walsh, C. *Nature* **2000**, *406*, 775–781.
102. Blount, K. F.; Zhao, F.; Hermann, T., et al. *J. Am. Chem. Soc.* **2005**, *127*, 9818–9829.
103. Bastida, A.; Hidalgo, A.; Chiara, J.L., et al. *J. Am. Chem. Soc.* **2006**, *128*, 100–116.
104. Fong, D. H.; Berghuis, A. M. *EMBO J.* **2002**, *21*, 2323–2331.
105. Vetting, M. W.; Hegde, S. S.; Javid-Majd, F., et al. *Nat. Struct. Biol.* **2002**, *9*, 653–658.
106. Vetting, M. W.; Magnet, S.; Nieves, E., et al. *Chem. Biol.* **2004**, *11*, 565–573.
107. Zhao, F.; Zhao, Q.; Blount, K. F., et al. *Angew. Chem. Int. Ed.* **2005**, *44*, 5329–5334.
108. Wang, X.; Migawa, M. T.; Sannes-Lowery, K. A., et al. *Bioorg. Med. Chem. Lett.* **2005**, *15*, 4919–4922.
109. Greenberg, W. A.; Priestley, E. S.; Sears, P. S., et al. *J. Am. Chem. Soc.* **1999**, *121*, 6527–6541.
110. Hanessian, S.; Tremblay, M.; Kornienko, A., et al. *Tetrahedron* **2001**, *57*, 3255–3265.
111. Haddad, J.; Kotra, L. P.; Llano-Sotelo, B., et al. *J. Am. Chem. Soc.* **2002**, *124*, 3229–3237.
112. Ding, Y.; Hofstadler, S. A.; Swayze, E. E., et al. *Org. Lett.* **2001**, *3*, 1621–1623.
113. Ding, Y.; Hofstadler, S. A.; Swayze, E. E., et al. *Angew. Chem. Int. Ed.* **2003**, *42*, 3409–3412.
114. Shuker, S. B.; Hajduk, P. J.; Meadows, R. P., et al. *Science.* **1996**, *274*, 1531–1534.
115. Yu, L.; Oost, T. K.; Schkeryantz, J. M., et al. *J. Am. Chem. Soc.* **2003**, *125*, 4444–4450.
116. Mayer, M.; James, T.L. *J. Am. Chem. Soc.* **2004**, *126*, 4453–4460.
117. Mayer, M.; Meyer, B. *Angew. Chem. Int. Ed.* **1999**, *38*, 1784–1788.
118. Tjandra, N.; Bax, A. *Science* **1997**, *278*, 1111–1114.

6

STRUCTURAL COMPARISONS BETWEEN PROKARYOTIC AND EUKARYOTIC RIBOSOMAL DECODING A SITES FREE AND COMPLEXED WITH AMINOGLYCOSIDES

JIRO KONDO AND ERIC WESTHOF

Institut de Biologie Moléculaire et Cellulaire, UPR9002 CNRS, Université Louis Pasteur, 67084 Strasbourg, France

Aminoglycoside Antibiotics: From Chemical Biology to Drug Discovery, Second Edition, by Dev P. Arya

6.1. INTRODUCTION

The A-site is an asymmetric internal loop framed by two G–C base pairs. In prokaryotes, the internal loop is most generally constituted of only adenines (A1408 on one strand, and the two A1492 and A1493 on the other strand). In cytoplasmic ribosomes of eukaryotes, the single A1408 is generally replaced by a G1408. The other two adenines are universally conserved. It has been shown by Venki Ramakrishnan and co-workers[1–4] that the two adenines when bulged out of the helix decode the first two base pairs of the codon–anticodon minihelix by recognizing their shallow/minor grooves and forming A-minor contacts as shown in Figures 6.1 and 6.2. In the free state, the two adenines adopt various conformations and the A-site is conformationally dynamic (for discussion see reference[5]). We call this resting state "off." On the other hand, in the state where the two adenines are bulging out and able to recognize the codon–anticodon helix, a single conformation of the internal loop is generally observed. We call this decoding state, the "on" state. As shown in Figure 6.3, in prokaryotes, aminoglycosides bind to and stabilize the "on" state by forming several invariant contacts between the neamine core and conserved residues (A1408, A1492, A1493, G1494, and U1495) (for comparative descriptions see reference[6]).

6.1.1. The Decoding "On" State of the Prokaryotic A-Site

The "on" state of the prokaryotic A-site without aminoglycosides has been observed in the crystal structure of *Thermus thermophilus* 30S ribosomal particle (see Figure 6.4a and Table 6.1).[1] At the both ends of the internal loop of the prokaryotic A-site, five canonical Watson–Crick pairs—C1404–G1497, G1405–C1496, C1407–G1494, C1409–G1491 and A1410–U1490—are formed. The U1406 residue forms a bifurcated base pair with the U1495 residue by using the Watson–Crick edge. The A1408 residue, which is universally conserved in all prokaryotic A-sites, does not form any base

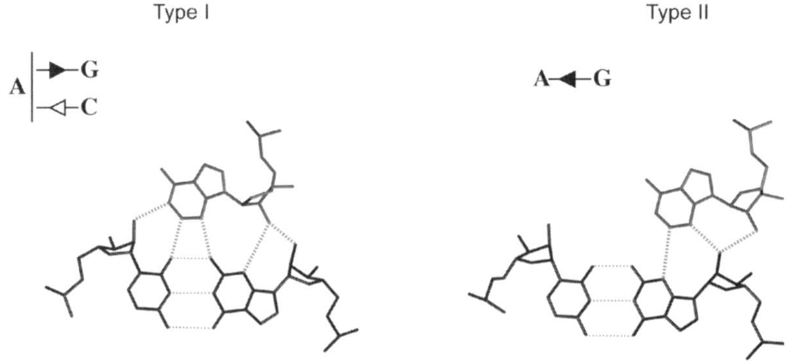

Figure 6.1. The two main types of A-minor interactions, type I and type II, according to reference[7], or sugar–sugar *trans* and sugar–sugar *cis* according to reference.[8]

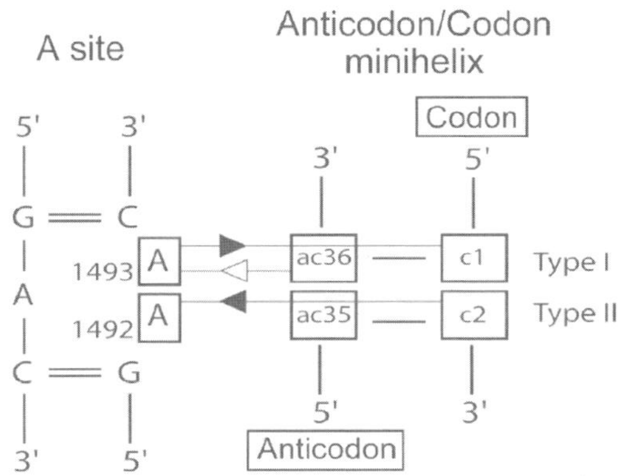

Figure 6.2. The mode of recognition between the ribosomal A-site and the codon–anticodon minihelix formed between the messenger RNA codon and the triplet of the anticodon loop of the cognate transfer RNA according to references[1–4]. The first two nucleotides of the codon triplet (C1, C2) paired with the last two nucleotides of the anticodon triplet (ac36, ac35) are recognized by A1493 and A1492, respectively, through A-minor interactions as described in Figure 6.1.

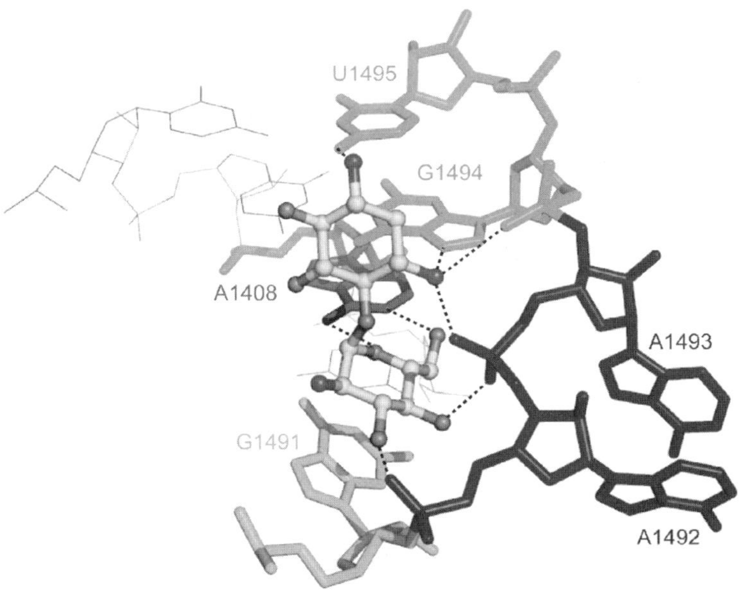

Figure 6.3. Conserved interactions between the prokaryotic A-site and the neamine core of 4,5- and 4,6-disubstituted aminoglycosides as deduced from crystal structures.[1,3,6,10,12–20] See color plates.

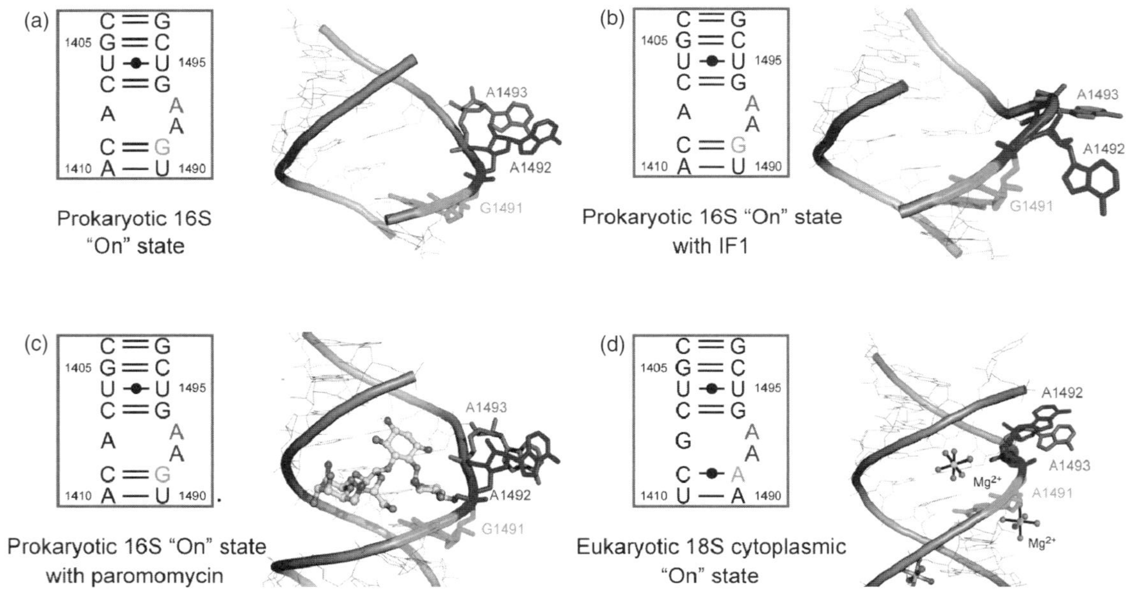

Figure 6.4. The observed structures of the decoding "on" state of the prokaryotic (a (reference[1]), b (reference[11]), c (reference[15])) and the eukaryotic cytoplasmic A-site (d (reference[5])).

TABLE 6.1. The "On" States of the Prokaryotic and the Eukaryotic A-Sites Found in Crystal Structures[a]

	Substrates	Aminoglycosides	PDB ID	Resolution (Å)	Reference
Prokaryotic "On" State					
T. thermophilus 30S ribosome	Cognate tRNA-mRNA		1IBM	3.31	1
T. thermophilus 30S ribosome	Initiation factor IF1		1HR0	3.20	11
Model RNA A-site[c]			1T0D	2.20	9
Model RNA A-site[c]			1T0E	1.70	9
Model RNA A-site			1ZX7	2.15	10
Prokaryotic "On" State with Aminoglycoside					
T. thermophilus 30S ribosome	Cognate tRNA–mRNA	Paromomycin	1IBL	3.11	1
T. thermophilus 30S ribosome	Cognate tRNA–mRNA	Paromomycin	1XMO	3.25	12
T. thermophilus 30S ribosome	Cognate tRNA–mRNA	Paromomycin	1XMQ	3.00	12
T. thermophilus 30S ribosome	Near-cognate tRNA–mRNA	Paromomycin	1IBK	3.31	1
T. thermophilus 30S ribosome[b]	Near-cognate tRNA–mRNA	Paromomycin	1FJG	3.00	13
T. thermophilus 30S ribosome	Near-cognate tRNA–mRNA	Paromomycin	1N32	3.00	3
T. thermophilus 30S ribosome	Near-cognate tRNA–mRNA	Paromomycin	1N33	3.35	3
T. thermophilus 30S ribosome	Near-cognate tRNA–mRNA	Paromomycin	1XNQ	3.05	14
T. thermophilus 30S ribosome	Near-cognate tRNA–mRNA	Paromomycin	1XNR	3.10	14
Model RNA A site		Paromomycin	1J7T	2.50	15

(*continued overleaf*)

TABLE 6.1. (*continued*)

	Substrates	Aminoglycosides	PDB ID	Resolution (Å)	Reference
Model RNA A site		Tobramycin	1LC4	2.54	16
Model RNA A site		Geneticin	1MWL	2.40	17
Model RNA A site		Apramycin	1YRJ	2.70	18
Model RNA A site		Neomycin derivative	1ZZ5	3.00	10
Model RNA A site		Neomycin	2A04	2.95	10
Model RNA A site		Paromomycin derivative	2BE0	2.63	19
Model RNA A site		Paromomycin derivative	2BEE	2.60	19
Model RNA A site		Kanamycin	2ESI	3.00	6
Model RNA A site		Lividomycin	2ESJ	2.20	6
Model RNA A site		Gentamicin C1A	2ET3	2.80	6
Model RNA A site		Neomycin B	2ET4	2.40	6
Model RNA A site		Ribostamycin	2ET5	2.20	6
Model RNA A site[c]		Neamine	2ET8	2.50	6
Model RNA A site		Neamine	2F4S	2.80	20
Model RNA A site[c]		Neamine derivative	2F4T	3.00	20
Model RNA A site[c]		Neamine derivative	2F4U	2.60	20
Eukaryotic Cytoplasmic "On" State					
Model RNA A site[c]			2FQN	2.30	5

[a]Crystal structures listed in this table have resolutions higher than 3.35 Å.
[b]Crystal structure contains other aminoglycosides, which do not bind to the A-site.
[c]Crystal structure also contains the "off" state of the A-site.

pair and is stacked between C1407 and C1409. Two adenine residues, A1492 and A1493, are fully bulged out and interact with shallow/minor groove of the cognate codon–anticodon stem of mRNA–tRNA complex through the A-minor motifs. Although resolution is not high enough, two magnesium ions are found in the deep/major groove of the A-site. Almost the same conformation has been observed also in crystal structures of model oligonucleotides containing an A-site.[9,10] In this case, the U1406 residue forms a symmetrical base pair with U1495 instead of a bifurcated one, and several water molecules or half-occupancy of sulfate ion are found in the deep/major groove of the A-site. Two bulged-out adenines make the A-minor contacts with a Watson–Crick stem of a neighboring duplex, which mimics the cognate codon–anticodon stem. The conformation of the A-site found in the crystal structure of the complex between the *T. thermophilus* 30S particle and the initiation factor IF1, which forces the initiator tRNA to bind in the P-site, is the other example of the "on" state of the prokaryotic A-site (see Figure 6.4b and Table 6.1).[11] The bulged-out A1492 residue is inserted into the cavity formed at the interface between IF1 and ribosomal protein S12, whereas the bulged-out A1493 residue is buried in the pocket on the surface of IF1. In both cases, each adenine residue is stabilized by stacking interaction with conserved arginine residues Arg^{41} and Arg^{46}, respectively.

6.1.2. The Decoding "On" State of the Prokaryotic A-Site Complexed with Aminoglycosides

It has been reported that 4,5- and 4,6-disubstituted aminoglycosides and the aminocyclitol containing apramycin, which is also classified as an aminoglycoside owing to its physicochemical properties, specifically bind to the "on" state of the prokaryotic A-site with several conserved contacts (see Figures 6.3 and 6.4c and Table 6.1).[1,3,6,10,12–20] First, the puckered sugar ring I is inserted into the A-site helix by stacking on the G1491 residue and forms a pseudo-pair with the Watson–Crick sites of the universally conserved A1408 residue. Second, the central ring II (2-deoxystreptamine; 2-DOS) makes four hydrogen bonds through N1-H. . .O4(U1495), N3-H. . .N7(G1494), N3-H. . .O2P(G1494), and N3-H. . .O1P(A1493). These common interactions force to maintain the "on" state with two bulged-out adenines regardless of whether the codon–anticodon stem is cognate or near-cognate type, which eventually induces misreading of the codon. In the case of the prokaryotic "on" state complexed with neamine and its several derivatives,[6,20] U1406 does not form a base pair with U1495, but bulges out of the A-site. The place normally occupied by U1406 is filled with A1492 coming from a neighboring molecule which forms a *cis* Hoogsteen/Watson–Crick A1492oU1495 base pair (where A1492 is in the *syn* conformation).

6.1.3. The Decoding "On" State of the Eukaryotic Cytoplasmic A-Site

The "on" state of the eukaryotic cytoplasmic A-site has been recently solved by X-ray analysis of model oligoribonucleotide containing two A-sites.[5] Its overall conformation is very similar to that observed for the prokaryotic A-site (see Figure 6.4d and Table 6.1). The eukaryotic cytoplasmic A site has the Watson–Crick C14090A1491 with a single H-bond N4-H...N1 and U1410–A1490 base pairs instead of the Watson-Crick C1409–G1491 and A1410–U1490 base pairs in the prokaryotic A site. The effect of the base change at position 1408 from adenine to guanine is not significant on the overall conformation of the "on" state of the A-site. Two adenine residues in the internal loop of the cytoplasmic A-site, A1492 and A1493, are fully bulged out and make contacts with the shallow/minor groove of a neighboring duplex. Two hexa-hydrated magnesium ions are observed as bound to the "on" state of the eukaryotic cytoplasmic A-site. One of them interacts with the Hoogsteen edge of G1494 of the C1407–G1494 base pair and O4 of U1406 of the symmetrical U1406–U1495 base pair. In addition, the hydrated ion bridges the Watson–Crick edge of the G1408 residue and the phosphate backbone of A1492 and A1493 through second shell water-mediated hydrogen bonds. The other hexa-hydrated magnesium ion directly binds to the shallow/minor groove of A1491 and U1410 of the Watson–Crick C14090A1491 and U1410–A1490 base pairs.

6.1.4. The Resting "Off" State of the Prokaryotic A-Site

While the "on" state has been observed in only a single conformation, four conformations of "off" states have been reported for the prokaryotic A-site so far.[3,5,6,9,20–25] In all "off" states, the five canonical Watson–Crick base pairs and the UoU (bifurcated or symmetrical) base pair as found in the "on" state are conserved. Differences are evidently observed as conformational versatility around the two adenine residues. A first example is the "off" state with bulged-in A1492 and A1493, which is found in crystal structures of the *T. thermophilus* 30S ribosomal particle (see Figure 6.5a and Table 6.2).[22,24] Only A1493 forms a base pair with A1408 in the Watson–Crick geometry, whereas A1492 is stacked between A1493 and G1491. Also in the crystal structures of the *Escherichia coli* 30S subunit,[21] the *E. coli* 70S ribosome[25] and one of the *T. thermophilus* 30S ribosomal particles,[23] the two A1492 and A1493 are bulged in, but they do not form any hydrogen-bonding contact within the internal loop of the A-site (see Figure 6.5b and Table 6.2). Although stacking interactions between G1491–A1492–A1493 are observed, the two adenines have poor densities or are disordered. The second and third examples of the "off" state of the prokaryotic A-site present one bulged-out adenine and one bulged-in adenine (see Figure 6.5c–e and Table 6.2). In the "off" state with bulged-out A1492, as found in the crystal structures of the model RNA A-sites,[6,20] A1493 has a *syn* conformation around the glycoside bond and forms a *cis* Hoogsteen/Watson–Crick A1493oA1408 base pair (see Figure 6.5c). On the other hand, in the "off" state with bulged-out A1493 observed in crystal structures of the *E. coli* 70S ribosome[25] and the model RNA A sites,[9] the A1492 residue stays inside of the helix and forms a *trans* Hoogsteen/Watson–Crick and a Watson–Crick A1492oA1408 base pairs, respectively (see Figure 6.5d,e). Another "off" state with two bulged-out adenines has also been found in crystal

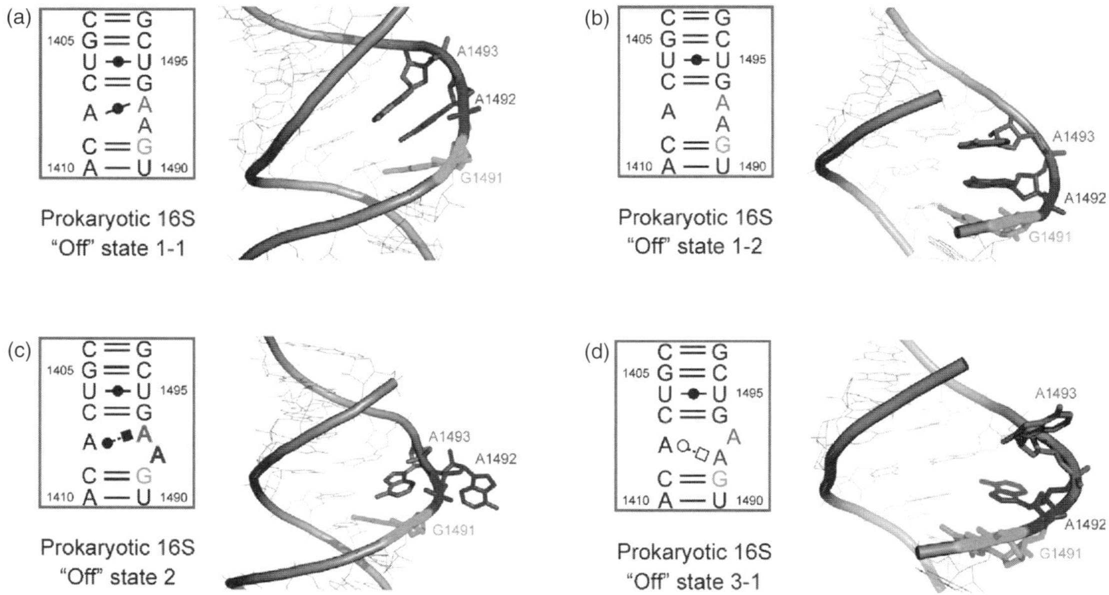

Figure 6.5. The observed structures of the resting "off" state of the prokaryotic (a (references[24]), b (reference[25]), c (reference[6]), d (reference[25]), e (reference[9]), f (reference[3]), and the eukaryotic (g (reference[5]), h (reference[26])) cytoplasmic A-site.

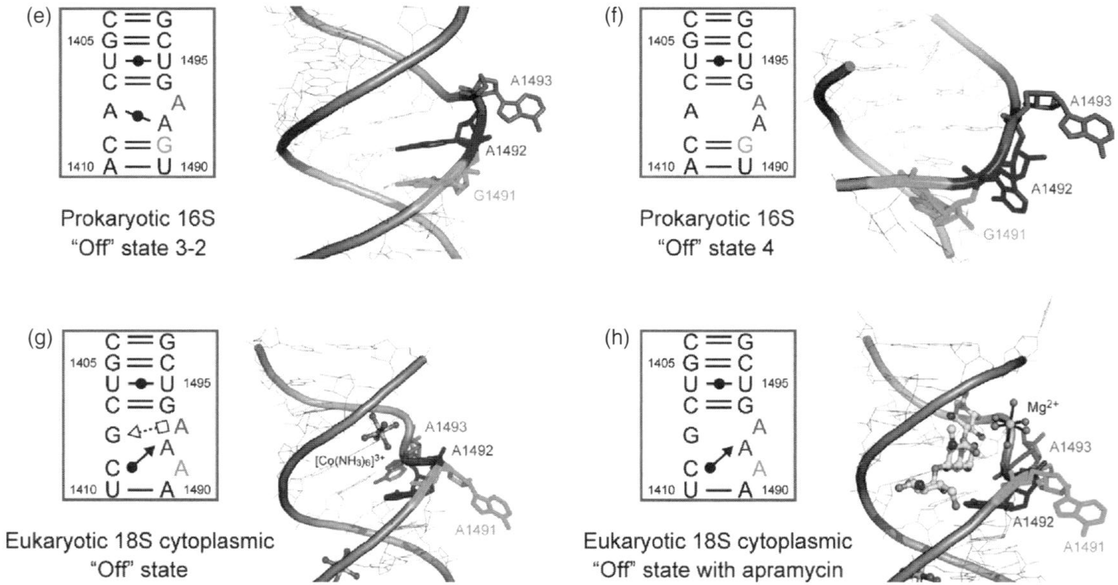

Figure 6.5. *(continued)*

TABLE 6.2. The "Off" States of the Prokaryotic and the Eukaryotic A-Sites Found in Crystal Structures[a]

	Substrates	Aminoglycosides	PDB ID	Resolution (Å)	Reference
Prokaryotic "Off" State 1 (Bulged-In Two As)					
E. coli 30S ribosome			1FKA	3.30	21
T. thermophilus 30S ribosome[b]			1HNW	3.40	22
T. thermophilus 30S ribosome[b]			1HNX	3.40	22
T. thermophilus 30S ribosome[b]			1HNZ	3.30	22
T. thermophilus 30S ribosome			1I94	3.20	23
T. thermophilus 30S ribosome			1J5E	3.05	24
E. coli 70S Ribosome			2AVY	3.46	25
Prokaryotic "Off" State 2 (Bulged-Out A1492)					
Model RNA A-site[c]			2ET8	2.50	6
Model RNA A-site[c]			2F4U	2.60	20
Prokaryotic "Off" State 3 (Bulged-Out A1493)					
E. coli 70S ribosome			2AW7	3.46	25
Model RNA A-site[c]			1T0D	2.20	9
Model RNA A-site[c]			1T0E	1.70	9
Prokaryotic "Off" State 4 (Bulged-Out Two As)					
T. thermophilus 30S ribosome	Near-cognate tRNA–mRNA		1N34	3.80	3
T. thermophilus 30S ribosome	Near-cognate tRNA–mRNA		1N36	3.80	3
Model RNA A-site[c]			2F4S	2.80	20
Model RNA A-site[c]			2F4T	3.00	20
Eukaryotic Cytoplasmic "Off" State					
Model RNA A-site[c]			2FQN	2.30	5
Eukaryotic cytoplasmic "Off" State with Aminoglycoside					
Model RNA A-site		Apramycin	2G5K	2.80	26

[a]Crystal structures listed in this table have resolutions higher than 3.80 Å.
[b]Crystal structure contains other aminoglycosides, which do not bind to the A-site.
[c]Crystal structure also contains the "on" state of the A-site.

structures of the *T. thermophilus* 30S ribosomal particles and the model RNA A-sites (see Figure 6.5f and Table 6.2).[3,20] In this case, however, A1492 and A1493 have poor densities or high B-factors.

6.1.5. The Resting "Off" State of the Eukaryotic Cytoplasmic A-Site

The "off" state of the eukaryotic cytoplasmic A-site has been found in the crystal structure of model oligoribonucleotide containing two A-sites (see Figure 6.5g and Table 6.2), in which one of the two A-site forms the "on" state discussed above.[5] Base-pair geometries at both ends of the internal loop are the same as those in the "on" state of the eukaryotic cytoplasmic A-site. On the other hand, the conformation of the internal loop of the A-site is completely different from that observed for the "on" state. In addition, it is important to note that the conformation of the "off" state of the eukaryotic cytoplasmic A-site is completely different from any of the four "off" states observed for the prokaryotic A-site. The G1408 residue forms a *trans* Sugar-edge/Hoogsteen base pair with A1493 through only one H-bond N2-H...N7, and it halfway bulges out to the solvent. The A1492 residue is maintained inside the helix by forming a *cis* Sugar-edge/Watson–Crick base pair with C1409 through three H-bonds, O2′–H...O2, O2′–H...N3, and N3...H-N4. The A1491 residue does not have any partner to form a base pair in the internal loop of the A-site, and it fully bulges out to the solvent. An hexammine cobalt cation binds to the deep/major groove of G1494 of the Watson–Crick C1407–G1494 base pair.

6.1.6. The Resting "Off" State of the Eukaryotic Cytoplasmic A-Site Complexed with the Aminoglycoside Apramycin

We have recently reported the crystal structure of the eukaryotic cytoplasmic A-site complexed with apramycin.[26] Unexpectedly and surprisingly, apramycin specifically binds to the "off" state of the eukaryotic cytoplasmic A-site (see Figure 6.5h). Only a single difference from the eukaryotic cytoplasmic "off" state without apramycin is observed on the A1493 residue, which forms the *trans* Hoogsteen/Sugar-edge base pair with G1408 in the native A-site and is free from any base-pair interaction in the RNA/apramycin complex. Ring II of apramycin interacts with C1404–G1497, G1405–C1496, and C1407–G1494. The central bicyclic ring I forms pseudo-pair with G1408. Ring III recognizes the Watson–Crick/Sugar-edge C1409–A1492 base pair from the deep/major groove.

6.2. CONCLUSIONS

A-minor motifs are very frequent in RNA–RNA contacts and several RNA motifs allow for the formation of such contacts.[27] Thus, the A-minor contacts formed by the decoding "on" state of the A-site are identical to those formed by GNRA tetraloops and helically stacked Watson–Crick base pairs. The asymmetrical A-site internal loop is, however, rare in RNA secondary structures. Comparisons between the published structures of ribosomal RNAs or oligoribonucleotides containing the A-site show that, whatever the environment and despite sequence variations, one conformational state is geometrically invariant in all structures of the A-site, the state we have called "on"—that is, the one with two bulging adenines and one internal purine residue. In contrast, the state we have called resting or "off" displays several states in both the prokaryotic and eukaryotic sequences.

The observed conformational similarity between the prokaryotic and cytoplasmic A-sites in the decoding "on" state—that is, the state used in prokaryote for recognition of mRNA–tRNA complex—supports the view that a similar recognition mechanism is used in eukaryotes. On the contrary, the resting "off" state, or the state adopted by a ribosome with an empty A-site, adopts a great variety of conformations in crystal structures containing the prokaryotic or eukaryotic A-site.[5] Thus, it appears that in the absence of the intermolecular stabilizing interactions made by the bulging adenines, the A-site adopts disordered or various conformers. When an aa-tRNA enters the A-site and forms cognate codon–anticodon stem with mRNA template, the A-site changes its conformation to an identical 'on' state. Indeed, the two bulged-out adenines, A1493 and A1492, have to bind to and recognize the first and second Watson–Crick base pair of codon–anticodon, respectively. It is natural to consider that aminoglycoside binding to the eukaryotic cytoplasmic A-site contributes to their toxicity. Further structural determinations and analysis of structures of complexes should allow the design of new aminoglycoside antibiotics with higher selectivity for the prokaryotic ribosome and less toxicity to mammals.

ACKNOWLEDGMENT

During part of this work, Dr. J. Kondo was supported by the Japan Society for the Promotion of Science.

REFERENCES

1. Ogle, J. M.; Brodersen, D. E.; Clemons, W. M. Jr.; Tarry, M. J.; Carter, A. P.; Ramakrishnan, V. Recognition of cognate transfer RNA by the 30S ribosomal subunit. *Science* **2001**, *292*, 897–902.
2. Ogle, J. M.; Carter, A. P.; Ramakrishnan, V. Insights into the decoding mechanism from recent ribosome structures. *Trends. Biochem. Sci.* **2003**, *28*, 259–266.
3. Ogle, J. M.; Murphy, F. V.; Tarry, M. J.; Ramakrishnan, V. Selection of tRNA by the ribosome requires a transition from an open to a closed form. *Cell* **2002**, *111*, 721–732.
4. Ogle, J. M.; Ramakrishnan, V. Structural insights into translational fidelity. *Annu. Rev. Biochem.* **2005**, *74*, 129–177.

5. Kondo, J.; Urzhumtsev, A.; Westhof, E. Two conformational states in the crystal structure of the *Homo sapiens* cytoplasmic ribosomal decoding A site. *Nucleic Acids Res.* **2006**, *34*, 676–685.
6. François, B.; Russell, R. J. M.; Murray, J. B.; Aboul-ela, F.; Masquida, B.; Vicens, Q.; Westhof, E. Crystal structures of complexes between aminoglycosides and decoding A site oligonucleotides: Role of the number of rings and positive charges in the specific binding leading to miscoding. *Nucleic Acids Res.* **2005**, *33*, 5677–5690.
7. Klein, D. J.; Moore, P. B.; Steitz, T. A. The roles of ribosomal proteins in the structure assembly, and evolution of the large ribosomal subunit. *J. Mol. Biol.* **2004**, *340*, 141–177.
8. Leontis, N. B.; Westhof, E. Geometric nomenclature and classification of RNA base pairs. *RNA* **2001**, *7*, 499–512.
9. Shandrick, S.; Zhao, Q.; Han, Q.; Ayida, B. K.; Takahashi, M.; Winters, G. C.; Simonsen, K. B.; Vourloumis, D.; Hermann, T. Monitoring molecular recognition of the ribosomal decoding site. *Angew. Chem. Int. Ed. Engl.* **2004**, *43*, 3177–3182.
10. Zhao, F.; Zhao, Q.; Blount, K. F.; Han, Q.; Tor, Y.; Hermann, T. Molecular recognition of RNA by neomycin and a restricted neomycin derivative. *Angew. Chem. Int. Ed. Engl.* **2005**, *44*, 5329–5334.
11. Carter, A. P.; Clemons, W. M. Jr.; Brodersen, D. E.; Morgan-Warren, R. J.; Hartsch, T.; Wimberly, B. T.; Ramakrishnan, V. Crystal structure of an initiation factor bound to the 30S ribosomal subunit. *Science* **2001**, *291*, 498–501.
12. Murphy, F. V., 4th; Ramakrishnan, V.; Malkiewicz, A.; Agris, P. F. The role of modifications in codon discrimination by tRNA(Lys)UUU. *Nat. Struct. Mol. Biol.* **2004**, *11*, 1186–1191.
13. Carter, A. P.; Clemons W. M.; Brodersen, D. E.; Morgan-Warren, R. J.; Wimberly, B. T.; Ramakrishnan, V. Functional insights from the structure of the 30S ribosomal subunit and its interactions with antibiotics. *Nature* **2000**, *407*, 340–348.
14. Murphy, F. V., 4th; Ramakrishnan, V. Structure of a purine-purine wobble base pair in the decoding center of the ribosome. *Nat. Struct. Mol. Biol.* **2004**, *11*, 1251–1252.
15. Vicens, Q.; Westhof, E. Crystal structure of paromomycin docked into the eubacterial ribosomal decoding A site. *Structure* **2001**, *9*, 647–658.
16. Vicens, Q.; Westhof, E. Crystal structure of a complex between the aminoglycoside tobramycin and an oligonucleotide containing the ribosomal decoding a site. *Chem. Biol.* **2002**, *9*, 747–755.
17. Vicens, Q.; Westhof, E. Crystal structure of geneticin bound to a bacterial 16S ribosomal RNA A site oligonucleotide. *J. Mol. Biol.* **2003**, *326*, 1175–1188.
18. Han, Q.; Zhao, Q.; Fish, S.; Simonsen, K. B.; Vourloumis, D.; Froelich, J. M.; Wall, D.; Hermann, T. Molecular recognition by glycoside pseudo base pairs and triples in an apramycin–RNA complex. *Angew. Chem. Int. Ed. Engl.* **2005**, *44*, 2694–2700.
19. François, B.; Szychowski, J.; Adhikari, S. S.; Pachamuthu, K.; Swayze, E. E.; Griffey, R. H.; Migawa, M. T.; Westhof, E.; Hanessian, S. Antibacterial aminoglycosides with a modified mode of binding to the ribosomal-RNA decoding site. *Angew. Chem. Int. Ed. Engl.* **2004**, *43*, 6735–6738.
20. Murray, J. B.; Meroueh, S. O.; Russell, R. J.; Lentzen, G.; Haddad, J.; Mobashery, S. Interactions of designer antibiotics and the bacterial ribosomal aminoacyl-tRNA site. *Chem. Biol.* **2006**, *13*, 129–138.

21. Schluenzen, F.; Tocilj, A.; Zarivach, R.; Harms, J.; Gluehmann, M.; Janell, D.; Bashan, A.; Bartels, H.; Agmon, I.; Franceschi, F.; Yonath, A. Structure of functionally activated small ribosomal subunit at 3.3 angstroms resolution. *Cell* **2000**, *102*, 615–623.
22. Brodersen, D. E.; Clemons, W. M., Jr.; Carter, A. P.; Morgan-Warren, R. J.; Wimberly, B. T.; Ramakrishnan, V. The structural basis for the action of the antibiotics tetracycline, pactamycin, and hygromycin B on the 30S ribosomal subunit. *Cell* **2000**, *103*, 1143–1154.
23. Pioletti, M.; Schlunzen, F.; Harms, J.; Zarivach, R.; Gluhmann, M.; Avila, H.; Bashan, A.; Bartels, H.; Auerbach, T.; Jacobi, C.; Hartsch, T.; Yonath, A.; Franceschi, F. Crystal structures of complexes of the small ribosomal subunit with tetracycline, edeine and IF3. *EMBO J.* **2001**, *20*, 1829–1839.
24. Wimberly, B. T.; Brodersen, D. E.; Clemons, W. M., Jr.; Morgan-Warren, R. J.; Carter, A. P.; Vonrhein, C.; Hartsch, T.; Ramakrishnan, V. Structure of the 30S ribosomal subunit. *Nature* **2000**, *407*, 327–339.
25. Schuwirth, B. S.; Borovinskaya, M. A.; Hau, C. W.; Zhang, W.; Vila-Sanjurjo, A.; Holton, J. M.; Cate, J. H. D. Structures of the bacterial ribosome at 3.5 Å resolution. *Science* **2005**, *310*, 827–834.
26. Kondo, J.; François, B.; Urzhumtsev, A.; Westhof, E. Crystal structure of the *Homo sapiens* cytoplasmic ribosomal decoding site complexed with apramycin. *Angew. Chem. Int. Ed. Engl.* **2006**, *45*, 3310–3314.
27. Leontis, N. B.; Lescoute, A.; Westhof, E. The building blocks and motifs of RNA architecture. *Curr. Opin. Struct. Biol.* **2006**, *16*, 279–287.

7

BINDING OF ANTIBIOTICS TO THE AMINOACYL-TRNA SITE OF BACTERIAL RIBOSOME

DALE KLING
Department of Chemistry and Biochemistry, University of Notre Dame, Notre Dame, IN 46556

CHRISTINE CHOW
Department of Chemistry, Wayne State University, Detroit, MI 48202

SHAHRIAR MOBASHERY
Department of Chemistry and Biochemistry, University of Notre Dame, Notre Dame, IN 46556

7.1. INTRODUCTION

Proteins are synthesized by the ribosomal assembly in living organisms, details of which have been the subject of excellent recent reviews.[1–4] The ribosome is a marvelously complex system of three large RNA molecules and over 50 proteins. This assembly translates the information from mRNA into protein sequence with a fidelity of one mistake per each 10^4 elongation cycles.[5]

Codon–anticodon recognition between mRNA and the stem loop of tRNA on the ribosomal surface brings one amino acid at a time for the formation of amide bonds at the top of ribosomal helix 44, a site known as the aminoacyl-tRNA site (A-site). It is not surprising that the A-site is also the binding site for

Aminoglycoside Antibiotics: From Chemical Biology to Drug Discovery, Second Edition, by Dev P. Arya

certain antibiotics. Binding of antibiotics to this specific site interferes with the translation events, leading to bacterial death.

Of central importance to this discussion is the class of aminoglycoside antibiotics. Structural evidence aimed at identifying the specific interactions between aminoglycosides and 16S ribosomal RNA has been accumulating steadily for nearly a decade. Puglisi and co-workers[6] described binding of paromomycin (**1**) to a 27-nucleotide RNA fragment using nuclear magnetic resonance spectroscopy (NMR). These studies revealed that paromomycin binds an asymmetric loop (3:4) near the region of bases A^{1492} and A^{1493}, resulting in a bulging of A^{1492} away from helix 44. Interestingly, the vacancy created by this bulge is filled by one of the rings of paromomycin, an aminoglycoside, which assumes a base-stacking role with G^{1491}. This study revealed for the first time the interactions that anchor this antibiotic class to the A site. Support for these findings was demonstrated a year later through a binding comparison with the related aminoglycoside neomycin, a 6′-amino derivative of paromomycin.[7]

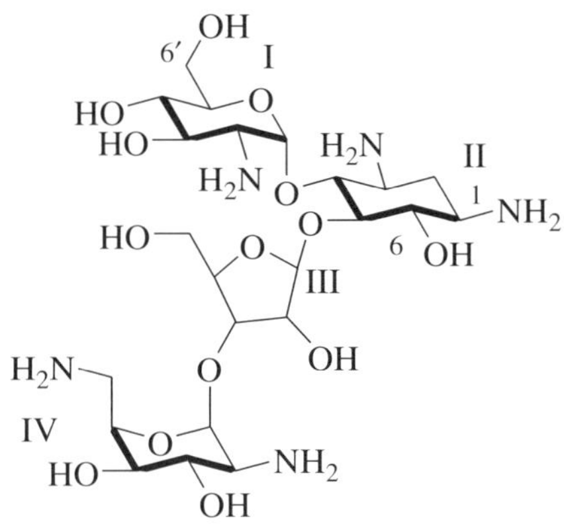

1

Naturally, the question of whether or not a 27-nucleotide fragment could faithfully mimic the complete 30S subunit had to be answered. Are there nuances to this short stem of RNA binding with paromomycin that do not exist when bound to the intact subunit? What role, if any, do neighboring RNA and protein moieties of the ribosome play in binding when the 30S subunit is whole? Answers to these questions have become much clearer recently with the outpouring of X-ray crystal structures containing aminoglycosides bound to the complete 30S ribosomal subunit.[8–14]

Ramakrishnan and co-workers' 3-Å crystal structures of two different aminoglycosides (paromomycin and streptomycin) bound to the 30S subunit laid to rest much of the debate concerning the binding mode of aminoglycoside antibiotics.[13] Paromomycin was found bound to the major groove of helix 44 (H44), confirming mutagenesis studies that had been carried out previously.[15]

Furthermore, ring I of paromomycin was found to imitate a nucleotide base and stack with G^{1491} as Puglisi had described by NMR. Of additional note was the finding that ring I is involved in hydrogen bonding interactions with A^{1408}, an interaction that was not well-defined by NMR. It was also observed by NMR that binding of paromomycin to the short RNA fragment resulted in bulging of base A^{1492} away from helix 44. This observation further support in the X-ray crystal structure, with the determination that ring I of paromomycin is tightly bound to the phosphate backbone of A^{1493} and actually helps "lock" two bases, A^{1492} and A^{1493}, into an extrahelical conformation. In contrast, each base appears to be dislocated from the helix to an even greater extent than NMR studies had revealed.

The X-ray crystal structure of paromomycin bound to the 30S subunit showed A^{1492} and A^{1493} pointing directly toward the minor groove of the A-site codon–anticodon helix, a significant conformational change compared to their original base-stacked positions within helix 44. To complement this finding, both X-ray crystallography and methyl substitution studies have exposed the identity of two hydrogen bonds related to this base "flipping"; these bonds have been hypothesized to be keys in binding of tRNA to the A-site. First, Puglisi and co-workers[16] described the effects of N1 methylation on these two adenine bases, making the observation that such modifications interfered with the binding of tRNA to the A-site in the absence of any drug. Moreover, mutation of either base was lethal in strains of *Escherichia coli*, suggesting a functional role for this dynamic region of 16S rRNA.[12]

Furthermore, Ramakrishnan discovered that complexation of the 30S ribosomal subunit and the stem loop of tRNA also locks A^{1492} and A^{1493} in an extrahelical position. This represented the first structure showing the extrahelical positions of each base, arguing that their mobility was not merely due to the presence of antibiotics within the A-site. It also revealed two specific hydrogen-bonding interactions between A^{1492} and A^{1493} with two OH groups on each side of the codon–anticodon helix when in the "flipped out" position, an event that might have significance for the translation event. Since the distance between these OH groups is directly related to the base pairing geometry of codon and anticodon, each hydrogen bond would be sensitive to improper pairing, hence they might discriminate against the near cognate or noncognate interactions.

That aminoglycoside binding interferes with translation has been known for at least four decades.[17] The mode of action of these antibiotics decreases the fidelity of translation, such that the organism biosynthesizes proteins that are defective. The presence of defective or nonfunctional proteins leads to the demise of bacteria. Consequently, paromomycin's bactericidal mode of action may come from its ability to lock the A-site into a conformation in which discrimination of cognate and near cognate tRNA is significantly diminished.

The work of Puglisi, Ramakrishnan, and co-workers further shaped a growing understanding of how aminoglycosides bind the bacterial ribosome. Not surprisingly, this work led to a series of subsequent investigations aimed at identifying interactions made between other aminoglycosides and the bacterial ribosome, including a higher-resolution structure of paromomycin bound to an RNA fragment.[10,11,14,17] The accumulation of these data solidified some of the broad binding trends that had been reported in the past. For instance, many of the multiple amino groups found on all aminoglycosides become protonated at physiological pH, functioning as polycationic binders to accessible regions of polyanionic 16S rRNA. The ammonium groups physically displace common biological cations (such as Mg^{2+} and Ca^{2+}) that are normally bound within these regions of high anionic character. Furthermore, rings I and II of the paromomycin and neomycin families appear to be more critical for binding than either rings III or IV. Finally, the bactericidal activity of both natural and synthetic aminoglycosides has been traced to a single region of the ribosome, the bacterial decoding A-site.

Complementing the strictly physical nature of NMR and X-ray crystallography, a number of computational efforts have also come forward to describe the behavior of aminoglycoside binding to different RNA motifs.[8] Such studies have not only been instrumental in modeling the interactions between aminoglycosides with RNA, they have also played key roles in bridging the gap between a strong understanding of target structure and the potential for rational design of new drug candidates.

Hermann and Westof[18] were among the first to publish molecular modeling results based on the interaction of cationic antibiotics with RNA. Westof's statement that "The mere presence of positive charges in aminoglycosides does not explain how they are able, unlike many other oligocations, to discriminate between different RNA folding motifs" drove modeling studies of aminoglycosides bound to the hammerhead ribozyme.[19] This analysis led to the proposition that aminoglycosides offer a three-dimensional framework of ammonium groups that orient themselves in such a way that they exploit multiple ion (Mg^{2+}) binding sites, and therefore the origin of aminoglycoside binding affinity likely arises from a complementarity between the orientation of these ammonium groups (governed by the rigid architecture of the aminoglycoside rings) and their unique fit to polyanionic three-dimensional folds in RNA.

In order to substantiate these initial claims, Hermann and Westof would later use Brownian dynamics (BD) simulations to map the electrostatic field gradient of different RNA molecules known to bind aminoglycosides.[20] Localized areas of defined negative charge were then compared to regions known to actually bind aminoglycosides. Within the free (no drug) 16S rRNA oligonucleotide, two negatively charged spherical pockets were immediately located. One of these pockets contained all the nucleotides, except for A^{1492}, necessary to make the intermolecular contacts with paromomycin reported in the NMR structure. Remarkably though, only rings III and IV of paromomycin were found occupying this area of high electron density. In contrast, rings I and II, which have three ammonium groups all together, were found in regions of relatively low electron density proximal to bases G^{1491}–C^{1496} pointing to a very different binding mode for each pair of rings. While electrostatics appear to dominate interactions of the third and fourth rings, shape-sensitive van der Waals contacts control binding of the first and second ring. This is very much in line with Puglisi's NMR studies in that rings I and II impart specificity of antibiotic binding, whereas the other two rings make only weak contributions.

Building on the foundation of these studies, Haddad et al.[21] used rings I and II of paromomycin (called paromomine) as the minimum structural motif in a computational search of over 273,000 compounds that might bind an A-site template. Results of this search were then narrowed based on steric and energetic demands. Several compounds were subsequently synthesized and shown to bind to an RNA model of the A-site, with some of the compounds (e.g., **2**) exhibiting broad-spectrum activity in bacteria. X-ray structures with the RNA models and with the 30S ribosome vindicated the design principles.[8,9] The compounds do bind to the A-site and they do flip the two bases mentioned previously to the extrahelical position, as do other antibiotics of this class.

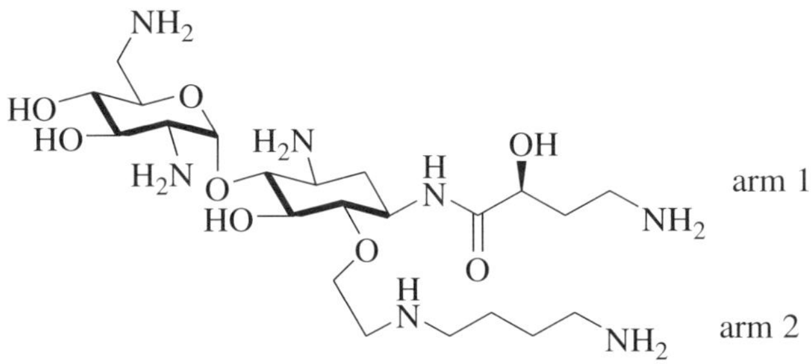

2

Numerous other design concepts have come forward in response to the growing structural understanding of the ribosomal A-site.[22–25] Simonsen and co-workers, for example, considered modifications at the primary alcohol position (6′) of ring I of paromomine in an effort to explore potential hydrogen bonding interactions of the shallow groove. However, none of these derivatives showed stronger activity in biochemical assays. Indeed even the most basic modifications, such as the addition of a single methylene unit (**3**), resulted in a 40-fold decrease in potency compared to paromomine, the parental molecular template. Addition of a second hydroxyl group (**4**) to this methylene unit did not restore biological activity either.

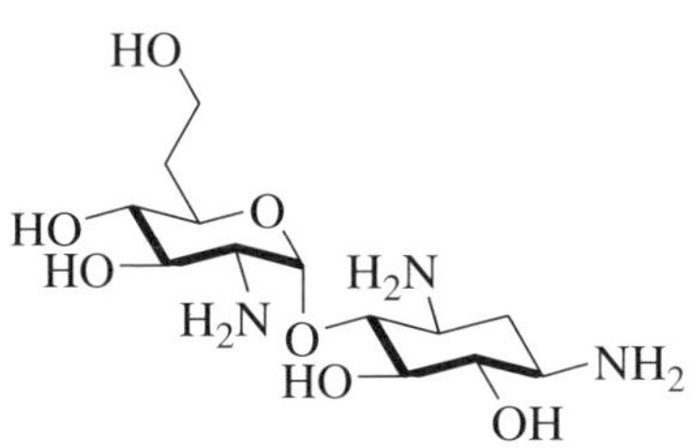

3

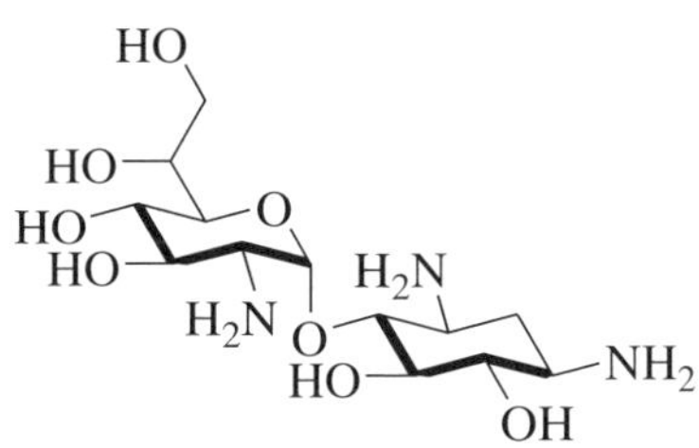

4

Conversely, Wong and co-workers took to surveying the major groove through dimerization of neamine with either a polyglycol or polymethylene linker. The length of this linker was varied to exploit possible contacts within the major groove of the RNA helix. Within the polyglycol class (**5**) binding constants for the bacterial ribosome improved in comparison to neamine nearly 10-fold while minimum inhibitory concentrations (MICs) hovered around 100–250 μM. Alternatively, those within the polymethylene group (**6**) showed an approximate twofold increase in binding constants, but took a significant hit in activity (>250 μM). Further work to improve binding specificity as well as binding affinity in this class of neamine analogs is ongoing.

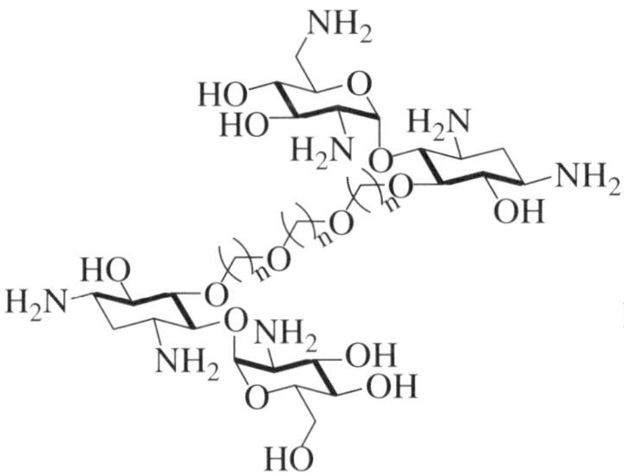

5

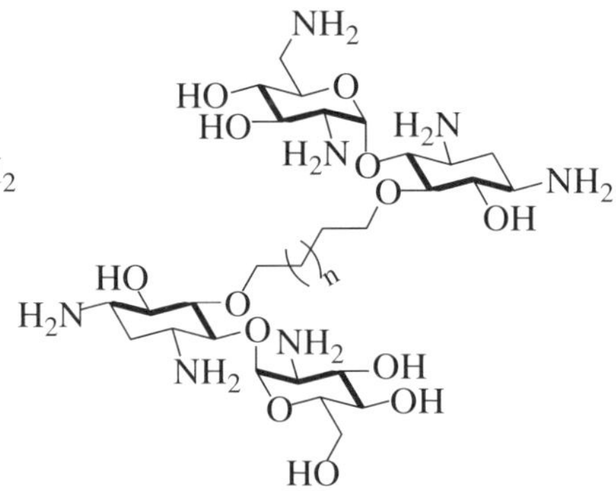

6

Lastly, a series of investigations have been carried out recently to look at the role of aminoglycoside conformation in binding 16S rRNA.[26–29] Part of the effectiveness of aminoglycosides from a therapeutic standpoint comes from their ability to adopt multiple conformations; this is a result of the pseudoglycosidic bonds, which unite most of their rings. The side effect of this freedom is that aminoglycosides tend to lack high RNA target selectivity, binding not only to 16S rRNA but also medicinally relevant ribozymes and the HIV-1 RNAs known as RRE and TAR.[30] In addition, bacterial resistance enzymes that target these compounds take advantage of their conformational adaptability, binding aminoglycosides in a conformationally distinct matter.[26] Such differences in binding mode led to the assumption that covalent linkage of individual rings may reduce the number of available conformations to such a degree that target selectivity could be altered.

Paromomycin and neomycin adopt a compact L-shaped conformation when bound to the A-site.[31,32] In contrast, these same two compounds adopt a more extended conformation when bound to TAR and a surprisingly high-energy conformation when bound to an antibiotic resistance enzyme ANT(4′).[33] With the knowledge of these conformational differences, two groups set out to covalently link rings I and III of neomycin (**7**). Molecular modeling was used to assess which RNA binding site the constrained neomycin analog would favor based on the known orientation of its unrestricted neomycin counterpart when bound to 16S rRNA, TAR, and ANT(4′). These simulations revealed that the restricted derivative was energetically disfavored to bind either TAR or the resistance enzyme

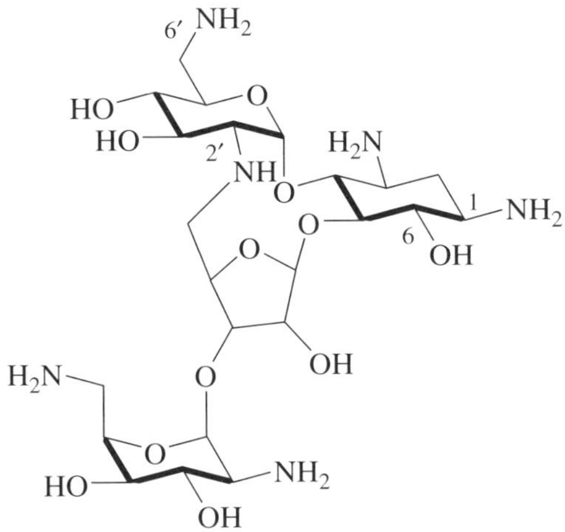

7

ANT(4′), but that the A-site RNA was still a viable target.[29] Upon completing the synthesis of the constrained neomycin, its biological activity was tested in comparison with unmodified neomycin. In the case of enzymatic modification, ANT(4′) adenylated the OH group at the 4′ position of neomycin within two minutes under the conditions used in the experiment. A completely different behavior was exhibited by the constrained derivative however. After more than two hours with the enzyme, an excess of 95% remained unmodified. With regard to target promiscuity, the results were not as promising. Binding of restricted neomycin to TAR showed only an 11-fold decrease in binding affinity compared to neomycin itself. Unfortunately, this was accompanied by a 22-fold decrease in binding affinity for the A-site, leading the Tor group to conclude that RNA binders may not be as nonselective as previously thought.[27]

The issue of selectivity of aminoglycosides to a given RNA is also pertinent with regard to the prokaryotic versus eukaryotic ribosomes as well. The fact that aminoglycosides have antibacterial activity is not merely an indication of the fact that these compounds bind to the RNA. In the case of prokaryotic cells there is an adventitious ATP-driven transport process, which is absent in eukarotic cells.[34] However, when direct comparison is made between the prokaryotic and eukaryotic ribosomes, aminoglycoside antibiotics indeed discriminate between prokaryotic and eukaryotic ribosomes.[35] In eukaryotes, the A•A base pair is disrupted because of the presence of a G in place of one of the adenines and there is different mispairing at positions 1408–1493, preventing the formation of the antibiotic (ring I) binding pocket.[6,36] The specificity of recognition of bacterial over human A-site RNA by aminoglycoside antibiotics has been examined in several different model systems.[37–39] Neomycin, paromomycin, and kanamycin A display a 30- to 50-fold increased affinity for the prokaryotic 16S rRNA compared to eukaryotic 18S rRNA. Lividomycin, neamine, and many neamine derivatives bind 16S rRNA with only slightly higher (~5-fold) or lower affinity than that of 18S rRNA.[39] In contrast, neamine derivative **2** has a >45-fold greater specificity for 16S over 18S rRNA. Thus, neamine derivative **2** is a promising new antibiotic due to the greater specificity that it shows for the bacterial A-site in comparison with the human A-site. Binding studies of novel antibiotics to the human ribosomal A-site RNA would be useful in the evaluation of toxicity of potential drugs.

REFERENCES

1. Poehlsgaard, J.; Douthwaite, S. *Nat. Rev. Microbiol.* **2005**, *3*, 870–881.
2. Paola, L. *FEMS Microbiol. Rev.* **2005**, *29*, 185–200.
3. Nilsson, B.; Soellner, M.; Raines, R. *Annu. Rev. Biophys. Biomol. Struct.* **2005**, *34*, 91–118.
4. Wilson, D.; Neirhaus, K. *Angew. Chem. Int. Ed.* **2003**, *42*, 3464–3486.
5. Rodnina, M.; Wintermeyer, W. *Trends Biochem. Sci.* **2001**, *26*, 124–130.
6. Fourmy, D.; Recht, M.; Blanchard, S.; Puglisi, J. *Science* **1996**, *274*, 1367–1371.
7. Fourmy, D.; Recht, M.; Puglisi, J. *J. Mol. Biol.* **1998**, *277*, 347–362.
8. Murray, J.; Meroueh, S.; Russell, R.; Lentzen, G.; Haddad, J.; Mobashery, S. *Chem. Biol.* **2006**, *13*, 129–138.
9. Russell, R.; Murray, J.; Lentzen, G.; Haddad, J.; Mobashery, S. *J. Am. Chem. Soc.* **2003**, *125*, 3410–3411.
10. Vicens, Q.; Westof, E. *Chemistry Biol.* **2002**, *9*, 747–755.
11. Vicens, Q.; Westof, E. *Structure* **2001**, *9*, 647–658.
12. Wimberly, B.; Brodersen, D.; Clemons, W.; Morgan-Warren, R.; Carter, A.; Vonrhein, C.; Hartsch, T.; Ramakrishnan, V. *Nature* **2000**, *407*, 327–339.
13. Carter, A.; Clemons, W.; Brodersen, D.; Morgan-Warren, R.; Wimberly, B.; Ramakrishnan, V. *Nature* **2000**, *407*, 340–348.
14. Brodersen, D.; Clemons, W., Jr.; Carter, A.; Morgan-Warren, R.; Wimberly, B.; Ramakrishnan, V. *Cell* **2000**, *103*, 1143–1154.
15. Spahn, C.; Prescott, C. *J. Mol. Med.* **1996**, *74*, 423–439.
16. Yoshizawa, S.; Fourmy, D.; Puglisi, J. *Science* **1999**, *285*, 1722–1725.
17. Magnet, S.; Blanchard, J. *Chem. Rev.* **2005**, *105*, 477–497.
18. Hermann, T.; Westof, E. *J. Mol. Biol.* **1998**, *276*, 903–912.
19. Hermann, T.; Westof, E. *J. Med. Chem.* **1999**, *42*, 1250–1261.
20. Hermann, T.; Westof, E. *Structure* **1998**, *6*, 1303–1314.
21. Haddad, J.; Kotra, L.; Llano-Sotelo, B.; Kim, C.; Azucena, E., Jr.; Liu, M.; Vakulenko, S.; Chow, C.; Mobashery, S. *J. Am. Chem. Soc.* **2002**, *124*, 3229–3237.
22. Vourloumis, D.; Winters, G.; Simonsen, K.; Takahashi, M.; Ayida, B.; Shandrick, S.; Zhao, Q.; Han, Q.; Hermann, T. *ChemBioChem* **2005**, *6*, 58–65.
23. Simonsen, K.; Ayida, B.; Vourloumis, D.; Takahashi, M.; Winters, G.; Barluenga, S.; Qamar, S.; Shandrick, S.; Zhao, Q. and Hermann, T. *ChemBioChem* **2002**, *3*, 1223–1228.

24. Liang, C.; Romero, A.; Rabuka, D.; Sgarbi, P.; Marby, K.; Duffield, J.; Yao, S.; Cheng, M.; Ichikawa, Y.; Sears, P. *Bioorg. Med. Chem. Lett.* **2005**, *15*, 2123–2128.
25. Agnelli, F.; Sucheck, S.; Marby, K.; Rabuka, D.; Yao, S.; Sears, P.; Liang, F.; Wong, C. *Angew. Chem. Int. Ed.* **2004**, *43*, 1562–1566.
26. Bastida, A.; Hidalgo, A.; Chiara, J.; Torrado, M.; Corzana, F.; Pérez-Cañadillas, J.; Groves, P.; Garcia-Junceda, E.; Gonzalez, C.; Jimenez-Barbero, J.; Asensio, J. *J. Am. Chem. Soc.* **2006**, *128*, 100–116.
27. Zhao, F.; Zhao, Q.; Blount, K.; Han, Q.; Tor, Y.; Hermann, T. *Angew. Chem. Int. Ed.* **2005**, *44*, 5329–5334.
28. Blount, K.; Zhao, F.; Hermann, T.; Tor, Y. *J. Am. Chem. Soc.* **2005**, *127*, 9818–9829.
29. Asensio, J.; Hidalgo, A.; Bastida, A.; Torrado, M.; Corzana, F.; Chiara, J.; García-Junceda, E.; Cañada, J.; Jiménez-Barbero, J. *J. Am. Chem. Soc.* **2005**, *127*, 8278–8279.
30. Chow, C.; Bogdan, F. *Chem. Rev.* **1997**, *97*, 1489–1513.
31. Szilaghi, R.; Shahzad-ul-Hussan, S.; Weimar, T. *ChemBioChem* **2005**, *6*, 1270–1276.
32. François, B.; Szychowski, J.; Adhikari, S.; Pachamuthu, K.; Swayze, E.; Griffey, R.; Migawa, M.; Westhof, E.; Hanessian, S. *Angew. Chem. Int. Ed.* **2004**, *43*, 6735–6738.
33. Faber, C.; Sticht, H.; Schweimer, K.; Rösch, P. *J. Biol. Chem.* **2000**, *275*, 20660–20666.
34. Wright, G.; Berghuis, A.; Mobashery, S. *Adv. Exp. Med. Biol.* **1998**, *456*, 27–69.
35. Recht, M.; Douthwaite, S.; Puglisi, J. *EMBO J.* **1999**, *18*, 3133–3138.
36. Lynch, S.; Puglisi, J. *J. Mol. Biol.* **2001**, *306*, 1023–1035.
37. Kaul, M.; Barbieri, C.; Pilch, D. *J. Am. Chem. Soc.* **2004**, *126*, 3447–3453.
38. Lynch, S.; Puglisi, J. *J. Mol. Biol.* **2001**, *306*, 1037–1058.
39. Llano-Sotelo, B. Ph.D. Thesis, Wayne State University, 2001.

8

METALLOAMINOGLYCOSIDES: CHEMISTRY AND BIOLOGICAL RELEVANCE

NIKHIL GOKHALE, ANJALI PATWARDHAN, AND J. A. COWAN
Department of Chemistry, Ohio State University, Columbus, OH 43210

Aminoglycoside Antibiotics: From Chemical Biology to Drug Discovery, Second Edition, by Dev P. Arya

8.1. INTRODUCTION

Aminoglycosides are a large family of antibiotics that appear to function by disruption of protein translation following binding to ribosomal RNA.[1] Aminoglycosides interact with seemingly unrelated families of RNA molecules including 16S rRNA,[2,3] RNase P,[4] self-splicing group I interons,[5] hammerhead ribozymes,[6,7] REV response element (RRE),[8] and trans-activating response element (TAR) of the human immunodeficiency virus HIV.[9–11] NMR solution structures for a variety of RNA–aminoglycoside complexes have already been reported.[12–22] Previous studies have shown that basic aminoglycosides also bind to DNA through electrostatic interactions at the negatively charged phosphate groups.[23] While widely used as consequence of low cost, aminoglycosides have been associated with ototoxicity. The underlying mechanisms remain uncertain, but possibilities include disruption of ribosomal translation of key proteins, or radical induced damage at the site of hearing caused by adventitious complex formation with redox active trace transition metals. While such redox chemistry poses obvious biological complications, it could also be harnessed in a therapeutically beneficial manner if an appropriate aminoglycoside could direct damaging chemistry toward a therapeutically relevant target molecule. As noted earlier, the potential for aminoglycosides to target a variety of structured RNAs supports the idea that suitable metalloaminoglycosides might promote therapeutically beneficial chemistry, while minimizing toxic side effects. Molecules that target such RNAs should possess a hydrolytic, photoactivatable, or redox active center, along with a recognition motif.[24,25] These requirements are inherent in metalloaminoglycosides. Accordingly, this chapter will review (a) the basic coordination chemistry of aminoglycosides to provide a working basis for the understanding of metal-promoted chemistry that underlies both biological toxicity and (b) the potential for therapeutic development. Every effort has been made to faithfully represent the published literature in this area.

The ability of aminoglycosides to coordinate metal ions is primarily governed by sugar ring substitutions on the 2-deoxystreptamine ring (ring B in **1**). Vicinal amine and hydroxyl groups can form a potential metal chelating motif, as found within the 2-deoxystreptamine ring (1-amine and 6-hydroxyl). The simplest 4-substituted aminoglycoside is neamine (**1**), which has a 2′-amine and 3′-hydroxyl juxtaposed to each other (ring A in **1**) and thus can effectively form a metal chelate in addition to the vicinal hydroxy-amine motif in ring B.

More complex aminoglycosides are obtained from substitutions at positions 4 and 5 (of the 2-deoxystreptamine ring) with amino sugar rings, which results in the neomycin class of compounds, whereas substitutions at the 4 and 6 positions result in the kanamycin class of aminoglycosides. The position-5 substituent in the neomycin class of compounds (ring D, neomycin B, **2**) offers one arrangement of vicinal amine and hydroxyl groups thus expanding its metal coordinating ability to rings A, B, and D (neomycin B, paromomycin, **2** and **3**). On the other hand, the position-6 substituent of the kanamycin class of aminoglycosides has 3″-amine and 4″-hydroxyl groups that can contribute toward metal coordination (ring C, kanamycin A, **4**) with position-4 substituent (ring A, **4** and **6**) contributing none toward metal binding, with the exception of kanamycin B (**5**), which has both ring A (2′-amine and 3′-hydroxyl) and ring C (3″-amine and 4″-hydroxyl) arrangements. Amikacin (**7**) derived from the semisynthetic modification of kanamycin A (by acylation of the 1-amino of ring B, with 4-amino-2-hydroxybutyric acid)

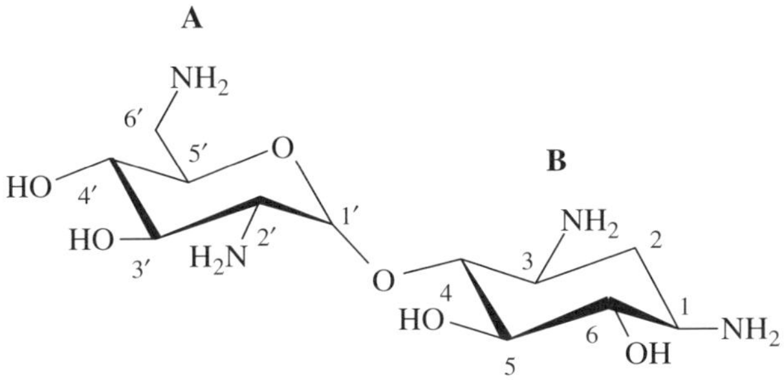

(**1**) neamine (4-substituted 2-deoxystreptamine ring)

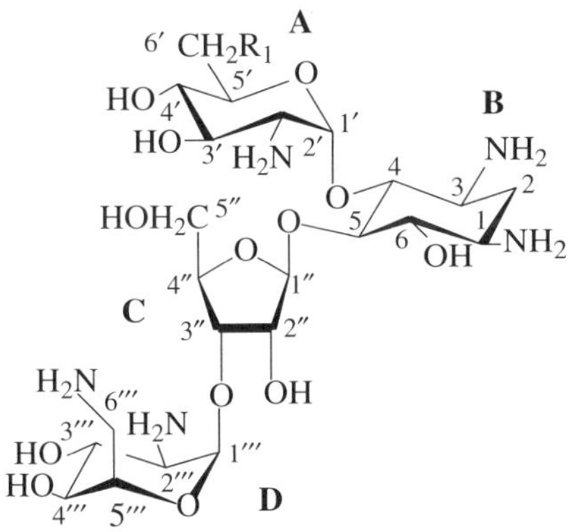

(**2**) neomycin B (R1 = NH2)

(**3**) paromomycin (R1 = OH)

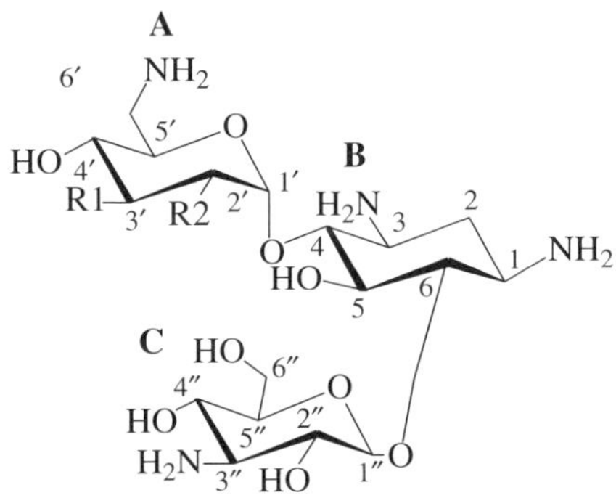

(**4**) kanamycin A (R1, R2 = OH)

(**5**) kanamycin B (R1= OH, R2 = NH2)

(**6**) tobramycin (R1 = H, R2 = NH2)

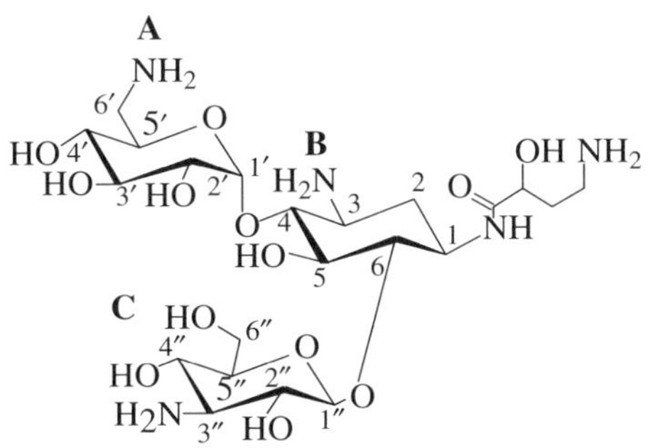

(**7**) Amikacin

has a potential metal binding motif through the 1-amidated nitrogen substituent along with the usual 3″-amine and 4″-hydroxyl functional groups.

Gentamicins form another class of aminoglycosides belonging to the kanamycin A class of compounds (gentamicin C1a**8**, gentamicin C1 **9**, gentamicin C2 10) and also possess metal-chelating donor atoms in ring C.

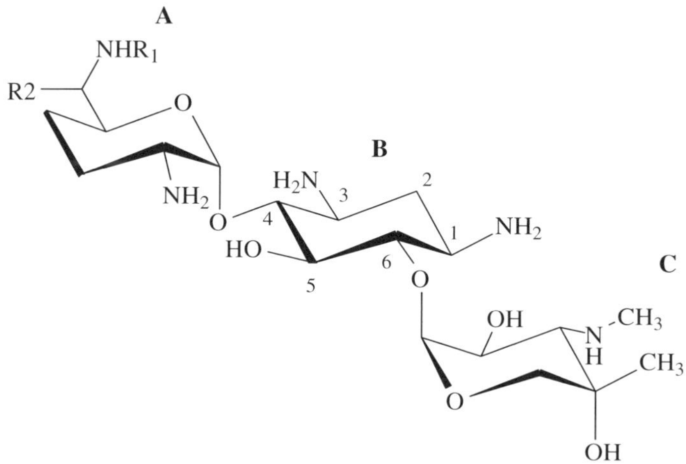

(**8**) gentamicin C1a (R1, R2 = H)

(**9**) gentamicin C1 (R1, R2 = CH3)

(**10**) gentamicin C2 (R1 = H, R2 = CH3)

The amino groups of the sugar rings exhibit variable pK_a values that range from 5.6 to 9.5. The aminoglycosides thus carry a net positive charge under physiological conditions (pH 7.4).[26] The amine groups on ring A are the most basic (pK_a ∼ 9.6), whereas those on ring B (pK_a ∼ 5.6 to ∼ 8.0) and ring C (pK_a ∼ 7.5) are closer to physiological pH. The total charge on the metalloaminoglycosides therefore depends on the oxidation state of the metal ion and protonation state of the coordinating hydroxyl group, in addition to the protonation states of amines at given pH.

Interaction of metal ions with aminoglycosides has previously been studied in light of preferential N-protection to generate acyl derivatives.[27] These studies utilized the fact that a metal ion bound to an amine group can be further stabilized by chelation via a vicinal hydroxyl group (protonated or deprotonated). These chelates are stable, and their solution structure characterization has been extensively reported utilizing UV–vis, EPR, XAS, EXAFS, and CD spectroscopy as recently reviewed by Kozlowaski et al.[26] The effectiveness of binding is dependant on the presence of a vicinal group to generate a chelate. Comparative studies with some amino–sugar Cu^{2+} complexes have shown that monodentate complexes involving metal coordination through the amine alone undergo hydrolysis above pH 7.0 and are not stable.[28] In all of these studies, copper binding to aminoglycosides results in the formation of a 1:1 metal–ligand complex. There are multiple vicinal —OH and —NH_2 groups (differing in pK_a values) on the various rings, which may lead to an increase in coordination number with increase in the pH. The aminoglycosides within a certain class differ from each other in subtle ways. These changes can have large effects as far as Cu^{2+} binding is concerned, which is often regarded as the principal pathway for aminoglycoside-related pharmacological activity and toxicity.[29] For example, the metal-chelating abilities of aminoglycosides can result in adverse effects under physiological conditions, where complex formation with transition metal ions can lead to deleterious oxidative reactions promoted by aminoglycoside-bound metal ions.[30–34]

A large number of published studies of the coordination chemistry of aminoglycosides are focused on their chemistry with iron, and especially copper. Given the biological significance of complexes of aminoglycosides and physiologically relevant redox active transition metal ions such as iron and copper, the chemistry of aminoglycosides with these metals and others (including, nickel, cobalt, and redox inactive zinc) has been investigated in the context of cellular toxicity.[35,36] Studies of iron complexes of neomycin and gentamicin[31,34,37] have provided molecular insight on potential pathways underlying induced ototoxicity.

In the following sections we will review the copper complexes of the various classes of aminoglycosides described earlier. Discussion will focus on complexes that are formed at physiological pH (7.4) but will take into account the differing modes of coordination at other pH values, as well as accounting for the distinct binding modes of other transitions metal ions.

8.2. COPPER(II)-AMINOGLYCOSIDES

Copper(II)-aminoglycoside complexes are often isolated as monomeric species over a wide pH range. The coordination complex is usually formed in 1:1 metal: ligand stoichiometry, and 1:2 metal:ligand stoichiometric complexes are rare.[26]

8.2.1. Copper(II)-Neamine

Neamine is a derivative of neomycin B following acid-catalyzed hydrolysis.[38,39] It represents the simplest of aminoglycosides with two potential sites of metal chelation (namely the 1,2-hydroxy-amine motif in either ring A or B). A 1:1 stoichiometric reaction of neamine with $Cu(OAc)_2$ in MeOH, under refluxing conditions for 48 h, results in formation of a blue 1:1 Cu^{2+}-neamine complex ($\lambda_{max} \sim 712$ nm, $\varepsilon \sim 65\ M^{-1}\ cm^{-1}$).[40] Extensive C-13 and H-1 NMR relaxation studies support binding of Cu^{2+} to ring A in neamine at pH levels close to physiological pH (7.5).[40] Based on UV–vis, EPR, and C-13 and H-1 NMR experiments the Cu^{2+}-neamine solution structure at pH 7.5 is best described as Cu^{2+} coordinated to the 1,2-hydroxyamine motif in ring A (**11**) with formation of a five-membered chelate ring involving 2′-NH_2 and 3′-OH of ring A in neamine. Water molecules that complete an octahedral geometry occupy the remaining coordination sites around copper.

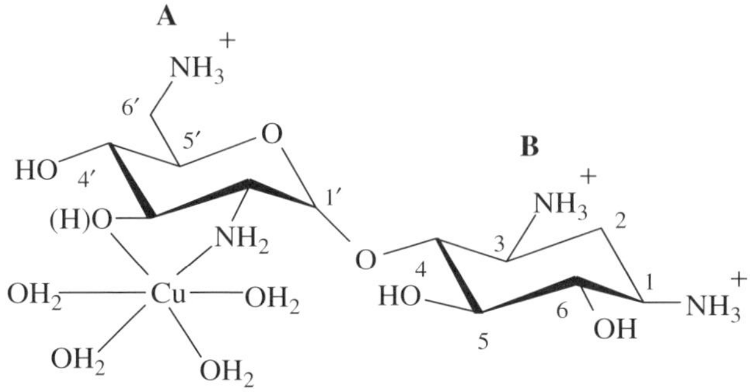

(**11**) Cu^{2+}-neamine at pH 7.5.[40]

8.2.2. Copper(II)-Kanamycin A

Cowan and co-workers[40,41] have extensively characterized the Cu^{2+} complexes of two different classes of aminoglycosides—namely, kanamycin A (**4**) and neomycin B (**2**)—in addition to the neamine (**1**) class of aminoglycosides. Their findings, based upon the interaction of $CuSO_4$ with kanamycin A and neomycin B, identified a blue monomeric 1:1 complex of Cu^{2+}-kanamycin A as well as a dark green 1:2 complex Cu^{2+}-(kanamycin A)$_2$ and a 1:1 Cu^{2+}-neomycin B complex, although the latter two were found to be less stable in aqueous solution with observable precipitation within 4 h. The Cu^{2+}-kanamycin A complex was found to be most stable in aqueous solutions over a period of one day.[42] Based on C-13 and H-1 NMR relaxation experiments, the Cu^{2+}-kanamycin A solution structure at pH = 7.9 (**12**) is described as Cu^{2+} coordinated to the

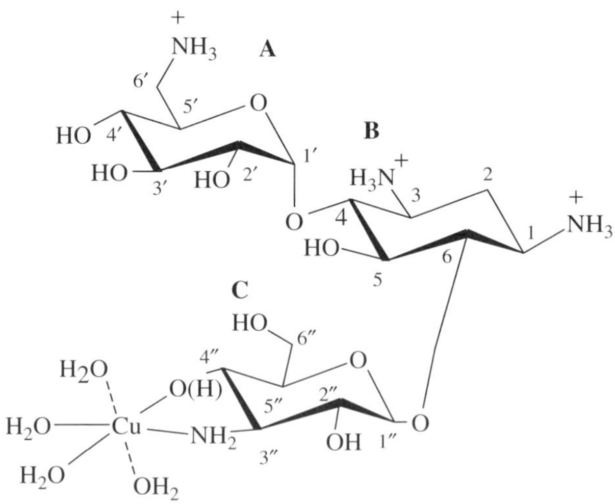

(**12**) Cu^{2+}- kanamycin A at pH 7.9.[40]

1,2-hydroxy-amine motif in ring C with formation of a five-membered chelate ring involving 3″-NH_2 and 4″-OH of ring C in kanamycin A.

Characterization of Cu^{2+}-kanamycin A at basic pH values has been studied by Grapsas et al.[43] coworkers (at pH 9.5 or 10.2) and recently by D'Amelio et al.[44] at three different pH values (5.5, 7.4, and 12.0). Each of the studies implicate involvement of ring C of kanamycin A in copper coordination up to pH 10.2. On the basis of NMR and EPR spectroscopic studies D'Amelio and coworkers identified involvement by the amino group of ring A (6′-NH_2) in metal coordination at higher pH (>10), which is accompanied by a switch in conformation of ring C. At pH 12.0, the 6″-hydroxyl group (or the ring oxygen) was observed to take part in copper coordination, combined with the 3″-amino nitrogen of ring C. The 5-hydroxyl group of ring B and 6′-amino of ring A complete the N_2O_2 coordination around the copper center (**13**).

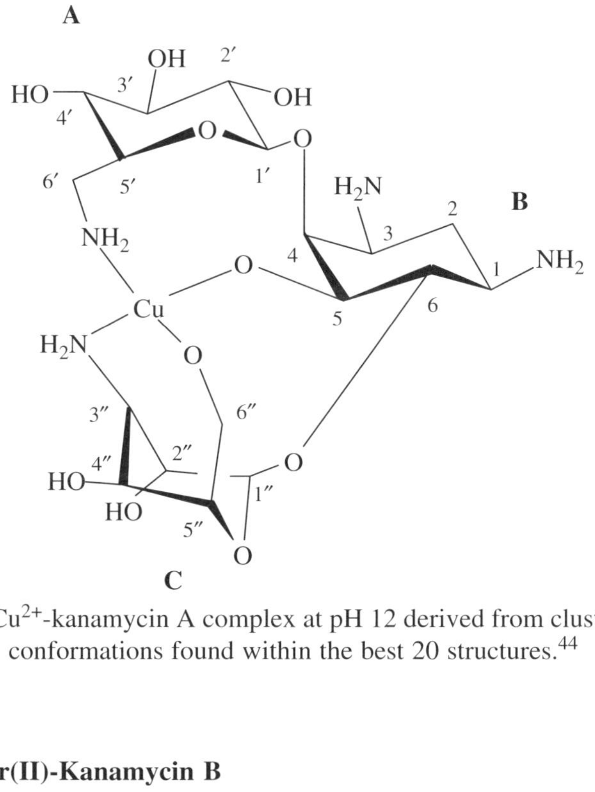

(**13**) Cu^{2+}-kanamycin A complex at pH 12 derived from clustered conformations found within the best 20 structures.[44]

8.2.3. Copper(II)-Kanamycin B

Jezowska-Bojczuk et al.[46] studied the interaction of Cu^{2+} ion with kanamycin B (**5**), which has an additional 2′-amino group as compared to kanamycin A (**4**), and in particular explored the involvement of this group in metal coordination by use of potentiometric, UV–vis, CD, and EPR methods. The presence of five amino groups in kanamycin B results in a significant lowering of the pK_a value of one of its amino nitrogens (pK_a = 5.74) with the pK_a values for other amino

TABLE 8.1. Stability Constants (log β and pK_a Values)[a#] of Kanamycin B, Amikacin and Tobramycin Complexes with Cu^{2+}

Species	Cu^{2+}-Kanamycin B		Cu^{2+}-Amikacin		Cu^{2+}-Tobramycin	
	log β	pK_a	log β	pK_a	log β	pK_a
CuH_3L	28.85±3	—	29.68±9	—	29.80±1	—
CuH_2L	23.06±1	5.79	23.94±1	5.74	23.45±1	6.35
CuHL	16.45±1	6.61	17.77±1	6.17	16.66±1	6.79
CuL	8.81±1	7.66	10.20±1	7.57	9.22±1	7.44
$CuH_{-1}L$	0.70±1	8.11	1.61±1	8.59	0.63±1	8.59
$CuH_{-2}L$	−8.56±1	9.26	−7.83±1	9.44	−9.31±1	9.94

[a]At 25°C, $I = 0.1$ M KNO_3, $\beta[MH_nL] = [MH_nL]/[M]\,[L]\,[H^+]^n$, where L is the aminoglycoside ligand.[45]

nitrogens falling in the normal range (pK_a = 6.9 to 9.1). Potentiometric and spectroscopic studies indicated a range of monomeric copper complexes from CuH_3L to $CuH_{-2}L$ with 1:1 metal:ligand stoichiometry (Table 8.1, where L is kanamycin B). It was evident from these studies that only ring C contributes to metal coordination, while the ring A amino group does not participate in binding the metal ion at physiological pH. At higher pH (>9), complex CuH_2L predominates, with additional axial coordination from the 3′-hydroxyl of ring A of kanamycin, which deprotonates with a pK_a 9.26 and maintains the coordination set.

8.2.4. Copper(II)-Neomycin B

Jezowska-Bojczuk et al.[47] have thoroughly investigated the coordination chemistry of neomycin B (**2**), the only active ingredient of neomycin analogues (neomycin A, B, and C) produced by *Streptomyces fradiei*. Based on their NMR, UV–vis, CD, EPR, XAS spectroscopic, and potentiometric and mass spectrometric studies, both mono-and dinuclear complexes of Cu^{2+}-neomycin B have been reported. Unlike the usual coordination sites encountered in the kanamycin class of compounds, Cu^{2+} was shown to coordinate the —NH_2 group of ring B above pH 5.0. Interestingly, other donor groups for the metal complexes corresponding to CuH_2L species predominate over pH range 6–7 (Table 8.2, where L is neomycin B), in particular, a second —NH_2 group in ring B, and the —OH group in ring A were involved in forming an N_2O donor set. From C-13 paramagnetic relaxation rates and potentiometric data the involvement of 1-NH_2, 3-NH_2 groups from ring B and the 4′-O^- (deprotonated hydroxyl group) of ring A as coordinating ligands to Cu^{2+} was suggested, which was further supported by EPR parameters (A$\|$ and g$\|$ of 184 Gs and 2.24, respectively, confirming involvement of two nitrogen donors) and UV–vis (616 nm, 356 nm) and CD spectroscopy (CT bands at 300 nm and 370 nm assigned to $NH_2 \rightarrow Cu^{2+}$ and $O^- \rightarrow Cu^{2+}$ LMCT transitions, respectively).[47]

TABLE 8.2. Stability Constants (log β and pK_a Values)^a of Neomycin B in 1:1 Complex Formation with Copper Ion

Species	log β	pK_a
CuH_3L	31.80±4	—
CuH_2L	26.24±1	5.56
CuHL	19.02±1	7.22
CuL	10.94±1	8.08
$CuH_{-1}L$	1.97±1	8.97
$CuH_{-2}L$	−7.99±1	9.96

^{a}At 25°C, $I = 0.1$ M KNO_3, $\beta[MH_nL] = [MH_nL]/[M][L][H^+]^n$, where L is neomycin B.[47]

TABLE 8.3. Stability Constants (log β)^a of 1:1 and 2:1 Complexes of Copper Ion and Neomycin B

Species	log β
CuH_5L	43.73±3
CuH_4L	37.64±6
Cu_2HL	23.90±3
Cu_2L	17.62±2
$Cu_2H_{-1}L$	10.89 ± 1
$Cu_2H_{-2}L$	2.65±6
$Cu_2H_{-3}L$	−7.19 ± 2

^{a}At 25°C, $I = 0.1$ M KNO_3, $\beta[MH_nL] = [MH_nL]/[M][L][H^+]^n$, where L is neomycin B.[47]

Uniquely, the Cu^{2+}-neomycin B system has been shown to form two-centered copper complexes at pH values below 6.5 if the metal ion is present in excess (Table 8.3) with less than 15% (relative to total Cu^{2+} present) of mononuclear species detected at pH values in the range of 4 to 6. At a 2:1 Cu^{2+} : neomycin B stoichiometric ratio it was shown that only the dinuclear species exists in solution at pH 7.5. The second binding site in the dinuclear complex was proposed to be in ring D, however, this unique feature is only observed in the case of the copper complex of neomycin B.

8.2.5. Copper(II)-Tobramycin

Jezowska-Bojczuk et al.[48] compared the coordination modes of tobramycin (**6**) with Cu^{2+} with respect to kanamycin B (tobramycin possesses a 3′-H as opposed to the 3′-OH functionality encountered in kanamycin B). The copper complexes of tobramycin formed at various solution pH values were observed to closely resemble those of kanamycin B (Table 8.1). However, the UV–vis spectrum of the Cu^{2+}-tobramycin complex of type CuHL (where L is tobramycin) lacks a characteristic charge transfer band at 350 nm, which was a prominent feature of the absorption spectra of kanamycin B,[46] originating from the 3′-OH hydroxyl binding to Cu^{2+}. Some degree of involvement of ring A in copper coordination was, however, proposed by evoking second chelate ring formation involving the 4′-OH group of ring A, which is less suitably positioned for coordination, forming a more relaxed and labile six-membered chelate ring. The unfavorable orientation of the 4′-OH was suggested to result in a high degree of flexibility for the second chelate ring, which results in a diminished charge transfer band at ~320 nm. The proposed structure of the Cu^{2+}-tobramycin complex at physiological pH is schematically shown here (**14**), with the central Cu^{2+} metal ion in an N_2O_2 coordination sphere comprising 3″-NH_2, 4″-O^-, 4′-OH, and 6′-NH_2 from rings A and C.

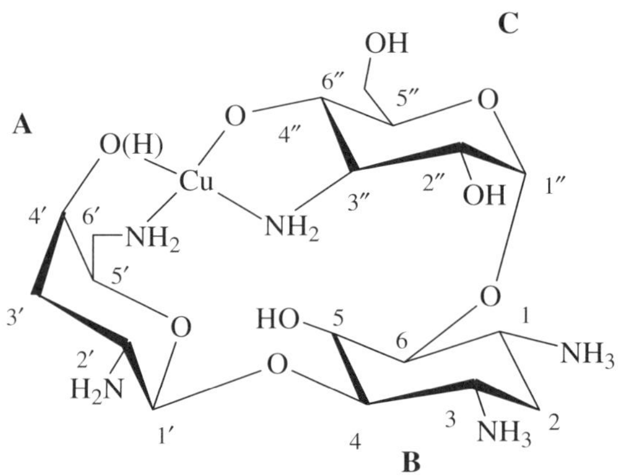

(**14**) Cu^{2+}-tobramycin complex at physiological pH.[48]

8.2.6. Copper(II)-Amikacin

Amikacin (**7**) is derived from kanamycin A by acylation of the 1-amino of 2-deoxystreptamine ring (ring B) with 4-amino-2-hydroxybutyric acid. These potential donor groups are arranged so that they can all participate in an alternative metal binding site, to the characteristic 1,2-hydroxy-amine motif (3″-amine and 4″-hydroxyl arrangement) of aminoglycosides. However, physical and spectroscopic studies (potentiometry, UV–vis, CD, NMR and EPR) of the Cu^{2+}-amikacin complex at physiological pH (7.4) inferred that a monomeric 1:1 copper complex (**15**) at pH 7.4 has all of the donor groups from ring B (2-deoxystreptamine ring).[49] The stability constants derived from the potentiometric studies, however, show close resemblance to those of kanamycin A and tobramycin (Table 8.1) complexes. The coordination of amikacin to copper thus involves 3-amino, 5-hydroxyl, and 1-amide nitrogen centers that generate a 10-member chelate ring. Furthermore, an equilibrium between monomeric

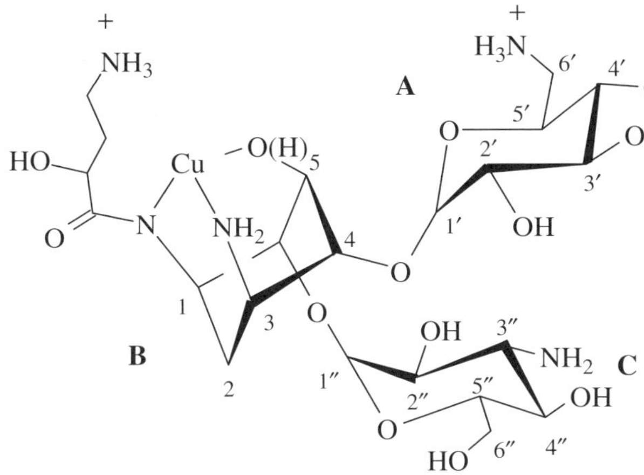

(**15**) Cu^{2+}-amikacin complex at pH 7.4 (the protonated and deprotonated species at B5 oxygen exist in equimolar equilibrium).[49]

and dimeric species was detected at complex concentrations above 5 mM (log $K_{dim} = 1.84 \pm 0.03$).[50] A strongly coupled Cu^{2+}-amikacin dimer was observed to form at concentrations higher than 10 mM of divalent copper and amikacin. EXAFS studies of this dimeric complex indicated a Cu–Cu distance of 3.86 A.[50] The dimeric complex (**16**) essentially retains the same Cu^{2+} coordination sphere as that of the monomeric species with the C5-oxygen of ring B bridging the two copper centers. The dimeric complex was shown to undergo facile but irreversible oxidation (+0.5 to +0.66 V, reduction potentials) to a Cu(III) species at a platinum electrode (0.1 M K_2SO_4, 25°C scan rate 25 mV/s).[50] The monomeric complex was able to catalyze a hydrogen peroxide disproportionation reaction at pH 7.4 in a multistep process, mediated by hydroxyl radicals and involving both Cu^{2+}/Cu^{1+} and Cu^{3+}/Cu^{2+} redox pairs.[50]

8.2.7. Copper(II)-Geneticin

Jezowska-Bojczuk et al.[45] further studied the geneticin class of aminoglycosides, which closely resemble to the gentamicin class (**8–10**) of aminoglycosides. Geneticin contains three amino and one methylamino group. The primary mode of Cu^{2+} coordination is through 3″-$NHCH_3$ and 4″-OH of ring C (corresponding to CuH_3L and CuH_2L species, Table 8.4), which is a typical binding mode for Cu^{2+} to the kanamycin A class of compounds at physiological pH (**17**). The presence of the 3″-$NHCH_3$ group in geneticin slightly lowers the binding affinity for Cu^{2+}.[45] Above pH 7.0, ring A of geneticin participates in coordination with its 2′-NH_2 and 3′-hydroxyl groups as shown schematically in (**18**). The species, which are not seen in any other class of aminoglycosides, correspond to $CuH_{-3}L/CuH_{-3}L_2$ products and are seen in the case of Cu^{2+}-geneticin complexes above pH 9–10.

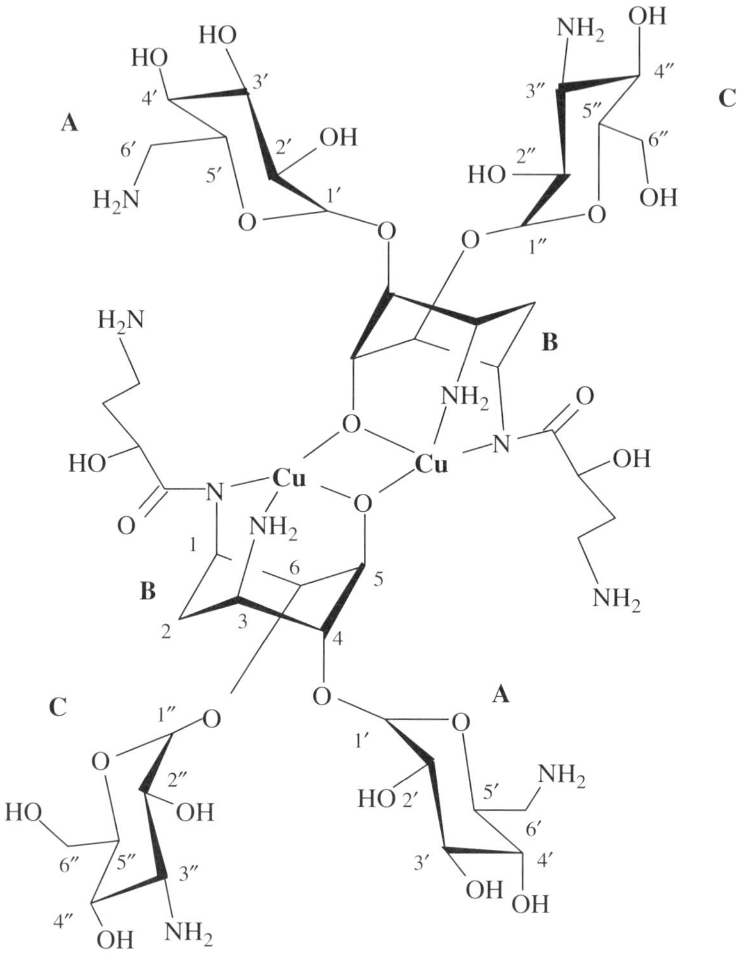

(**16**) Proposed dimer of Cu^{2+}-amikacin complex based on XAS data.[26,50]

These species occur with lowering of the Cu^{2+} binding symmetry and formation of 1:2 Cu^{2+}: gemeticin is proposed to include all amino binding to Cu^{2+} with two molecules of geneticin binding through either A or C rings.

8.2.8. Copper(II)-Gentamicin

Pecoraro, Schacht, and co-workers[29] explored the solution chemistry of Cu^{2+} complexes of the gentamicin class of aminoglycosides using various spectroscopic techniques, including UV–vis, EPR, and CD along with potentiometry. While the studies particularly focused on the Cu^{2+}-gentamicin C1a complex, each of the gentamicin C1a, C2, and C1 ligands were found to coordinate Cu^{2+} at ring C, while differences in the 6′-substituent in ring A do not involve metal

TABLE 8.4. Stability Constants (log β and pK_a Values)[a] of Geneticin Complexes with Copper Ion

Species	log β	pK_a
CuH_3L	—	—
CuH_2L	21.13±1	6.30
CuHL	14.83±1	6.87
CuL	7.96±1	7.65
$CuH_{-1}L$	0.31±1	8.59
$CuH_{-2}L$	−8.28±1	9–10
$CuH_{-3}L_2$	−15.44±3	9–10

[a] At 25°C, $I = 0.1$ M KNO_3, $\beta[MH_nL] = [MH_nL]/[M]\,[L]\,[H^+]^n$, where L is geneticin.[45]

TABLE 8.5. Stability Constants (log β and pK_a Values)[a] of Gentamicin C1a, C2, and C1 Complexes with Copper Ion

	Cu^{2+}-Gentamicin C1a		Cu^{2+}-Gentamicin C2		Cu^{2+}-Gentamicin C1	
Species	log β	pK_a	log β	pK_a	log β	pK_a
CuH_3L	30.68±1	—	30.92±1	—	30.73±1	—
CuH_2L	24.25±1	6.43	24.34±1	6.58	24.15±1	6.58
CuHL	17.03±1	7.22	17.15±1	7.19	16.77±1	7.38
CuL	9.25±1	7.78	9.45±1	7.7	8.93±1	7.84
$CuH_{-1}L$	0.48±1	8.77	0.56±1	8.89	−0.13±1	9.06
$CuH_{-2}L$	−9.51±11	9.99	−9.50±1	10.60	−10.15±1	10.28

[a] At 25°C, $I = 0.1$ M KNO_3, $\beta[MH_nL] = [MH_nL]/[M]\,[L]\,[H^+]^n$, where L is the aminoglycoside.[29]

coordination. The UV–vis and CD spectroscopic parameters of CuH_3L and CuH_2L complexes (Table 8.5) were consistent with coordination to copper by 3″-NH_2 and 4″-O^- groups, and an energy minimized model was proposed for CuH_3L species based on the spectroscopic evidences,[29] which is shown here schematically (**19**). Based on the EPR parameters recorded at pH 7.5 and 8.5 (where CuHL and CuL species are predominant, and L is gentamicin C1a), the species exhibited N_2O_2 equatorial coordination to Cu^{2+}. The resulting structural species represented by (**19**) are converted to $CuH_{-1}L$ and $CuH_{-2}L$ species at pH 9.5, represented by (**20**); the most likely reason for this distortion is the involvement of axial binding of an oxygen donor from the 2-deoxystreptamine ring (**20**). The binding of this 5-hydroxyl to the copper center was evoked in comparison to that of geneticin complex (**18**),[45] wherein the 3′-OH coordinates to the metal ion. Gentamicin C1a lacks this hydroxyl group; however, spectroscopic evidence is in favor of 5-hydroxyl binding with a decrease in the intensity of d–d bands, suggesting distortion and the presence of an $O^- \rightarrow Cu^{2+}$ CT transition at 251 nm in the CD spectrum. The energy-minimized structure of the $CuH_{-1}L$ complex is schematically shown here (**20**).

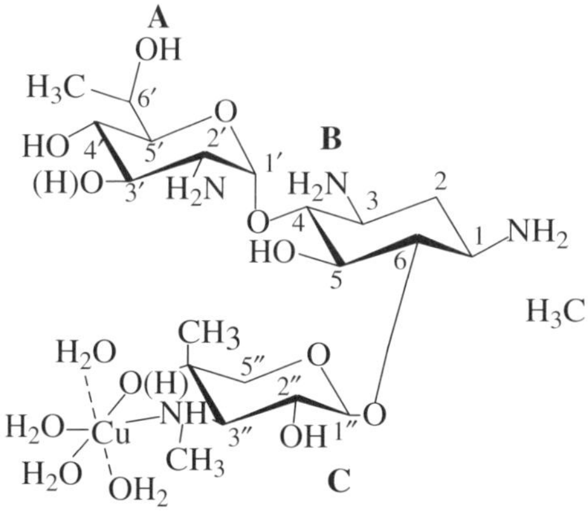

(**17**) Cu(II)-geneticin complex (CuH_2L species).

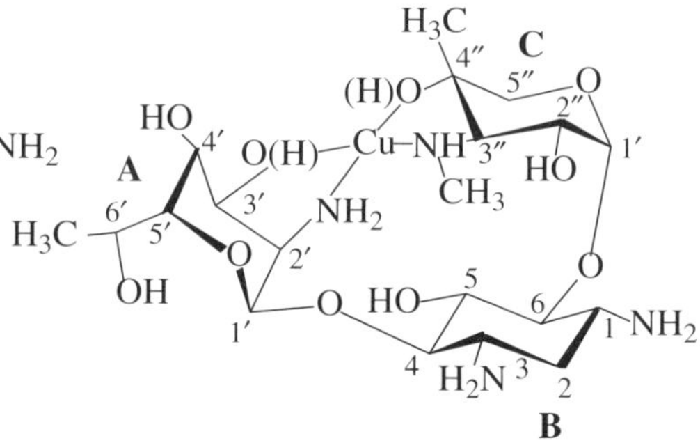

(**18**) Schematic representation of a Cu(II)-geneticin complex[45] ($CuH_{-2}L$ species).

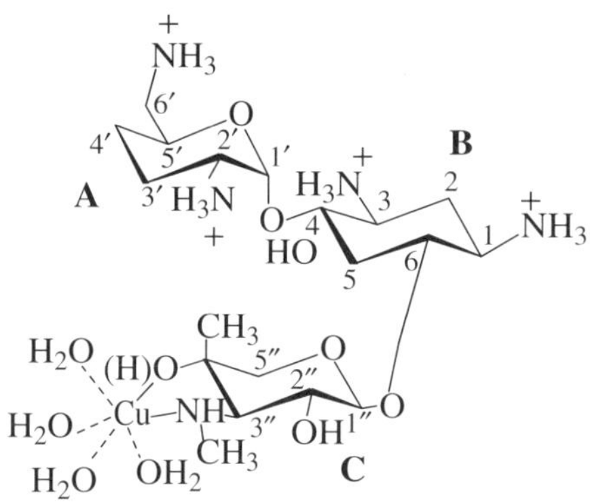

(**19**) Cu^{2+}-gentamicin C1a complex (corresponding to CuH_3L).

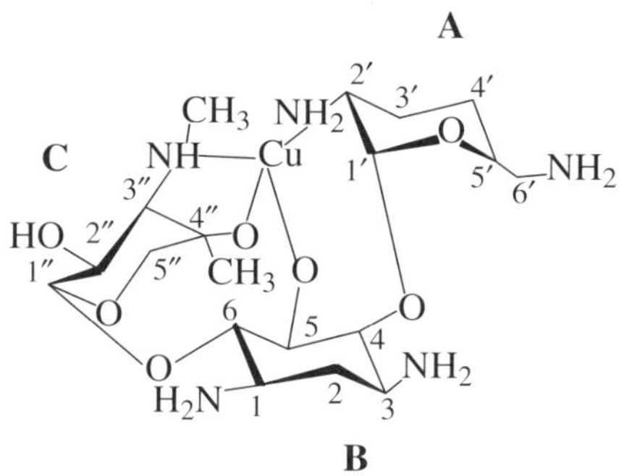

(**20**) Cu^{2+}-gentamicin C1a complex (corresponding to $CuH_{-1}L$).

8.3. ZINC(II)-AMINOGLYCOSIDES

In addition to Cu^{2+} complexes of the gentamicin class of compounds Pecoraro, Schacht, and co-workers[29] have also explored the Zn^{2+} binding ability of this class of aminoglycosides. The stability constants for Zn^{2+} with gentamicin C1a (Table 8.6) were, however, at least 3 log units lower than those of the corresponding Cu^{2+}-gentamicin C1a complexes (Table 8.5), thus eliminating a plausible biological role for Zn^{2+}-aminoglycosides.

TABLE 8.6. Stability Constants (log β and pK_a Values)[a] of Gentamicin C1a Complexes with Zinc Ion

Species	log β	pK_a
ZnHL	14.05±1	—
ZnL	5.57±1	8.48
$ZnH_{-1}L$	−3.19±1	8.76
$ZnH_{-2}L$	−13.22±1	10.03

[a]At 25°C, $I = 0.1$ M KNO_3, $\beta[MH_nL] = [MH_nL]/[M]\ [L]\ [H^+]^n$, where L is gentamicin C1a.[29]

8.4. IRON(III/II)-AMINOGLYCOSIDES

While much of attention has been focused on the synthesis, characterization, and solution properties of Cu-aminoglycoside complexes, Fe-aminoglycosides have also been proposed as equally important for understanding aminoglycoside-related toxicity and pharmacological activity. The underlying mechanism of toxicity of the gentamicin class of compounds was first postulated to be related to gentamicin-induced free radical formation, presumably via Fe^{3+}-gentamicin complex.[30,31] Extending on this hypothesis a competitive iron chelation therapy was also established.[34,51]

8.4.1. Iron(III)-Gentamicin

Pecoraro and co-workers[30] have reported both ferrous and ferric complexes of gentamicin. The solution structures and binding modes of gentamicin to ferrous/ferric metal ions were evaluated based on the H-1 NMR methods. The Fe^{3+}-gentamicin complex was generated *in situ* by air oxidation of Fe^{2+}-gentamicin at pH 4.5. The magnetic susceptibility of the Fe^{3+}-gentamicin complex was found to be intermediate ($\mu_{eff} = 4.46$) between, high-spin Fe^{3+} ($S = 5/2$, $\mu_{eff} = 5.92$) and low-spin Fe^{3+} ($S = 3/2$, $\mu_{eff} = 3.87$), suggesting more than one species in solution.[30]

8.4.2. Iron(II)-(Gentamicin)$_2$

Similarly, Fe^{2+} binding to gentamicin was proposed based on H-1 NMR resonances and NMR magnetic susceptibility experiments (Evans method).[52] Two binding constants were determined (high affinity 0.02 M^2, low affinity 0.2 M^2) at pH 6.5.[30] Formation of a 2:1 gentamicin:Fe^{2+} (low spin) complex was proposed, with two gentamicin molecules coordinating to Fe^{2+} (low spin) ($S = 0$). Two possible coordination isomers were proposed fac–fac [(**21**) (facial = fac) and mer–mer (**22**) (meridonial = mer)], consistent with the finding that there are two binding constants for the complex. The binding modes proposed were based on the evidence that low-spin iron complexes require at least four or more ligating nitrogens and that a 1:1 Fe^{2+}-gentamicin complex would generate sterically unfavorable arrangements in the gentamicin molecule. This conclusion was further supported by the biphasic nature of the concentration dependence of complex formation with added Fe^{2+} and gentamicin. The Fe^{2+}-(gentamicin)$_2$ represents one of very few examples of metalloaminoglycosides with 2:1 aminoglycoside:metal stoichiometry. The proposed isomers for these complexes are schematically shown (**21, 22**).

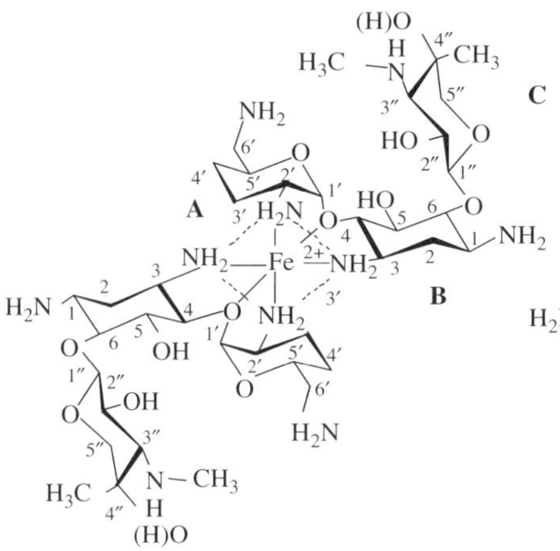

(**21**) Fac–Fac

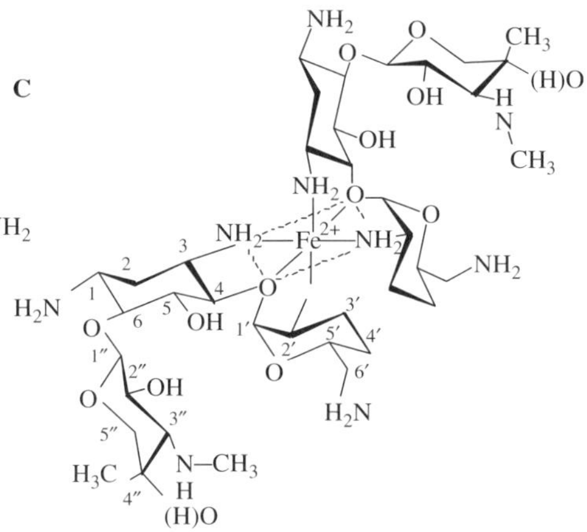

(**22**) Mer–Mer

8.5. BIOLOGICAL PERTINENT CHEMISTRY OF METALLOAMINOGLYCOSIDES

8.5.1. Oxidative Chemistry of Metalloaminoglycosides

Numerous studies have been reported that describe the redox chemistry of copper aminoglycoside complexes. Cyclic voltammetry investigations, carried out over a wide pH range, indicate that electrochemical reduction of aminoglycoside-bound Cu^{2+} results in complex decomposition and formation of metallic copper (Cu^0)[53], suggesting that the $Cu^{2+/1+}$ redox couple is irreversible under physiological conditions if the Cu^{1+} state is not oxidized prior to dismutation. On the other hand, oxidation of Cu^{2+}-aminiglycoside species has been shown to proceed via formation of Cu^{3+} species, as shown for Cu^{2+}-kanamycin A and Cu^{2+}-amikacin complexes.[50,53] As a result of the flexibility of oxidation state, Cu^{2+}-aminoglycosides have been shown to disproportionate hydrogen peroxide to reactive oxygen species such as superoxide and hydroxyl radicals.[50,53–55] Such reactivity may be related to recent findings that implicate Cu^{2+}-aminoglycosides in inducing an immunological response in host cells,[56] thereby activating cytokines such as interferon (IFN), tumor necrosis factor (TNF), and interleukin-10 (IL-10).

A wide range of Cu^{2+}-aminoglycosides (Cu^{2+}-amikacin, Cu^{2+}-kanamycin A, Cu^{2+}-kanamycin B, Cu^{2+}-tobramycin, Cu^{2+}-geneticin, Cu^{2+}-gentamicin and Cu^{2+}-neomycin B complexes) have been shown to mediate disproportionation of H_2O_2 at physiological pH involving Cu^{2+}/Cu^{1+} and Cu^{3+}/Cu^{2+} redox pairs.[26] These complexes have been shown to promote oxidation of 2′-deoxyguanosine to 7,8-dihydro-8-oxo-2′-deoxyguanosine, double-stranded DNA cleavage[54,57] and cleavage of $tRNA^{Phe}$ at the anticodon specific site[58] under both hydrolytic and oxidative conditions. Cu^{2+}-kanamycin A and Cu^{2+}-neomycin B complexes have been shown to cleave a variety of nucleic acids in the presence of H_2O_2 or O_2/ascorbic acid, in particular promoting highly specific DNA strand scissions by the mechanism of C- 4′ hydrogen abstraction from the deoxyribose motif.[59] In these reactions that mediate the nucleic acid cleavage reactions by generation of reactive oxygen species (ROS), unwanted side effects such as lipid peroxidation[29,60] and free radical induced toxicity may limit their use as has been postulated in case of $Fe^{3+/2+}$-gentamicin complexes.[30]

Cu^{2+}-kanamycin A and Cu^{2+}-neamine, however, have shown promising characteristics by being the most active DNA nucleases reported to date, operating through a hydrolytic mechanisms.[40,41] The hydrolytic DNA-cleaving ability of Cu^{2+}-neamine complex at micromolar levels has surpassed all other synthetic nucleases, with several-million-fold enhancement over uncatalyzed DNA cleavage (rate of hydrolysis of inactivated phosphodiester 6×10^{-14} min^{-1}, rate of DNA cleavage by Cu^{2+}-neamine complex $= k_{cat} = 5.95 \times 10^{-4}$ min^{-1}).[40] Furthermore, the Cu^{2+}-neamine complex exhibits enzyme-like nuclease activity with Michaelis–Menten and multiturnover behavior. The utility of Cu^{2+}-kanamycin A complexes over free kanamycin A have been extended further to specifically cleave RNA within the bacterial cells (*in vivo*), suggesting again its potential as a metal-based drug.[61]

8.6. CONCLUSIONS

Aminoglycoside antibiotics are widely used for the treatment of infections caused by gram negative and some gram-positive organisms.[3,62] The mechanism by which these antibiotics kill bacteria is related primarily toward their inhibitory action on the translational step in microbial protein synthesis. Despite their nephrotoxicity and ototoxicity, these antibiotics are a preferred choice for antibiotic treatment. The interaction of aminoglycosides with metal ions—in particular, transition metal ions—has been related to their pharmacological activity and toxicity, through formation of reactive oxygen species by redox active metal centers (Cu^{2+} and Fe^{2+}). The enhanced biological activities of the metalloaminoglycosides over free aminoglycosides suggest a potential therapeutic application for metalloaminoglycosides. The Cu^{2+}-aminoglycosides in particular have already shown such promise.

ACKNOWLEDGMENT

The authors gratefully acknowledge the support of the NIH (GM063740) in facilitating the preparation of this manuscript as well as related research activities.

REFERENCES

1. Cowan, J. A.; Ohyama, T.; Wang, D.; Natarajan, K. *Nucleic Acids Res.* **2000**, *28*, 2935–2942.
2. Purohit, P.; Stern, S. *Nature* **1994**, *370*, 659–662.
3. Moazed, D.; Noller, H. F. *Nature* **1987**, *327*, 389–394.
4. Mikkelsen, N. E.; Brannvall, M.; Virtanen, A.; Kirsebom, L. A. *Proc. Natl. Acad. Sci. USA* **1999**, *96*, 6155–6160.
5. von Ahsen, U.; Davies, J.; Schroeder, R. *J. Mol. Biol.* **1992**, *226*, 935–941.
6. Stage, T. K.; Hertel, K. J.; Uhlenbeck, O. C. *RNA* **1995**, *1*, 95–101.
7. Clouet-d'Orval, B.; Stage, T. K.; Uhlenbeck, O. C. *Biochem.* **1995**, *34*, 11186–11190.
8. Zapp, M. L.; Stern, S.; Green, M. R. *Cell* **1993**, *74*, 969–978.
9. Litovchick, A.; Evdokimov, A. G.; Lapidot, A. *FEBS Lett.* **1999**, *445*, 73–79.
10. Litovchick, A.; Evdokimov, A. G.; Lapidot, A. *Biochem.* **2000**, *39*, 2838–2852.
11. Mei, H.-Y.; Galan, A. A.; Halim, N. S.; Mack, D. P.; Moreland, D. W.; Sanders, K. B.; Truong, H. N.; Czarnik, A. W. *Bioorg. Med. Chem. Lett.* **1995**, *5*, 2755–2760.
12. Lynch, S. R.; Recht, M. I.; Puglisi, J. D. *Methods Enzymol.* **2000**, *317*, 240–261.
13. Fourmy, D.; Recht, M. I.; Blanchard, S. C.; Puglisi, J. D. *Science* **1996**, *274*, 1367–1371.
14. Fourmy, D.; Recht, M. I.; Puglisi, J. D. *J. Mol. Biol.* **1998**, *277*, 347–362.
15. Fourmy, D.; Yoshizawa, S.; Puglisi, J. D. *J. Mol. Biol.* **1998**, *277*, 333–345.
16. Battiste, J. L.; Tan, R.; Frankel, A. D.; Williamson, J. R. *J. Biomol. NMR* **1995**, *6*, 375–389.
17. Batey, R. T.; Battiste, J. L.; Williamson, J. R. *Methods Enzymol.* **1995**, *261*, 300–322.
18. Varani, G.; Tinoco, I., Jr. *Q. Rev. Biophys.* **1991**, *24*, 479–532.
19. Peterson, R. D.; Feigon, J. *J. Mol. Biol.* **1996**, *264*, 863–877.
20. Peterson, R. D.; Bartel, D. P.; Szostak, J. W.; Horvath, S. J.; Feigon, J. *Biochemistry* **1994**, *33*, 5357–5366.
21. Battiste, J. L.; Mao, H.; Rao, N. S.; Tan, R.; Muhandiram, D. R.; Kay, L. E.; Frankel, A. D.; Williamson, J. R. *Science* **1996**, *273*, 1547–1551.
22. Yoshizawa, S.; Puglisi, J. D. *NATO ASI Series, Ser. A: Life Sci.* **1998**, *301*, 173–182.
23. Zimmer, C.; Triebel, H.; Thrum, H. *Biochim. Biophys. Acta* **1967**, *145*, 742–751.
24. Billadeau, M.; Morrison, H. *Metal Ions Biol. Syst.* **1996**, *33*, 269.
25. Sigman, D. S., Landgraf R., Perrin, D. M., and L. Pearson *Metal Ions Biol. Syst.* **1996**, *33*, 485.
26. Kozlowski, H.; Kowalik-Jankowska, T.; Jezowska-Bojczuk, M. *Coord. Chem. Rev.* **2005**, *249*, 2323–2334.
27. Hanessian, S.; Patil, G. *Tetrahedron Lett.* **1978**, *12*, 1035–1038.

28. Jezowska-Bojczuk, M.; Lamotte, S.; Trnka, T. *J. Inorg. Biochem.* **1996**, *61*, 213–219.
29. Lesniak, W.; Harris, W. R.; Kravitz, J. Y.; Schacht, J.; Pecoraro, V. L. *Inorg. Chem.* **2003**, *42*, 1420–1429.
30. Priuska, E. M.; Clark-Baldwin, K.; Pecoraro, V. L.; Schacht, J. *Inorg. Chim. Acta.* **1998**, *273*, 85–91.
31. Priuska, E. M.; Schacht, J. *Biochem. Pharmacol.* **1995**, *50*, 1749–1752.
32. Sha, S. H.; Schacht, J. *Free Radical Biol. Med.* **1999**, *26*, 341–347.
33. Sha, S. H.; Zajic, G.; Epstein, C. J.; Schacht, J. *Audiol. Neuro-Otol* **2001**, *6*, 117–123.
34. Song, B. B.; Sha, S. H.; Schacht, J. *Free Radical Biol. Med.* **1998**, *25*, 189–195.
35. Abu-El-Wafa, S. M.; El-Ries, M. A.; Abou-Attia, F. M.; Issa, R. M. *Anal. Lett.* **1989**, *22*, 2703–2716.
36. Chohan, Z. H.; Ansari, M.-U. H. *Pakistan J. Pharmacol.* **1989**, *6*, 21–26.
37. Conlon, B. J.; Perry, B. P.; Smith, D. W. *Laryngoscope* **1988**, *108*, 284–287.
38. Roestamadji, J.; Grapsas, I.; Mobashery, S. *J. Am. Chem. Soc.* **1995**, *117*, 11060–11069.
39. Grapsas, I.; Cho, Y. J.; Mobashery, S. *J. Org. Chem.* **1994**, *59*, 1918–1922.
40. Sreedhara, A.; Freed, J. D.; Cowan, J. A. *J. Am. Chem. Soc.* **2000**, *122*, 8814–8824.
41. Sreedhara, A.; Cowan, J. A. *Chem. Commun.* **1998**, 1737–1738.
42. Sreedhara, A.; Patwardhan, A.; Cowan, J. A. *Chem. Commun.* **1999**, 1147–1148.
43. Grapsas, I.; Massova, I.; Mobashery, S. *Tetrahedron* **1998**, *54*, 7705–7720.
44. D'Amelio, N.; Gaggelli, E.; Gaggelli, N.; Molteni, E.; Baratto, M. C.; Valensin, G.; Jezowska-Bojczuk, M.; Szczepanik, W. *Dalton Trans.* **2004**, 363–368.
45. Jezowska-Bojczuk, M.; Karaczyn, A.; Bal, W. *J. Inorg. Biochem.* **1998**, *71*, 129–134.
46. Jezowska-Bojczuk, M.; Bal, W.; Kozlowski, H. *Inorg. Chim. Acta* **1998**, *275–276*, 541–545.
47. Jezowska-Bojczuk, M.; Szczepanik, W.; Mangani, S.; Gaggelli, E.; Gaggelli, N.; Valensin, G. *Eur. J. Inorg. Chem.* **2005**, 3063–3071.
48. Jezowska-Bojczuk, M.; Karaczyn, A.; Kozlowski, H. *Carbohydrate Res.* **1998**, *313*, 265–269.
49. Jezowska-Bojczuk, M.; Bal, W. *J. Chem. Soc., Dalton Trans.* **1998**, 153–160.
50. Jezowska-Bojczuk, M.-l.; Lesniak, W.; Bal, W.; Koz-lowski, H.; Gatner, K.; Jezierski, A.; Sobczak, J.-l.; Mangani, S.; Meyer-Klaucke, W. *Chem. Res. Toxicol.* **2001**, *14*, 1353–1362.
51. Song, B. B.; Schacht, J. *Hear. Res.* **1996**, *94*, 87–93.
52. Evans, D. F. *J. Chem. Soc.* **1959**, 2003.
53. Szczepanik, W.; Kaczmarek, P.; Sobczak, J.; Bal, W.; Gatner, K.; Jezowska-Bojczuk, M. *New J. Chem.* **2002**, *26*, 1507–1514.
54. Szczepanik, W.; Dworniczek, E.; Ciesiolka, J.; Wrzesinski, J.; Skala, J.; Jezowska-Bojczuk, M. *J. Inorg. Biochem.* **2003**, *94*, 355–364.
55. Jezowska-Bojczuk, M.; Lesniak, W. *J. Inorg. Biochem.* **2001**, *85*, 99–105.
56. Szczepanik, W.; Czarny, A.; Zaczynska, E.; Jezowska-Bojczuk, M. *J. Inorg. Biochem.* **2004**, *98*, 245–253.
57. Jezowska-Bojczuk, M.; Szczepanik, W.; Lesniak, W.; Ciesiolka, J.; Wrzesinski, J.; Bal, W. *Eur. J. Biochem.* **2002**, *269*, 5547–5556.

58. Szczepanik, W.; Ciesiolka, J.; Wrzesinski, J.; Skala, J.; Jezowska-Bojczuk, M. *Dalton Trans.* **2003**, 1488–1494.
59. Patwardhan, A.; Cowan, J. A. *Chemical Commun.* **2001**, 1490–1491.
60. Lesniak, W.; Pecoraro, V. L.; Schacht, J. *Chem. Res. Toxicol.* **2005**, *18*, 357–364.
61. Chen, C.; Cowan, J. A. *Chem. Commun.* **2002**, 196–197.
62. Coates, A.; Hu, Y.; Bax, R.; Page, C. *Nature Rev. (Drug Discovery)* **2002**, *1*, 895–910.

9

ADVERSE EFFECTS OF AMINOGLYCOSIDE THERAPY

ANDRA E. TALASKA AND JOCHEN SCHACHT
Kresge Hearing Research Institute, Department of Otolaryngology, University of Michigan, Ann Arbor, MI 48109

9.1. INTRODUCTION

The introduction of streptomycin as an antibiotic in 1944[1] was welcomed as the long-sought solution to combat tuberculosis and other intractable gram-negative infections. The first clinical trial, however, already highlighted the side effects

Aminoglycoside Antibiotics: From Chemical Biology to Drug Discovery, Second Edition, by Dev P. Arya

of this class of drugs, namely ototoxicity (i.e., damage to the inner ear) and nephrotoxicity (i.e., damage to the kidney).[2] The subsequent development of other natural or semisynthetic aminoglycosides (including neomycin, tobramycin, gentamicin, kanamycin, and amikacin) created a useful armamentarium of highly efficacious antibiotics, but ototoxic and nephrotoxic potentials remained associated with all of them. Despite these adverse effects, the aminoglycoside antibiotics must be considered one of the major success stories of chemotherapy in the twentieth century.

During the last 20 years, aminoglycosides have been largely replaced by new generations of antibiotics including the cephalosporins, but they nevertheless still enjoy widespread use today. Aminoglycoside antibiotics are nearly ideal medications for emergencies because of their broad antibacterial spectrum and lack of allergic responses; they are the drugs of choice for inhalation therapy that cystic fibrosis patients must undergo periodically; and they are part of the drug cocktail recommended by the World Health Organization to combat multidrug-resistant tuberculosis. Furthermore, their use continues almost unabated in developing countries where aminoglycoside antibiotics in general and gentamicin in particular are frequently the only affordable medications. Gentamicin may be the most commonly used antibiotic today.

A major clinical distinction between the effects on the inner ear and the kidney is the fact that the renal effects are reversible while the effects on the inner ear are irreversible, leading to permanent loss of balance or auditory function. Furthermore, renal insults can more easily be monitored and thereby largely prevented, while monitoring of impending auditory or vestibular damage is not always possible. Ototoxic side effects frequently develop after cessation of aminoglycoside treatment, sometimes delayed by weeks. This review will therefore focus on the ototoxic side effects as a major unresolved issue in aminoglycoside toxicity.

9.2. PATHOLOGY AND PATHOPHYSIOLOGY

9.2.1. Target Tissues and Pharmacokinetics

The pathology of aminoglycoside damage was established quickly after the discovery of the clinical symptoms, while it took almost five decades to unravel the underlying molecular mechanisms. The ototoxic side effects of aminoglycosides include damage to the vestibular system, resulting in balance disorders, and damage to the cochlea, resulting in hearing loss.[3] Nephrotoxicity is associated with damage to the proximal tubules, potentially resulting in renal failure. Aminoglycoside antibiotics can express one or more of these toxicities both in patients and in animal models, but the tissue targets of these drugs are not predictable by any structure–activity relationship. For example, gentamicin is considered more vestibulotoxic than cochleotoxic in the human and is therefore frequently used for vestibular ablation in Ménière's disease.[4] In contrast, amikacin or neomycin may primarily target the cochlea. Most strikingly, streptomycin is almost exclusively vestibulotoxic while the closely related dihydrostreptomycin is almost exclusively cochleotoxic. Likewise, there is no correlation between the effects of a given aminoglycoside on the inner ear and on the kidney.

There is also no correlation between the differential predilection for the vestibular organ or the cochlea or the varying severity of toxicity and the pharmacokinetics of the drugs. The concentrations reached in the inner ear by different aminoglycosides do not correlate with the magnitude of their ototoxic potential[5] nor with their preferred site of action.[6] Aminoglycoside antibiotics enter the inner ear via the bloodstream and do so rather rapidly within minutes following a parenteral administration.[7] In contrast to the quick tissue penetration, aminoglycosides are less efficiently cleared. While the half-life of most aminoglycosides in serum is around 3–5 hours, half-life in inner ear tissues and fluids may exceed 30 days. Therefore, the drugs may still persist in the inner ear at a time when they have been essentially cleared from the bloodstream.

This difference in half-lives in serum and tissues was misinterpreted in the early literature as an "accumulation" of aminoglycosides in the inner ear and held responsible for their organ-specificity. Such an interpretation is, however, not tenable. Aminoglycoside antibiotics are present in the fluids (perilymph and endolymph) and the tissues of the inner ear at relatively low levels, typically at one-tenth of peak serum levels.[8] The reason for the differential sensitivity of inner ear sensory cells may then be based on the extreme persistence of the drugs or on an intrinsic susceptibility to their actions, notably to the generation of reactive oxygen species as described later.[9]

9.2.2. Incidence of Side Effects

Ototoxicity and nephrotoxicity are not invariable consequences of aminoglycoside treatment, and several factors influence their incidence. One determinant factor is the choice of drug since different aminoglycosides have different toxic potentials. For example, neomycin is considered more ototoxic than gentamicin or tobramycin, which in turn may be more toxic than netilmicin. Another factor is the dose and duration of treatment. Longer therapies, such as those associated with tuberculosis, lead to a higher incidence of hearing loss. Impaired kidney function or the nutritional and physiological state of the subject may also contribute to the development of adverse effects.[10] Within these variables, ototoxicity may affect up to 10–20% of patients,[11,12] and nephrotoxicity can likewise complicate 10–20% of therapeutic courses.[13]

There is, however, a notable exception to the observation that aminoglycoside ototoxicity usually develops during chronic drug administration. Clinical cases have been reported in which a single parenteral injection led to a profound hearing loss. This sensitivity was traced to a mutation in the mitochondrial genome, expressed in location 1555 of the mitochondrial ribosomal RNA where a guanosine to adenosine substitution occurs.[14] Seventeen percent of patients who sustain aminoglycoside-induced hearing loss may be carriers of this mutation, while the incidence of this mutation in the population overall is less than 1%.

9.2.3. Cellular Pathology

Despite the differences in tissue selectivity and severity of ototoxic potential, the primary cellular targets of the ototoxic actions of all aminoglycoside antibiotics are the sensory cells of the inner ear. These "hair cells" are essential for the transduction of auditory stimuli (inner and outer hair cells) and balance sensation (type I and type II vestibular hair cells). In the auditory system, aminoglycosides mostly affect the outer hair cells and damage them in a systematic fashion from the basal part of the cochlea, where high frequencies are being processed, to the apical part, where the low frequencies are being processed (Figure 9.1). Therefore, aminoglycoside-induced hearing loss is first manifested as a high-tone hearing loss.[15,16] The inability of the auditory hair cells to regenerate is behind the irreversibility of the loss of function.

In the vestibular organ, the initial hair cell damage occurs in the apex of the cristae and the striolar regions of the maculi. Hair cell loss then progresses toward the periphery of the vestibular receptor organ with type I hair cells affected earlier than the type II hair cells.[17,18] Vestibular disturbances from systemic aminoglycoside administration present as severe unsteadiness that becomes worse in the dark, but an objective clinical assessment is difficult. In contrast to the situation in the cochlea, regeneration of vestibular hair cells has been observed in mammalian species.[19]

Morphological damage by aminoglycosides is not limited to hair cells, and the cochlear sensory epithelium may become completely replaced by a squamous epithelium after prolonged drug treatment.[15] Changes also occur in the

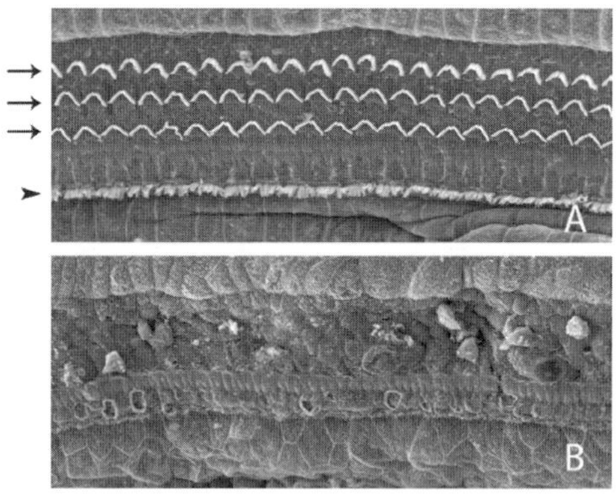

Figure 9.1. Aminoglycoside-induced damage to the cochlea. These scanning electron micrographs present a "surface view" of the apical ends of the cochlear sensory cells. Panel A is taken from an intact inner ear of a guinea pig, clearly outlining three rows of outer hair cells (arrows) and one row of inner hair cells (arrow head) with their distinctive tufts of stereocilia ("hairs"). Panel B is typical of the devastation that aminoglycosides can cause to this structure. The hair cells have completely disappeared and have been replaced by tissue scars formed by their surrounding supporting cells. (Photos courtesy of Drs. Masahiko Izumikawa and Yehoash Raphael, University of Michigan.)

stria vascularis, an accessory structure that maintains the endocochlear potential, usually evident as a thinning or degeneration of the tissue. The effect of aminoglycoside-induced hearing loss on the nerve fibers (spiral ganglion cells) is of special clinical interest. Deafened patients are candidates for a cochlear prosthesis (cochlear implant), and the survival of auditory neurons is of primary importance for the successful restoration of speech recognition. Progressive destruction of nerve fibers is a well-documented sequel to hair cell loss in experimental animals and humans,[20,21] and pathological changes may continue long after drug treatment has been terminated. However, neurotrophic factors can influence the integrity of nerve fibers and aid in the maintenance and regeneration of auditory neurons.[22,23] Vestibular nerve endings and ganglion cells will eventually also deteriorate, but like their cochlear counterparts, they may be protected by nerve growth factors.[24] Neurotrophins may therefore have a potential therapeutic value in the prevention of loss or promotion of re-growth of nerve fibers after ototraumatic insults.

9.3. BIOCHEMICAL AND MOLECULAR MECHANISM OF DRUG ACTION

9.3.1. Sundry Biochemical Actions

Knowledge of the pathology of aminoglycoside toxicity initially came from human inner ears. Mechanistic investigations, however, rely on animal models; fortunately, such models have proven to be reliable representations of ototoxicity. While the guinea pig has long held a traditional position in auditory investigations, recent developments in medical research are favoring the mouse because of the availability of genetic information, probes for the analysis of molecular biology, and transgenic or knock-out animals. An adult mouse model has been recently developed which demonstrates essentially identical pathology and pathophysiology of aminoglycoside ototoxicity as humans and other experimental animals.[25]

Aminoglycoside antibiotics are multifaceted drugs that can exert a number of actions on cells and tissues, not all of which relate to their chronic toxicity. Several of these actions are based on the positive charges that aminoglycosides carry at physiological pH and that enable them to bind to a variety of negatively charged cell components or displace cations from such binding sites. One of the earliest recognized pharmacological actions of aminoglycosides is an antagonism to calcium, discovered in connection with a neuromuscular block that could be achieved with these drugs.[26] Aminoglycoside–calcium interactions at the plasma membrane of the cell became a major focus of early research and made these drugs valuable tools in elucidating the pharmacology of different types of calcium channels.[27,28] Aminoglycoside antibiotics also bind with very high affinity and selectivity in an almost receptor–ligand-like fashion to polyphosphoinositides. This observation led to the use of aminoglycosides as probes into the biochemical mechanisms and physiological role of these membrane lipids which serve as precursors to second messenger mechanisms in most cells.[29,30]

A plethora of other effects on cell constituents from proteins to lipids to DNA is also documented in the early literature without, however, establishing any link between these reactions and the toxic consequences. Such an inability to deduce a toxic mechanism is not necessarily surprising. After all, aminoglycosides are highly aggressive drugs that kill cells and tissues, and a cell in the process of destruction will show a multitude of deranged biochemical and molecular pathways.

9.3.2. Formation of Reactive Oxygen Species

What eventually amounted to a breakthrough in establishing the mechanisms of aminoglycoside toxicity came from observations that aminoglycoside antibiotics elicited the formation of reactive oxygen species (ROS; free radicals) in isolated renal mitochondria[31] and that a radical scavenger was able to protect from the auditory side effects of kanamycin.[32] At the time, however, these observations were challenged by other studies failing to link free radicals to nephrotoxicity[33] or to confirm radical scavenger protection against aminoglycoside ototoxicity.[34] Although more evidence emerged that antioxidants could limit ototoxicity and that free radicals were generated in tissues exposed to aminoglycosides,[10,35,36] there appeared to be considerable reluctance to accept reactive oxygen species as causal agents in aminoglycoside toxicity. One of the stumbling blocks was the unresolved question of how aminoglycoside antibiotics could participate in single electron transfers and oxidation–reduction reactions. Clear evidence now exists for the potential of aminoglycosides to catalyze ROS formation both nonenzymatically and by stimulating enzymatic reactions.

A key to understanding the redox actions of aminoglycosides was the finding that gentamicin was able to accelerate iron-mediated formation of free radicals.[37] Since iron-catalyzed oxidations can be greatly enhanced by chelators,[38] it was postulated that gentamicin and iron may be able to form redox-active complexes. Coordination complexes of aminoglycosides with other metal ions including cobalt, nickel, copper, and calcium had been previously described,[39] but none of them was redox-active. NMR studies confirmed the existence of complexes with both ferric and ferrous irons with ligation through amine groups and the glycosidic oxygen[40] whereby a 1:1 gentamicin–iron complex would contain available coordination sites for oxygen-dependent reactions. Cu^{2+}–aminoglycoside complexes were also extensively investigated including Cu^{2+}–amikacin complexes which catalyze hydrogen peroxide disproportionation involving Cu^{1+}/Cu^{2+} and Cu^{2+}/Cu^{3+} redox pairs.[41]

While these results provided a potential theoretical background for ROS formation by aminoglycoside–transition metal complexes, there were severe practical constraints. The stability constants–for example, of the Cu–amikacin complexes–were such that gentamicin would have to be present in cells or tissues at impossibly high levels (around 100 M) before a significant fraction of Cu ions would be bound. Computer simulations also showed that in a physiological environment other small chelators such as histidine would replace aminoglycosides. Likewise, aminoglycoside–iron complexes would be in competition with other natural chelators from the low-molecular-weight (or "free") intracellular iron pool rendering the presence of simple aminoglycoside–metal complexes unlikely under physiological conditions.

Aminoglycoside–iron-mediated superoxide formation from molecular oxygen required a polyunsaturated fatty acid as an electron donor with arachidonic acid being a particularly suitable co-reactant.[42] The availability of free arachidonic acid is low in an intracellular environment where most of it is esterified to phospholipids, in particular to polyphosphoinositides. The earlier observations that aminoglycosides bind strongly to phosphatidyl inositol 4,5-bisphosphate[30] now suggested the possibility that the lipid itself could provide reactive electrons through its arachidonic acid content. This notion was strengthened by the demonstration that polyphosphoinositides can indeed simultaneously bind to both gentamicin and iron, and Electron Spray Ionization Mass Spectrometry confirmed the existence of ternary complexes between $Fe^{2+/3+}$, gentamicin, and arachidonic acid.[43] The binding mode of iron involves two oxygen atoms donated by the carboxyl group of arachidonic acid and two electron donors from the gentamicin molecule, namely the 3-NH_2 group and the glycosidic oxygen connecting the 2-deoxystreptamine and purpurosamine rings. An H_2O molecule as part of the coordination pattern can be readily displaced by molecular oxygen to initialize the formation of reactive oxygen species, leading to the propagation of arachidonic acid peroxidation. In analogy to established patterns of iron-dependent oxidations promoted by iron chelators,[44] the initial steps of free radical formation could then have the following form:

$$\text{Arachidonic acid–}Fe^{2+}\text{–gentamicin} + O_2 \leftrightarrow [\text{arachidonic acid–}Fe^{2+}\text{–}O_2\text{-gentamicin} \leftrightarrow \text{arachidonic acid–}Fe^{3+} - O_2^{\bullet -} - \text{gentamicin}] \leftrightarrow \text{arachidonic acid–}Fe^{3+}\text{–gentamicin} + O_2^{\bullet -}$$

Since aminoglycoside complexes with membrane phosphoinositides are extremely stable, they could well exist in an intracellular environment with concentrations of aminoglycosides that are achieved in chemotherapeutic treatment.

While a nonenzymatic catalysis of ROS formation may contribute or even cause toxicity, it may not be the only means by which aminoglycosides can promote the generation of free radicals. One of the initial reactions of aminoglycoside antibiotics *in vivo* is an activation of redox-dependent molecular signaling pathways linked to Rho-GTPases, which control the assembly and disassembly of the actin cytoskeleton in many cells.[45] The activity of Rac-1, a member of the family of Rho-GTPases, is enhanced by aminoglycosides, thereby leading to activation of the NADPH oxidase complex which enzymatically promotes the formation of the superoxide radicals. Thus, aminoglycosides may be additionally involved in enzymatic mechanisms of ROS production, which ultimately may be linked to the regulation of cell death pathways.

The question then arises whether the formation of ROS can indeed be causally related to cell death or whether it simply constitutes an epiphenomenon. The strongest evidence for a causal relationship comes from studies that demonstrate the ability of iron chelators and antioxidants to protect against aminoglycoside-induced ototoxicity. Effective protectants include chelators such as deferoxamine and dihydroxybenzoic acid and a variety of other antioxidant molecules including vitamin E, salicylate, or methionine.[46–48] In these experiments in animal models, an otherwise profound loss of hearing or balance by aminoglycoside treatment was effectively prevented by the concurrent administration of the protective agents. Since similar compounds also protect against aminoglycoside nephrotoxicity,[49] it may be assumed that similar mechanisms of ROS toxicity also operate in the kidney.

9.3.3. Cell Death Pathways

Pathways of cell death triggered by aminoglycosides (Figure 9.2) are complex and range from apoptosis to necrosis of the sensory cells in the inner ear.[50,51] Oxidative stress and a high load of ROS have many downstream consequences. ROS readily oxidize cellular components, including proteins, lipids, and DNA, by virtue of a highly reactive unpaired electron. If the oxidative imbalance is too great to be neutralized by cellular antioxidant systems, specific signaling pathways are invoked in order to maintain homeostasis or to prepare for cell death. The regulation of these pathways, in turn, requires the concerted efforts of cellular response elements to stress, second messenger systems, protein kinases, and transcription factors.

In aminoglycoside-treated animals, the cells can be led to canonical apoptotic death through activation of caspases. Caspase-9 forms an "apoptosome" complex with cytochrome c and APAF-1 and leads to apoptosis through activation of caspase-3. Aminoglycosides activate caspases in auditory structures[52]; conversely, inhibition of caspase activity successfully blocks neomycin-induced vestibulotoxicity.[53] In contrast, apoptotic markers were essentially absent in a mouse model of chronic kanamycin ototoxicity where death of auditory sensory cells ensued via cathepsins.[51] The activation of cathepsin D was accompanied by the nuclear translocation of endonuclease G, necrotic cleavage of PARP, and activation of μ-calpain, all facets of necrotic cell death.

Cell death pathways may be counteracted by survival signaling cascades in the early stages of intoxication or in cells that are resistant to the toxic actions of the drugs. Signaling via the transcription factor NF-κB may constitute such a homeostatic mechanism in the cochlea. Cells that survived aminoglycoside treatment in a model of chronic ototoxicity were marked by NF-κB activation.[45]

9.4. PROTECTION

Attempts to protect patients from adverse side effects date back to the early years of aminoglycoside chemotherapy. An amazing variety of pharmacological agents

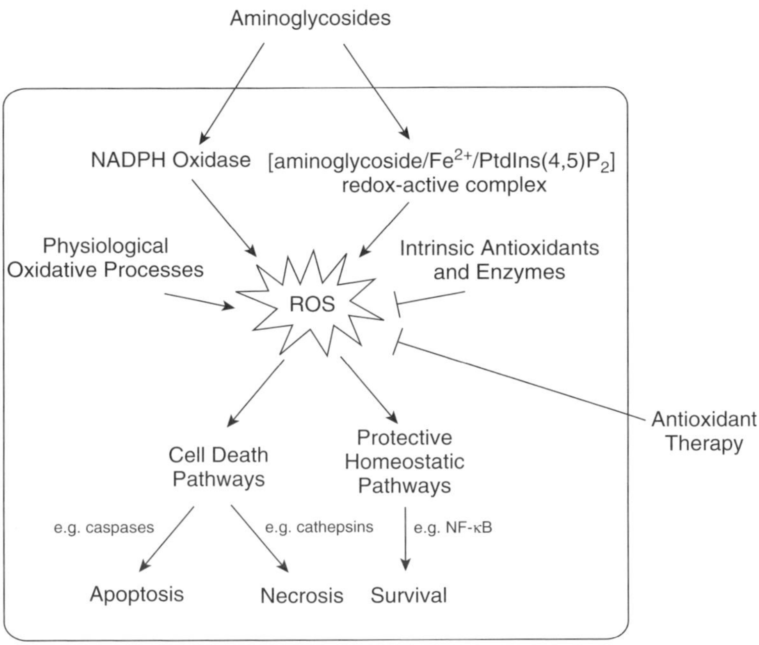

Figure 9.2. Mechanisms of aminoglycoside toxicity. This schematic representation summarizes the principles of aminoglycoside toxicity discussed in the text. Treatment with the drugs leads to the formation of reactive oxygen species through a redox-active complex with iron and unsaturated fatty acid or by triggering superoxide production by way of NADPH oxidase. An excess of reactive oxygen species, not balanced by intracellular antioxidant systems, will cause an oxidative imbalance potentially severe enough to initiate cell death pathways. Augmenting cellular defenses by antioxidant therapy can reverse the imbalance and restore homeostasis to protect the cell.

have been claimed at one time or another to attenuate ototoxicity in animals—for example, various vitamins, diverse amino acids, hormones, antibiotics, sulfhydryl compounds, or herbal concoctions. None of these claims, however, developed into a successful clinical treatment for the prevention of aminoglycoside-induced hearing loss.

The recent advances in the understanding of the mechanisms of ototoxicity have now led to the development of rational and successful protective therapies. Inhibitors of some of the many steps in the apoptotic cascade have proven effective in protecting hair cells in cell or organ culture. Such evidence, however, is only the first step to designing a clinical therapy, an end that may be difficult to reach since systemic intervention in crucial cell signaling pathways may have undesirable physiological consequences. Local gene therapy, the process of virally introducing a gene into the cochlea—for example, a gene coding for a neurotrophin or an antioxidant enzyme—has shown promise in alleviating ototoxicity in animals and may be clinically possible.[54]

A presently more practicable method of prevention may be antioxidant therapy, which has become a successful clinical approach to many of the pathologies that involve free radicals.[55,56] This type of intervention would act directly on the ROS, upstream of the ensuing cell death pathways, and therefore suppress toxic mechanisms at the very outset. Antioxidant therapy, indeed, seems indicated for aminoglycoside ototoxicity since antioxidants are effective against both the vestibular and cochlear side effects, regardless of which individual aminoglycoside is the causative agent.[46,47] Neither the serum levels of the aminoglycosides nor their antibacterial efficacy are compromised by this co-treatment. One of the more intriguing protective compounds is salicylate, the active principal of aspirin.[48] Recent clinical studies using this approach show promise that aminoglycoside ototoxicity may finally be preventable.[57]

9.5. OUTLOOK

The past decade has brought great advances in our understanding of the causes of the adverse effects of aminoglycoside chemotherapy. The basic observations, however, that the nephrotoxicity results from damage to the renal proximal tubules and that the ototoxicity stems from an insult on the sensory hair cells of the inner ear were never challenged. The evidence that oxidative stress is a major contributor to aminoglycoside-induced cell death should now drive the design of novel rational therapeutic interventions or perhaps of novel nontoxic aminoglycoside derivatives. The amelioration of ototoxicity and nephrotoxicity would have far-reaching implications for the safe use of a class of drugs whose chemotherapeutic value remains undisputed.

ACKNOWLEDGMENT

Dr. Schacht's research on drug-induced hearing loss is supported by grant DC-03685 from the National Institute on Deafness and Other Communication Disorders, National Institutes of Health.

REFERENCES

1. Schatz, A.; Bugie, E.; Waksman, S. A. *Proc. Soc. Exp. Biol. Med.* **1944**, *55*, 66–69.
2. Hinshaw, H. C.; Feldman, W. H. *Proc. Mayo Clin.* **1945**, *20*, 313–318.
3. Forge, A.; Schacht, *J. Audiol. Neuro-Otol.* **2000**, *5*, 3–22.
4. Blakley, B. W. *Am. J. Otol.* **1997**, *18*, 520–526.
5. Ohtsuki, K.; Ohtani, I.; Aikawa, T.; Sato, Y.; Anzai, T.; Ouchi, J.; Saito, *J. Ear Res. Jpn.* **1982**, *13*, 85–87.
6. Dulon, D.; Aran, J. M.; Zajic, G.; Schacht, *J. Antimicrob. Agents Chemother.* **1986**, *30*, 96–100.

7. Tran Ba Huy, P.; Bernard, P.; Schacht, J. *J Clin Invest.* **1986**, *77*, 1492–1500.
8. Henley, C. M.; Schacht, *J. Audiol.* **1988**, *27*, 137–146.
9. Sha, S-H.; Taylor, R.; Forge, A.; Schacht, J. *Hear. Res.* **2001**, *155*, 1–8.
10. Lautermann, J.; McLaren, J.; Schacht, J. *Hear. Res.* **1995**, *86*, 15–24.
11. Fee, W. E. *Laryngoscope* **1980**, *40*, 1–19.
12. Lerner, S.A.; Schmitt, B. A.; Seligsohn, R.; Matz, G. J. *Am J Med.* **1986**, *80*, 98–104.
13. Swan, S. K. *Sem Nephrol.* **1997**, *17*, 27–33.
14. Prezant, T. R.; Shohat, M.; Jaber, L.; Pressman, S.; Fischel-Ghodsian, N. *Am. J Med. Genet.* **1992**, *44*, 465–472.
15. Hawkins, J. E. In *Handbook of Sensory Physiology*, Vol. 5, part 3, Keidel, W. D.; Neff W. D., Eds.; Berlin: Springer-Verlag, 1976; pp. 707–748.
16. Fausti, S. A.; Rappaport, B. Z.; Schechter, M. A.; Frey, R. H.; War, T. T.; Brummett, R. E. *Am. J. Otolaryngol.* **1984**, *5*, 177–182.
17. Wersäll, J.; Lundquist, P-G.; Björkroth, B. *J. Infect. Dis.* **1969**, *119*, 410–416.
18. Wersäll, J.; Björkroth, B.; Flock, A.; Lundquist, P.-G. *Adv. Oto-Rhino-Laryng.* **1973**, *20*, 14–41.
19. Forge, A.; Li, L.; Corwin, J. T.; Nevill, G. *Science* **1993**, *259*, 1616–1619.
20. Hawkins, J. E.; Beger, V.; Aran, J. M. In *Sensorineural Hearing Processes and Disorders*; Graham, A. B., Ed.; Boston: Little Brown Co., 1967; pp. 411–425.
21. Johnsson, L. G.; Hawkins, J. E.; Kingsley, T. C., Black, F. O.; Matz, G. J. *Acta Otolaryngol.* **1981**, *S383*, 1–19.
22. Shah, S. B.; Gladstone, H. B.; Williams, H.; Hradek, G. T.; Schindler, R. A. *Am J Otol.* **1995**, *16*, 310–314.
23. Ernfors, P.; Duan, M. L.; ElShamy, W. M.; Canlon, B. *Nat. Med.* **1996**, *2*, 463–467.
24. Zheng, J. L.; Stewart, R. R.; Gao, W. Q. *J. Neurobiol.* **1995**, *28*, 330–440.
25. Wu, W.-J.; Sha, S.-H.; McLaren, J. D.; Kawamoto, K.; Raphael, Y.; Schacht, J. *Hear. Res.* **2001**, *158*, 165–178.
26. Vital-Brazil, O.; Corrado, A. P. *J Pharm Exp Ther.* **1957**, *120*, 452–459.
27. Keith, R. A.; Mangano, T. J.; Defeo, P. A.; Horn, M. B.; Salama, A. I. *J. Mol. Neurosci.* **1992**, *3*, 147–154.
28. Pichler, M.; Wang, Z.; Grabner-Weiss, C.; Reimer, D.; Hering, S.; Grabner, M.; Glossman, H.; Striessnig, *J. Biochemistry* **1996**, *35*, 14659–14664.
29. Schacht, J. *J. Neurochem.* **1976**, *27*, 1119–1124.
30. Schacht, *J. Arch. Otorhinolaryngol.* **1979**, *224*, 129–134.
31. Walker, P. D.; Shah, S. V. *Am. J. Physiol.* **1987**, *253*, C495–C499.
32. Pierson, M. G.; Møller, A. R. *Hear. Res.* **1981**, *4*, 79–87.
33. Ramsammy, L. S.; Josepovitz, C.; Ling, K.-Y.; Lane, B. P.; Kaloyanides, G. J. *Biochem. Pharmacol.* **1987**, *36*, 2125–2132.
34. Bock, G. R.; Yates, G. K.; Miller, J. J.; Moorjani, P. *Hear. Res.* **1983**, *9*, 255–262.
35. Clerici, W. J.; Hensley, K.; DiMartino, D. L.; Butterfield, D. A. *Hear. Res.* **1996**, *98*, 116–124.
36. Hirose, K.; Hockenberry, D. N.; Rubel, E. W. *Hear. Res.* **1997**, *104*, 1–14.
37. Priuska, E. M.; Schacht, J. *Biochem Pharmacol.* **1995**, *50*, 1749–1752.
38. Sutton, H. C. *Free Rad. Biol. Med.* **1985**, *1*, 195–202.

39. Barba-Behrens, N.; Bautista, J.; Ruiz, M.; Joseph-Nathan. P.; Flores-Parra, A.; Contreras, R. *J. Inorg Biochem.* **1990**, *40*, 201–215.
40. Priuska, E. M.; Clark, K.; Pecoraro, V.; Schacht, J. *Inorg. Chim. Acta* **1998**, *273*, 85–91.
41. Jezowska-Bojczuk, M.; Lesniak, W.; Bal, W.; Kozlowski, H.; Gatner, K.; Jezierski, A.; Sobczak, J.; Mangani, S.; Meyer-Klaucke, W. *Chem. Res. Toxicol.* **2001**, *14*, 1353–1362.
42. Sha, S.-H.; Schacht, J. *Free Rad. Biol. Med.* **1999**, *26*, 341–347.
43. Lesniak, W.; Pecoraro, V.-L.; Schacht, *J. Chem. Res. Toxicol.* **2005**, *18*, 357–364.
44. Stadtman, E. R. *Annu. Rev. Biochem.* **1993**, *62*, 797–821.
45. Jiang, H.; Sha, S.-H.; Schacht, J. *J. Neurosci. Res.* **2006**, *83*, 1544–1551.
46. Song, B.-B; Schacht, *J. Hear. Res.* **1996**, *94*, 87–93.
47. Song, B-B.; Sha, S.-H.; Schacht, *J. Free Rad. Biol. Med.* **1998**, *25*, 189–195.
48. Sha, S.H.; Schacht, *J. Lab Invest.* **1999**, *79*, 807–813.
49. Walker, P.D.; Shah, S.V. *J Clin Invest.* **1988**, *81*, 334–341.
50. Forge, A. *Hear Res.* **1985**, *19*, 171–182.
51. Jiang, H.; Sha, S.-H.; Forge, A.; Schacht, J. *Cell Death Diff.* **2006**, *13*, 20–30.
52. Cheng, A.G.; Cunningham, L.L.; Rubel, E.W.*J. Assoc. Res. Otolaryngol.* **2003**, *4*, 91–105.
53. Cunningham, L.L.; Cheng, A.G.; Rubel, E.W. *J Neurosci.* **2002**, *22*, 8532–8540.
54. Kawamoto, K.; Sha, S.-H.; Minoda, R.; Izumikawa, M.; Kuriyama, H.; Schacht, J.; Raphael, Y. *Mol. Ther.* **2004**, *9*, 173–181.
55. Hershko, C. *Mol. Asp. Med.* **1992**, *13*, 113–165.
56. Tanswell, A. K.; Freeman, B. A. *New Horizons* **1995**, *3*, 330–341.
57. Sha, S.-H.; Qiu, J.-H.; Schacht, J. *N Engl J. Med.* **2006**, *354*, 1856–1857.

10

TARGETING HIV-1 RNA WITH AMINOGLYCOSIDE ANTIBIOTICS AND THEIR DERIVATIVES

LEV ELSON-SCHWAB AND YITZHAK TOR
Department of Chemistry and Biochemistry, University of California, San Diego, La Jolla, CA 92093

10.1. INTRODUCTION

According to figures generated by the World Health Organization, approximately 40 million people worldwide currently live with HIV/AIDS.[1] Close to 5 million people have been newly infected in 2004, and nearly 3.1 million HIV/AIDS-related

Aminoglycoside Antibiotics: From Chemical Biology to Drug Discovery, Second Edition, by Dev P. Arya

deaths have been documented.[1] While it appears that the epidemic has essentially been controlled in large parts of the Western world, the global prognosis does not seem bright.[2] Projections suggest that the epidemic could at least double by the end of the decade. This forecast seems particularly bleak for heavily populated areas, including Eastern Europe, central Asia, and parts of China.

The development of antiretroviral therapy has been a major challenge since the discovery of the human immunodeficiency virus (HIV).[3,4] Early successes with nucleoside and non-nucleoside reverse transcriptase (RT) inhibitors, as well as the development of protease inhibitors have facilitated, in recent years, a highly active antiretroviral therapy (HAART), where a combination of drugs is simultaneously administered.[5,6] In spite of significant improvements in the morbidity and mortality of HIV-infected patients, the rapid appearance of resistant HIV-variants,[7] as well as adverse effects[8] and high cost of contemporary drugs necessitate the continuous development of independent classes of anti-HIV agents.[9,10]

In principle, various targets for therapeutic intervention exist. A successful drug can interact with viral receptors, virally encoded enzymes, viral structural components, viral genes or transcripts, or cellular factors required for viral replication. In recent years, attention has been given to small molecules that can target viral-specific RNA sites and prevent the formation of key RNA–protein and RNA–RNA complexes.[11,12] Aminoglycoside antibiotics have provided the impetus for this approach as discussed below.

10.2. HIV-1 REPLICATION AND RNA

As a retrovirus, HIV-1 utilizes RNA as its genetic material. Viral replication thus relies on first converting the viral genome into DNA and then transcribing it back into RNA. The elaborate life cycle of HIV-1 requires an ordered pattern of viral gene expression, and it can be divided into three major phases (Figure 10.1).[13,14] The first stage involves recognition of the host cell, fusion/uncoating, reverse transcription, and integration into the host cell chromosome. This establishes a permanent presence of the virus genetic material within the infected cell. The next major stage includes transcription of viral RNA, translation of viral proteins, and processing of RNA and proteins into their mature forms. This generates the necessary building blocks for the assembly of new viruses. In the final step, new viral particles are packaged and released from the cell to infect new cells.

Each stage described above involves at least one critical RNA–RNA and/or RNA–protein interaction. Initiation of reverse transcription involves the binding of a specific tRNA, which is packaged in the viral particle, to a short sequence of the viral RNA. Transcription and accumulation of viral RNA is dependent upon the sequence-specific interaction between two essential viral regulatory proteins, Tat and Rev, with their respective RNA sites, TAR and RRE (Figure 10.1).[15] Tat is a transcriptional activator,[16] whereas Rev acts post-transcriptionally to increase the cytoplasmic accumulation of the viral *gag-pol* and *env* messenger RNAs.[17] Viral assembly is initiated by formation of an RNA–RNA dimer where

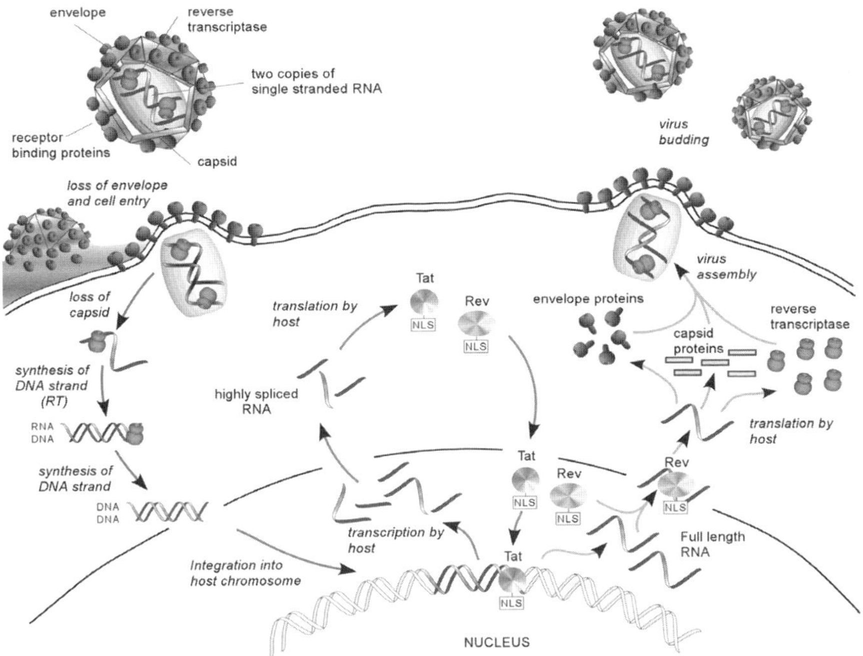

Figure 10.1. Schematic illustration of the life cycle of HIV-1. Following adhesion, fusion, and partial loss of capsid, the viral single-stranded RNA genome is reverse-transcribed to the corresponding double-stranded DNA that is then integrated into the host chromosome (red arrows). Transcription is rather ineffective in early stages, and short viral RNA transcripts are formed. These RNA sequences escape the nucleus and are responsible for the expression of a number of viral regulatory proteins, including Tat and Rev (blue arrows). Both Tat and Rev possess nuclear localization signals (NLS). Tat translocates into the nucleus and significantly increases the formation of the full-length viral RNA (blue arrows). Export of full-length and singly spliced viral RNA is mediated by the Rev protein (green arrows). The presence of viral RNA in the cytoplasm facilitates the biosynthesis of the necessary viral proteins as well as the packaging of the complete viral genomes in the new HIV particles (green arrows). (Reproduced with permission from Wiley-VCH.) See color plates.

the full genomic viral RNA dimerizes for packaging in new viral particles.[18] As viral RNA–RNA and RNA–protein interactions can be highly selective, small molecules that specifically disrupt such viral-specific biomolecular recognition events can have potential therapeutic utility. This chapter highlights some of the viral RNA sites that have emerged as potential targets for aminoglycoside antibiotics and related low-molecular-weight ligands.[19,20]

10.3. AMINOGLYCOSIDE ANTIBIOTICS AS RNA BINDERS

Aminoglycoside antibiotics represent a family of highly charged naturally occurring pseudo-oligosaccharides (Figure 10.2).[21] The common core of most

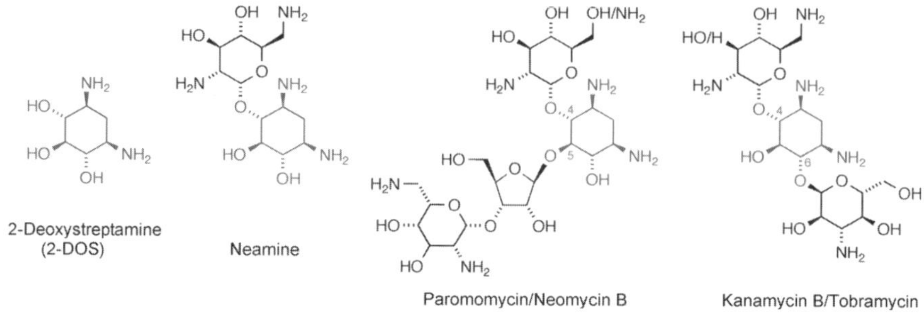

Figure 10.2. Structures of selected aminoglycosides and their fragments.

aminoglycosides is 2-deoxystreptamine (2-DOS), a highly functionalized aminocyclitol.[22] Glycosidation of the 2-DOS core with monosaccharides, typically at the 4 and 5 (or 4 and 6) positions, generates most aminoglycosides. These potent antibiotics have evolved to bind the bacterial ribosomal decoding site, also known as the A-site, and interfere with bacterial protein biosynthesis.[23] Despite significant adverse effects (e.g., nephro- and ototoxicity), aminoglycoside antibiotics remain clinically useful and important contemporary research tools.[24,25]

Moazed and Noller's 1987 article,[26] localizing the binding site of aminoglycosides to the decoding region on prokaryotic 16S rRNA, has essentially labeled these antibiotics as RNA binders and prompted further investigations. The popularity that aminoglycosides have experienced in recent years was driven by Schroeder and co-workers' 1991 report, demonstrating that these antibiotics can also inhibit splicing of group I introns.[27] Green and co-workers' 1993 observation that aminoglycosides can also inhibit the binding of the HIV-1 Rev protein to its RNA target, the Rev response element (RRE), by directly binding to the latter,[28] has focused the attention on aminoglycosides as potential viral RNA binders.

Stimulated by these intriguing observations suggesting aminoglycosides to be potent RNA binders, we and others started exploring this family of antibiotics for their potential to bind various other RNA targets.[12] Over the course of the past decade, numerous discoveries have established aminoglycosides as universal RNA binders. These antibiotics have been found to bind the hammerhead ribozyme,[29] human hepatitis delta virus ribozyme,[30] and RNase P,[31] as well as HIV-1 TAR[32] and even transfer RNA.[33] Additionally, multiple binding sites for aminoglycosides were discovered on small and large RNA molecules.[34] Aminoglycosides are therefore rather promiscuous RNA binders, a phenomenon that can be rationalized by a model suggesting "structural electrostatic complementarity."[35,36] While these naturally occurring antibiotics are unlikely to display the necessary selectivity to become therapeutically relevant antivirals,[37] they provide a solid foundation for the potential development of more potent and selective RNA binders. The following sections outline HIV RNA targets of therapeutic potential and summarize efforts to develop small molecules as potential inhibitors of specific stages in the complex life cycle of HIV-1.

10.4. INITIATION OF REVERSE TRANSCRIPTION AND THE PRIMER BINDING SITE

Initiation of reverse transcription in HIV-infected cells relies on a critical RNA–RNA interaction between $tRNA^{Lys}{}_3$, which is preferentially packaged into the viral particle,[38] and a specific viral RNA sequence. The 3′-terminal 18 nucleotides of $tRNA^{Lys_3}$ are complementary to the primer binding site (PBS) sequence located in the 5′-long terminal repeat (LTR) of the viral RNA genome (Figure 10.3). The UUU anticodon of the tRNA is complementary to and binds to an adenosine rich loop located 8 nucleotides upstream (5′) of the PBS. This RNA–RNA duplex which is formed when $tRNA^{Lys_3}$ binds to the PBS fits within the active site of HIV-1 reverse transcriptase,[39] but multiple interactions between the viral RNA and $tRNA^{Lys_3}$ are necessary for efficient initiation of reverse transcription.[38] This interaction nucleates the reverse transcription complex which contains viral RNA, reverse transcriptase, $tRNA^{Lys_3}$, $p17^{MA}$, nucleocapsid p7, and Vpr (Viral protein R), as well as multiple host factors.[40]

The A-rich loop plays a role in retroviral replication and likely pairs with the complementary $tRNA^{Lys_3}$ anticodon.[39] Structural analysis of this A-rich loop shows a U-turn motif, an RNA structure first seen in $tRNA^{Phe}$,[41] but which is a common structural motif in many tRNAs.[42] If, as evidence suggests, the four adenosines in the A-rich loop of the PBS pairing with the four uridines in the anticodon stem loop of $tRNA^{Lys_3}$ is a critical event in the HIV life cycle, this could provide a potential target for therapeutic intervention.

The therapeutic potential of this RNA target is, however, questionable due to the structural similarity of the A-rich loop and the anticodon stem loop of tRNA. Any potential drug must have not only reasonable affinity for its target, but must display a high degree of selectivity. Such discrimination might be difficult to achieve due to the prevalence of transfer RNAs in cells. Additionally, aminoglycosides would likely provide a poor framework for synthetic ligands targeting the A-rich loop as they are known to bind double-stranded RNA regions.[43,44] A molecule that preferentially binds the complex formed between the A-rich loop and $tRNA^{Lys_3}$ could potentially stabilize this interaction, thus promoting viral replication rather than hindering it. Based on these considerations, the PBS is

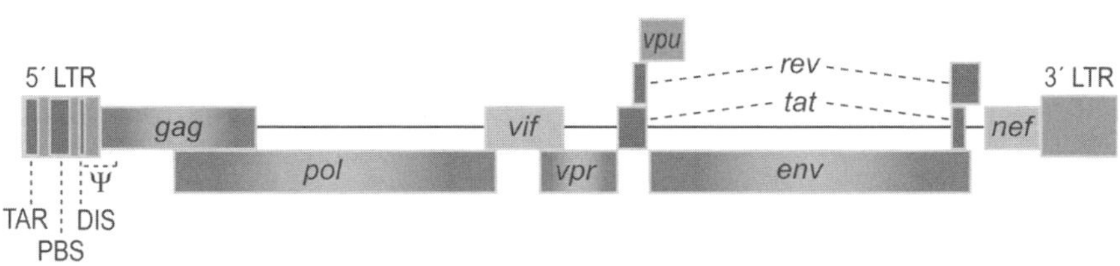

Figure 10.3. Schematic view of the HIV-1 genome, showing gene structure and splicing patterns, as well as important RNA sites localized in the 5′-long terminal repeat (LTR) and the ψ-site. See color plates.

likely to be classified as a poor target for modified aminoglycosides designed to interfere with reverse transcription.

10.5. REGULATION OF VIRAL TRANSCRIPTION AND THE TAR RNA

Once the viral RNA has been reverse-transcribed into DNA, it must enter the nucleus and be integrated into the host genome. The pre-integration complex (viral DNA, reverse transcriptase, $p17^{MA}$, Vpr, and host factors) forms and connects to host microtubules, which allow it to translocate into the nucleus.[45] Integration is promoted by integrase, a viral enzyme, as well as several host factors, and takes place at transcriptional "hot spots."[40,46] This process is largely RNA-independent, making it unlikely that aminoglycosides or related analogs, which preferentially bind to RNA, could provide useful inhibitors.

Once integrated into the host chromosome, the assembly of new viral particles necessitates the production of viral RNA transcripts and proteins. Initiation of viral transcription is also an RNA independent process where host transcription promoters and enhancer elements such as NF-κB bind to the 5′-LTR. The host transcriptional complex is then recruited and transcription commences.[47,48] Once transcription has been initiated, RNA and RNA–RNA interactions play a critical role in mediating the production of viral transcripts. The multiprotein transcription complex has a recognition factor for nonhost DNA and quickly releases from viral DNA, creating short, abortive transcripts. Processing and nuclear export of these transcripts leads to the translation of the HIV Tat protein,[49,50] a small early-phase viral protein (Figure 10.4) that plays a key role in the ultimate formation of full-length viral RNA transcripts.

HIV Tat can cross nuclear membranes and localize in the nucleus,[51,52] where it binds to TAR (transactivating response element), a stem-loop RNA found on

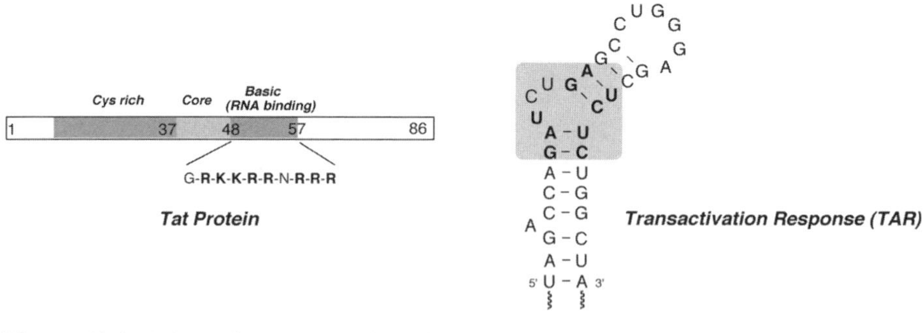

Figure 10.4. Schematic representation of the Tat protein and its functional regions, highlighting the basic RNA binding domain. The secondary structure of its RNA target, TAR, is shown. Critical residues for Tat binding within the recognition domain (highlighted) are shown in bold.

the 5′-LTR (Figures 10.3 and 10.4). The formation of this RNA–protein complex stimulates the transcription of full-length viral genome by preventing the release of the transcription complex from the nonhost viral DNA.[47,53] The Tat–TAR interaction is essential for transcription of full-length viral RNA. Tat, therefore, is involved in a positive feedback mechanism that ensures a high level of HIV transcription. Not surprisingly, the Tat–TAR interaction has been the subject of structural studies[54] and discovery assay development[55,56] and has been widely studied as a potential therapeutic target.[16,57]

About a decade ago, aminoglycoside antibiotics were shown to bind TAR and inhibit the formation of the Tat–TAR complex by Mei and co-workers.[31,58] Their observations stimulated additional explorations into aminoglycoside-related derivatives and other small molecules as TAR binders.[59] Lapidot and co-workers[60] demonstrated that conjugation of arginine to aminoglycosides generates potent TAR binders (Figure 10.5). Evidence was also provided for antiviral activity of these derivatives in infected cells.[61] Décout and co-workers[62] reported the conjugation of neamine to a TAR-targeted PNA sequences. The resulting conjugates demonstrated TAR cleavage *in vitro* and anti-HIV activity in infected cells. In addition, Yu and co-workers[63] reported TAR binding of several neomycin conjugates containing chloramphenicol and linezolid (Figure 10.5). The conjugates display approximately 10-fold higher affinity to TAR compared to the parent aminoglycoside.

Other families of small molecules have also been found to effectively compete with the Tat peptide for binding to the TAR RNA.[59] These include diphenylfuran derivatives developed by Wilson and Boykin.[64] Footprinting experiments suggested that the bulge region is critical for RNA binding, although TAR recognition by these derivatives appear to be shape- rather than sequence-specific.[64] In addition to small molecules such as the polyamine-acridine conjugates described by Bailly,[65,66] oligomeric peptides, peptidomimetic and oligoureas have been reported by Rana and co-workers[67] to bind TAR and serve as Tat antagonists.

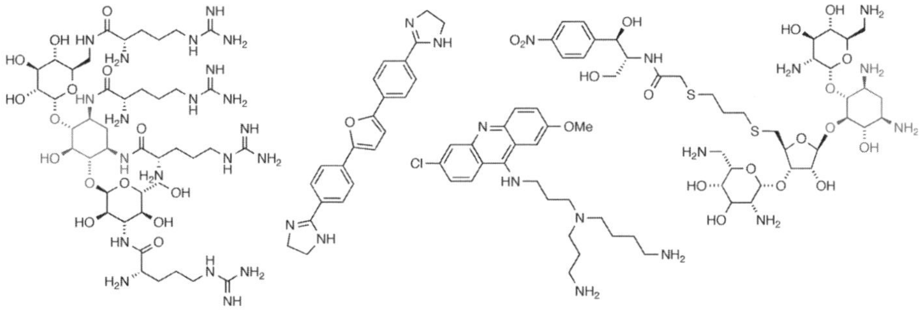

Figure 10.5. Selected examples of TAR binders. From left to right: Arginine–kanamycin A conjugate, diphenylfuran derivative, polyamine–acridine conjugate, and chloramphenicol–neomycin B conjugate.

To quantitatively determine the binding affinity and target selectivity of TAR ligands, a new assay has been developed and validated.[68] It is based on monitoring the fluorescence of a covalently incorporated pyrene fluorophore. Circular dichroism, enzymatic footprinting, and NMR experiments all indicate that the tertiary structure surrounding the bulged region of the TAR RNA changes upon binding of small molecules such as aminoglycosides.[69] Since the pyrene fluorophore is sensitive to its environment when incorporated into nucleic acids,[70] we reasoned that a pyrene in this region would accurately report these structural changes by either an increase or quenching of its fluorescence. Measuring the fluorescence of the pyrene-derivatized HIV-1 TAR RNA upon titration with aminoglycoside antibiotics yielded binding isotherms that fit well with a 1:1 interaction, generating accurate equilibrium dissociation constants. Reliable performance has been observed in the presence of competing nucleic acids (e.g., tRNA, dsDNA), thus providing a means for evaluating ligand selectivity.[68]

As discussed above, aminoglycosides are rather promiscuous RNA binders capable of binding numerous RNA sequences with comparable affinity. This lack of target selectivity results from both their electrostatically driven binding mode and their conformational adaptability. The inherent flexibility of the glycosidic bonds allows these natural products to assume a variety of conformations, permitting them to structurally adapt to diverse RNA targets.[36] This structural plasticity leads to the formation of aminoglycoside complexes with diverse RNA targets in which the antibiotics assume distinct conformations. Structural analysis of neomycin B and paromomycin bound to the HIV-1 TAR and prokaryotic ribosomal A-site, respectively, shows markedly different conformations. When bound to the A-site, paromomycin is compactly arranged with ring III in close proximity to the 2-DOS ring (Figure 10.6).[71,72] This is in contrast to the structure of neomycin B bound to the HIV-1 TAR where the aminoglycoside is arranged in a much more extended conformation.[69] We postulated that reducing the conformational flexibility of aminoglycosides could therefore alter their target selectivity.

To explore this possibility, conformationally constrained neomycin and paromomycin analogs were prepared (Figure 10.6).[73] These derivatives were designed to mimic the A-site bound aminoglycoside structure and disfavor the reported TAR-bound conformation. The results were illuminating. Not unexpectedly, locking the aminoglycosides into a rather compact structure had minimal impact on binding to the A-site.[74] Surprisingly, however, preorganizing these neomycin-class aminoglycoside antibiotics into a TAR-disfavored structure had no deleterious effect on binding to this HIV-1 RNA sequence.

These observations were rationalized by suggesting that the A-site and HIV TAR possess inherently different selectivities toward aminoglycoside antibiotics.[73] The inherent plasticity of the TAR RNA, coupled to the remaining flexibility within the conformationally constrained aminoglycoside analogs, makes this RNA site an accommodating target for such polycationic ligands. In contrast, the deeply encapsulating A-site is a more discriminating RNA target.

These observations suggest that future design of RNA-directed therapeutics must consider the inherent structural selectivity of the RNA target and not only

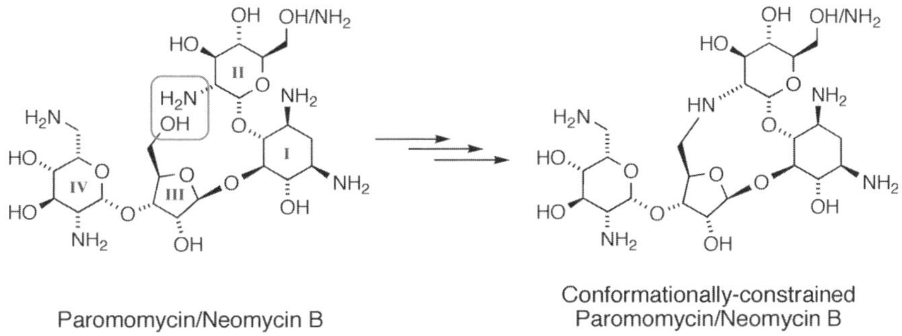

Figure 10.6. Conformationally constrained paromomycin and neomycin analogs.

the selectivity patterns displayed by the low-molecular-weight ligands. While the Tat–TAR interaction is critical for viral replication, it may be ultimately realized to be a rather poor target for low-molecular-weight binders because adequate *in vivo* selectivity might be difficult to attain.

10.6. REGULATION OF VIRAL REPLICATION: RNA EXPORT AND THE RRE

Full-length viral RNA transcripts are processed by the host machinery prior to nuclear export; however, if only processed viral RNA transcripts were exported from the nucleus, viral replication would halt. Both partially spliced and fully unspliced viral RNA transcripts must be exported from the nucleus to serve as open reading frames for proteins (Gag, Pol, and Env) that are essential for completing the viral life cycle (Figures 10.1 and 10.3).[75] Additionally, each new viral particle must contain two copies of the fully unspliced viral RNA, which serves as the primary genome for progeny virions.

The Rev protein serves to facilitate the nuclear export of viral RNA that has not been fully processed. Rev, a 116-amino- acid polypeptide, is expressed early during viral replication (Figure 10.7).[17] It is comprised of three functional domains: (a) an arginine-rich motif (aa 34–50) that functions as a nuclear localization signal (NLS) and RNA binding domain, (b) a leucine-rich nuclear export signal (NES, aa 75–83), that associates with CRM1, a cellular export receptor protein, and (c) a multimerization domain, flanking the RNA binding domain (Figure 10.7).[17] Rev binds to a unique high-affinity site on the RRE with low nanomolar affinity.[76,77] This binding event is followed by the multimerization of Rev on the entire RRE (a 350-nt-long RNA), a process that is facilitated by RNA–protein and protein–protein interactions. Once bound to the RRE of a partially or fully unspliced viral RNA transcript, the complex initiates nuclear export of the RNA. Since the arginine-rich motif of the Rev protein serves as both the RRE binding domain and the nuclear localization signal, it allows the protein to continuously shuttle back and forth between the cytoplasm and nucleus

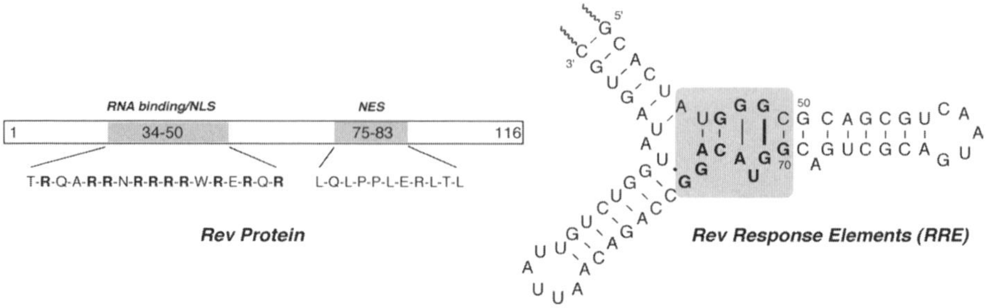

Figure 10.7. Schematic representation of the Rev protein, emphasizing its two key functional domains. The secondary structure of the RRE, highlighting the Rev biding site, is shown. Residues essential for RRE are in bold. The intervening bulge contains two non-Watson–Crick base pairs, G48:G71 and G47:A73, and a bulged base U72.[77]

as follows: Once translated, Rev can enter into the nucleus, bind to the RRE, and exit the nucleus with the RRE inclusive transcripts; after dissociation of the Rev–RRE complex in the cytoplasm, the Rev, whose NLS is now exposed, is free to reenter the nucleus and export additional viral RNA transcripts.

The RRE RNA sequence is an intriguing target for therapeutic intervention. This viral sequence serves both as the high-affinity Rev binding site and as part of the open reading frame for the gp41 "fusion domain" of the *env* protein (Figure 10.3).[78] This dual role is the likely cause for the unusually low mutation rate found at the primary Rev binding site on the RRE.[79] The nucleotides composing the primary Rev binding site are almost invariant between genetically diverse groups of HIV isolates. The primary Rev binding site on the RRE is, therefore, a highly attractive target for small-molecule therapeutics since (1) small molecules that bind to and inactivate this site should possess consistent activities among diverse HIV groups, and (2) due to the RRE's dual function, the evolution of resistant variants may be prevented or impeded.

Green's 1993 discovery, showing that aminoglycoside antibiotics, particularly neomycin and tobramycin, competitively inhibit the binding of the Rev protein to the RRE and interfere with Rev function in chronically infected cells,[28] has established a precedent for a novel and attractive antiviral strategy. The case has further been strengthened by Green and co-workers' report demonstrating that RRE variants that are highly neomycin B-resistant generally show decreased affinity for Rev.[80] This observation suggests that resistant HIV mutants harboring a neomycin B-resistant RRE variant, if evolved, are also likely to display compromised replication.

To further evaluate the binding of aminoglycosides to the RRE, Marino and co-workers have developed truncated RRE constructs (RRE IIB) where 2-aminopurine (2-AP), a fluorescent A isoster, replaces A at position 68 or U at position 72 (Figure 10.8).[81] The fluorescent constructs retain their ability to bind Rev and can be used to monitor ligand binding. By measuring fluorescence

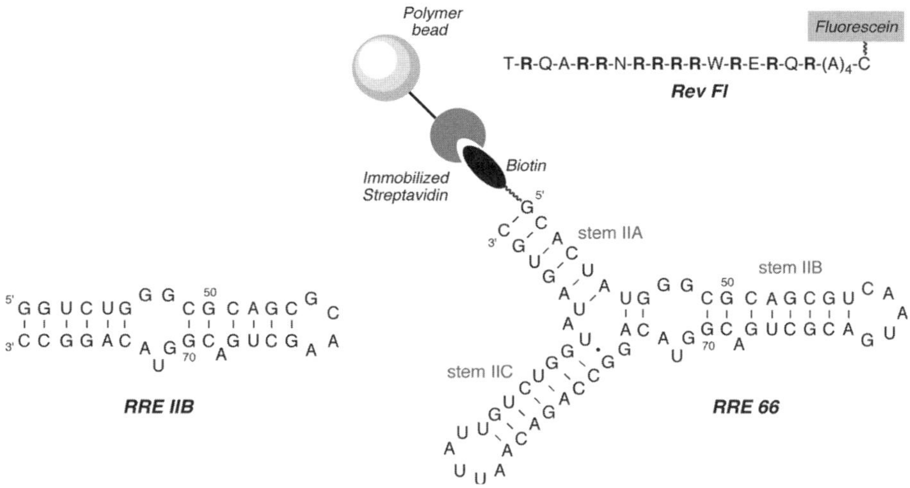

Figure 10.8. *Left*: The sequence of RRE IIB, a fragment used to study RRE–ligand binding. *Right*: The components of our solid-phase assay that utilize the native RRE sequence. See color plates.

changes upon neomycin titration, they determined the presence of a primary high-affinity binding site for neomycin ($K_d \sim 250$ nM) and one or more lower-affinity sites (high micromolar to low millimolar). Competition experiments suggested occupation of the high-affinity site is not disruptive to Rev binding.[81] We were able to reproduce this behavior with related constructs; however, studies with the naturally occurring RRE construct suggest that for this RNA construct, binding of neomycin to its highest-affinity binding site does competitively inhibit Rev binding. It is plausible that the neomycin binding site is altered by the truncations made to the native RRE sequence. Footprinting experiments reported by Green and co-workers in their original paper indicate that neomycin binds near the three-way junction of Stem IIA, IIB, and IIC of the RRE66.[28] This region is dramatically altered by truncation of the RRE where stems IIA and IIC are removed (Figure 10.8). This minimization of the RRE appears, therefore, to affect the mechanism by which neomycin displaces Rev from the RRE. Because the practice of truncating native RNA sequences into minimized "functional" domains is rather common, these conflicting observations should serve as a general lesson for investigators interested in RNA–ligand interactions.

While the binding of aminoglycosides to the RRE provides a proof of principle, their affinity and, in particular, selectivity traits need to be improved for true therapeutic utility. To facilitate the discovery of potent and selective RRE binders, we developed a solid-phase assay.[82] The components of this assembly include (a) insoluble agarose beads (or microtiter plates) covalently modified with streptavidin, (b) a biotinylated RRE fragment, and (c) a fluorescein-labeled Rev fragment (RevFl). Assembly of the three components generates an immobilized ternary complex whereby the biotinylated RRE binds to the beaded streptavidin, and the fluorescein-tagged Rev binds to the RRE (Figure 10.8). Notably, RevFl, where the amino terminus is succinylated and four helix-inducing Ala residues are incorporated between Rev_{34-50} and the fluorescently tagged C-terminus (Figure 10.8), binds to the immobilized RRE with the same affinity as measured in solution by fluorescence anisotropy ($K_d \cong 3$ nM).[83] To determine the inhibitory potency of RRE binders, the immobilized fluorescent Rev–RRE complex is challenged with increasing concentrations of the ligand. Measuring the amount of RevFl displaced generates the necessary information regarding a ligand's affinity. This robust assay reliably performs in the presence of large amounts of competing nucleic acids. Thus, to assess the selectivity traits of RRE binders, their potency is reevaluated in the presence of various competitors.[82] Taken together, these experiments yield indispensable information regarding potential inhibitors of the RRE–Rev interaction. Our studies with the solid-phase assay have been complemented by fluorescence anisotropy measurements where RRE–RevFl association and dissociation are followed in solution.[83] The combination of these tools has allowed us to discover and study several new families of RRE binders, including aminoglycoside dimers and conjugates,[84] as well as cyclized peptides[85] and octahedral Ru^{II} complexes.[86]

Several approaches have been explored in an attempt to increase the affinity and selectivity of aminoglycoside-based ligands to the RRE. Early work by Wong, where neamine was modified at the 5 position on the 2-DOS ring using Ugi-type multicomponent reactions, yielded numerous peptide conjugates with enhanced affinity compared to the aminoglycoside core.[87] Neamine was also modified at the 6′ primary aminomethyl group with various aromatic acids.[88,89] The pyrene derivative showed enhanced affinity to the RRE IIB compared to neamine. Note, however, that modification at this position neutralizes a residue that is likely to be essential for selective RNA recognition. Because selectivity studies were not reported, it is likely that binding is driven, at least partially, by the aromatic fragments rather than the aminoglycosidic moiety. This might result in poor RRE selectivity. A more successful approach has been reported by Yu and co-workers.[63] Conjugating chloramphenicol and linezolid to neomycin via a hydroxyl residue on the D-ribose residue generated potent and apparently more selective RRE binders (Figure 10.5). The same group has also reported the construction of a library of neomycin–dipeptide conjugates. Analogs with moderately improved RRE affinity and higher selectivity were obtained.[90]

The NMR structure of the arginine-rich RNA-binding domain of Rev complexed to a short RRE construct shows purine–purine pairing and a bulged-out pyrimidine residue (see Figure 10.7).[91] Because such noncanonical motifs may constitute favored intercalation sites,[92] we postulated that appending an intercalator to aminoglycosides may generate RNA binders with high RRE binding affinity. Indeed, a neomycin–acridine conjugate, trivially named neo-*S*-acridine (Figure 10.9), has an apparent K_i of 3 nM, which is approximately the same affinity as that of the Rev peptide.[93] Footprinting experiments have established that this novel derivative competitively inhibits Rev–RRE complex formation by binding the RRE in the same region as the arginine-rich Rev_{34-50} peptide.[93]

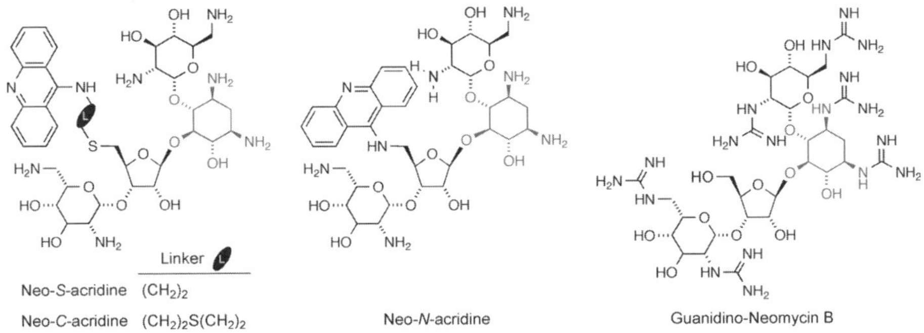

Figure 10.9. The structure of neomycin–acridine conjugates and guanidinoneomycin B.

This conjugate remains one of the strongest competitive inhibitors of Rev–RRE binding reported to date.

As might be expected, neo-*S*-acridine displays mediocre RRE selectivity as it binds competing nucleic acids including tRNA and DNA.[84] Modulating the selectivity of aminoglycoside–intercalator conjugates can be accomplished by varying the length of the linker (Figure 10.9). Thus neo-*C*-acridine, possessing a longer linker displays lower RRE selectivity, while neo-*N*-acridine, a ligand with a shorter linker, exhibits improved RRE selectivity. Compared to neomycin B, acridine conjugates based on tobramycin and kanamycin A have slightly lower RRE affinity, but improved RRE specificity.[84] These observations reflect a rather general trend where sacrificing affinity may improve ligand selectivity.[84]

While structural alterations, such as the ones described above, can improve RRE selectivity, additional RNA-selective scaffolds are essential. Inspiration came again from Nature. Numerous proteins utilize guanidinium groups, typically embedded within arginine-rich domains, for RNA recognition.[94] In contrast to ammonium groups, guanidinium groups are highly basic and planar. They are capable of both stacking and directed H-bonding interactions. We postulated that replacing ammonium with guanidinium groups could enhance the RRE affinity and selectivity of aminoglycosides and related ligands. A new family of RRE ligands, where all the ammonium groups of the natural aminoglycosides have been converted into guanidinium groups, was then synthesized and coined "guanidinoglycosides" (Figure 10.9).[95,96] A substantial increase in RRE affinity has been observed for all guanidinoglycosides when compared to their parent antibiotic.[95] Generally, the RRE specificity has been significantly improved, although it was found to depend on the total number of chargeable groups.[95] Guanidinylation of paromomycin (five basic groups), for example, increases its selectivity for the RRE, while guanidinylation of neomycin (six basic groups) diminishes its target selectivity. This observation supports the general trend outlined above, suggesting that increasing the affinity above a certain threshold may become detrimental to a ligand's RRE selectivity. Despite these challenges, the RRE and the RRE–Rev interaction remain important targets for future antiretroviral agents.

10.7. VIRAL PACKAGING AND RNA DIMERIZATION

Once the viral RNA has been translated and the large polyprotein has been processed to form individual viral proteins, new viral particles begin to form. Packaged in each virion are two copies of fully unspliced viral RNA. To be packaged into the virion, the RNA dimerizes in a highly orchestrated process involving a self-complementary stem-loop interaction. Such HIV-specific RNA events may prove useful for future therapeutic intervention, although they have so far received relatively little attention by medicinal chemists.

In mature virus particles, the viral genome is noncovalently dimerized near its 5′-end.[97] The initial events leading to dimerization have been confined to stem loop 1 (SL1 also denoted DIS for dimerization initiation site), the first out of four characterized stem loops found on the $\approx$ 120-nt-long HIV ψ-packaging site (Figure 10.10).[98] This stem-loop RNA structure initiates dimerization by forming a "kissing loop", where six complementary nucleotides from each loop associate via base pairing (Figure 10.10). This kissing loop complex can rearrange into a more stable extended duplex (Figure 10.10).[99] This process has been shown to be accelerated by the nucleocapsid protein (NCp).[100] Although RNA dimerization and packaging take place in late stages of the viral life cycle, the ability to disrupt the assembly of mature viral particles should constitute a viable antiretroviral therapeutic approach. Since changes in the DIS have been shown to dramatically diminish viral infectivity,[101] small molecules that modulate this essential process may exhibit activity across diverse strains.

The sophisticated and multifaceted process of viral RNA dimerization has been extensively studied, both by structural methods and by biophysical and biochemical approaches.[102] Much less, however, has been reported regarding the binding of low-molecular-weight ligands to this viral RNA. Goodisman and Dabrowiak have demonstrated the binding of aminoglycosides to the packaging region of HIV-1.[103] Footprinting experiments suggested that neomycin's primary binding site is found on SL1. Binding at this site and additional secondary sites induces conformational changes in the RNA. The dimerization initiation site (DIS), found on SL1, is of particular interest. A recent structural analysis of the solved crystal structure reported by Ehresmann and co-workers. [104] has shown the DIS to exhibit remarkable structural resemblance to the prokaryotic ribosomal A-site, the cognate target of aminoglycoside antibiotics. Similarity has been demonstrated not only in the primary, secondary, and tertiary structures, but also in the ability to bind certain aminoglycosides.[105] It would be of interest to explore the design and synthesis of modified aminoglycosides that can distinguish between the A-site and the dimerization initiation site.

10.8. SUMMARY

There is a great need to advance novel strategies for anti-HIV therapeutics. Small organic molecules that target unique viral RNA sites and that can prevent the

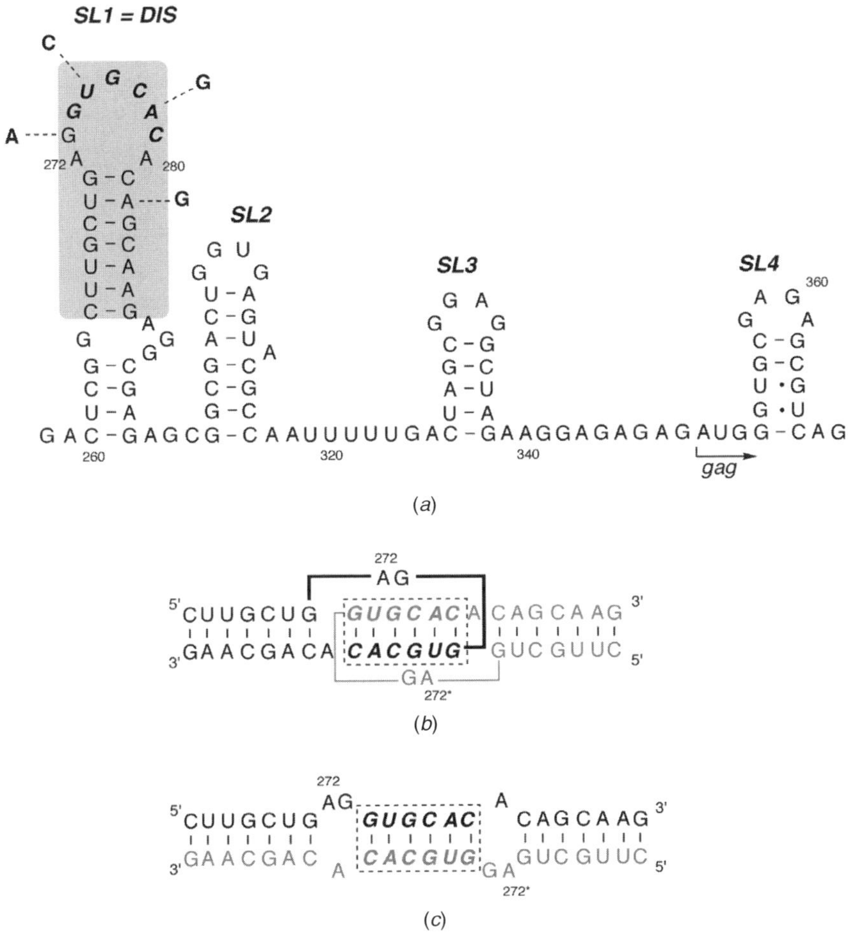

Figure 10.10. (a) Secondary structure of the HIV-1 encapsidation region (ψ-site) including the four stem loops. The main sequence is shown for HIV-1_{Mal} (type A), and the bold residues indicate the sequence found in HIV-1_{Lai} (type B). (b) A 23-nt-long secondary structure of the forming a kissing loop complex. (c) Secondary structure of the rearranged extended duplex.

formation of key regulatory RNA–protein complexes are promising candidates for drug discovery.[20,106] Ultimately, one would like to simultaneously target several RNA sequences that serve key roles in distinct stages of the viral life cycle. This is reminiscent of currently employed combination therapy where RT and protease inhibitors are co-administered.

Our understanding of the modes in which RNA is recognized by aminoglycosides and other low-molecular-weight ligands has advanced significantly in recent years.[107] Nevertheless, the discovery of RNA-specific binders has so far remained an empirical and challenging task.[108] Several reasons account for this: (a) Deep solvent-excluded clefts (resembling active sites in enzymes) are not abundant in most RNA folds. Many intermolecular contacts have to be established to gain substantial affinity, and maintain selectivity. (b) A major contribution by electrostatic interactions in many RNA–ligand complexes creates a delicate interplay between overall charge, affinity, and selectivity.[84] (c) RNA–ligand binding commonly involves a mutual induced-fit process where both the RNA host and the ligand undergo significant conformational changes. This phenomenon, referred to as co-folding or tertiary structure capture,[109] further complicates the design of high affinity and selective binders. The ability of many small molecules (e.g., aminoglycosides) to "remodel" according to the RNA topography and electrostatic potential, along with the utilization of water-mediated contacts, results in "plastic" and potentially promiscuous RNA binding.[36] (d) RNA has only recently been recognized as a valid therapeutic target, and significant structure–activity relationships (SARs) have yet to be established.

Despite the challenges, there is compelling evidence that low-molecular-weight ligands can regulate cellular process at the RNA level. Recent reports illustrating the specific interaction between low-molecular-weight metabolites and mRNAs related to their biosynthetic pathways are encouraging.[110] These observations strongly suggest specific regulatory roles for small molecules at the RNA level and hence the ability of low-molecular-weight RNA ligands to operate in a cellular environment.

A distinction needs to be made between ligands, which target RNA sites serving as riboswitches (such as the A-site), and RNA binders, which need to compete with high-affinity RNA binding proteins (such as Rev). While interfering with riboswitches clearly requires adequate selectivity, high-affinity might not always be essential to exert the biological outcome. In contrast, competing with high-affinity high-molecular-weight ligands imposes significant challenges in terms of achieving high affinity and selectivity. Nevertheless, the results discussed in this chapter demonstrate that (a) aminoglycosides and related molecules can effectively interfere with protein–RNA interactions, (b) synthetic ligands based on aminoglycosides can achieve very high RNA affinity, approaching that of the natural RNA-binding domains on proteins, and (c) the combination of different binding modes (e.g., ionic and intercalation) is a powerful approach for enhancing the RNA affinity of synthetic binders. In this particular case, aminoglycosides can function as RNA-selective scaffold to deliver additional recognition elements to a viral RNA target.[111] It is likely that aminoglycoside antibiotics will continue to serve as important research tools and useful scaffold for the developed of future RNA binders.

NOTES ADDED IN PROOFS

Crystal structures of a DIS construct bound to aminoglycosides have been published.[112] A fluorescence-based assay for evaluating ligand binding to DIS has been reported.[113]

ACKNOWLEDGMENTS

We thank the National Institutes of Health (AI 47673) for generous support, and we thank UARP for an IDEA grant (ID01–SD–027) and student fellowships.

REFERENCES

1. UNAIDS: AIDS epidemic update 2004. Can be downloaded from http://www.unaids.org/wad2004/report.html
2. Steinbrook, R. *N. Engl. J. Med.* **2002**, *347*, 553–554.
3. (a) Mitsuya, H.; Yachoan, R.; Broder, S. *Science* **1990**, *249*, 1533–1544. (b) Georgiev, V. St.; McGowan, J. J. *Ann. New York Acad. Sci.* **1990**, *616*, 1–10. (c) Mohan, P.; Baba, M. (Ed.). *Anti-AIDS Drug Development: Challenges, Strategies and Prospects*; Chur, Switzerland: Harwood Academic Publishers, 1995.
4. Emini, E. A. (Ed.). *The Human Immunodeficiency Virus: Biology, Immunology, and Therapy*; Princeton, N J: Princeton University Press, 2002.
5. Moore, R. D.; Chaisson, R. E. *AIDS*, **1999**, *13*, 1933–1942.
6. Baliant, G. A. *Pharm. Ther.* **2001**, *89*, 17–27.
7. Garcia-Lerma, J. C.; Heneine, W. *J. Clin. Virol.* **2001**, *21*, 197–212. Chen, R.; Quinones-Mateu, M. E.; Mansky, L. M. *Curr. Pharm. Design* **2004**, *10*, 4065–4070.
8. (a) Macchi, B.; Mastino, A. *Pharmacol. Res.* **2002**, *46*, 473–482. (b) Claessens, Y.-E.; Chiche, J.-D.; Mira, J.-P.; Cariou, A. *Critical Care* **2003**, *7*, 226–232.
9. Check, E. *Nature* **2003**, *424*, 361.
10. Gulick, R. M. *Clin. Microbiol. Infect.* **2003**, *9*, 186–196.
11. Hermann, T. *Angew. Chem. Int. Ed.* **2000**, *39*, 1891–1905.
12. Tor, Y. *ChemBioChem* **2003**, *4*, 998–1007 and references therein.
13. Vaishnav, Y. N.; Wong-Staal., F. *Annu. Rev. Biochem.* **1991**, *60*, 577–630. Emerman, M.; Malim, M. H. *Science* **1998**, *280*, 1880–1884.
14. Frankel, A. D.; Young, J. A. T. *Annu. Rev. Biochem.* **1998**, *67*, 1–25.
15. Rosen, C.A.; Pavlakis, G.N. *AIDS* **1990**, *4*, 499–509. Cullen, B. R. *Microbiol. Rev.* **1992**, *56*, 375–394. Green, M. R. In *AIDS Research Reviews III*; Koff, W. C.; Wong-Staal, F.; Kennedy, R. C., Eds.; New York: Marcel Dekker, 1993; pp. 41–55.
16. Karn, J. *J. Mol. Biol.* **1999**, *293*, 235–254. Rana, T.M.; Jeang, K.-T. *Arch. Biochem. Biophys.* **1999**, *365*, 175–185.
17. Pollard, V. W.; Malim, M. H. *Annu. Rev. Microbiol.* **1998**, *52*, 491–532. Hope, T. J. *Arch. Biochem. Biophys.* **1999**, *365*, 186–191.
18. Berkhout, B.; van Wamel, J. L. *J. Virol.* **1996**, *70*, 6723–6732.
19. For earlier reviews, see: Michael, K.; Tor, Y. *Chem. Eur. J.* **1998**, *4*, 2091–2098. Weizman, H.; Tor, Y. In *Carbohydrate-Based Drug Discovery*, Vol. 2; Wong, C.-H., Ed.; Weinheim: Wiley-VCH, 2003; pp. 661–683. Luedtke, N. W.; Tor, Y. In *Small Molecule DNA and RNA Binders: From Synthesis to Nucleic Acid Complexes*; Demeunynck, M.; Bailly, C.; Wilson, D., Eds.; Weinheim: Wiley-VCH, 2003; pp. 18–40.

20. Wilson W. D.; Ratmeyer, L.; Zhao, M.; Ding, D.; McConnaughie, A. W.; Kumar, A.; Boykin, D. W. *J. Mol. Recog.* **1996**, *9*, 187–196. Pearson, N. D.; Prescott, C. D. *Chem. Biol.* **1997**, *4*, 409–414. Hermann, T.; Westhof, E. *Curr. Opin. Biotech.* **1998**, *9*, 66–73. Cheng, A. C.; Calabro, V.; Frankel, A. D. *Curr. Opin. Struct. Biol.* **2001**, *11*, 478–484.
21. Hooper, I. R. In *Aminoglycosides Antibiotics. Handbook of Experimental Pharmacology*, Vol. 62; Umezawa, S.; Hooper, I. R., Eds.; New York: Springer-Verlag, **1982**; pp. 1–35.
22. Busscher, G. F.; Rutjes, F. P. J. T.; van Delft, F L. *Chem. Rev.* **2005**; *105*, 775–792.
23. Pilch, D. S.; Kaul, M.; Barbieri, C. M. *Topi. Curr. Chem.* **2005**, *253*, 179–204.
24. Zembower, T. R.; Noskin, G. A.; Postelnick, M. J.; Nguyen, C.; Peterson, L. R. *Int. J. Antimicrob. Agents* **1998**, *10*, 95–105.
25. Kotra, L. P.; Haddad, J.; Mobashery, S. *Antimicrob. Agents Chemother.* **2000**, *44*, 3249–3256.
26. Moazed, D.; Noller, H. F. *Nature* **1987**, *327*, 389–394.
27. von Ahsen, U.; Davies, J.; Schroeder, R. *Nature* **1991**, *353*, 368–370.
28. Zapp, M. L.; Stern, S.; Green, M. R. *Cell* **1993**, *74*, 969–978.
29. Stage, T. K.; Hertel, K. J.; Uhlenbeck, O. C. *RNA* **1995**, 1, 95–101.
30. Rogers, J.; Chang, A.H.; von Ahsen, U.; Schroeder, R.; Davies, J. *J. Mol. Biol.* **1996**, *259*, 916–925.
31. Mikkelsen N. E.; Brannvall, M.; Virtanen, A.; Kirsebom, L. A. *Proc. Natl. Acad. Sci. USA* **1999**, *96*, 6155–6160.
32. Mei, H.-Y.; Sanders, K. B.; Galan, A. A.; Mack, D. P.; Halim, N. S.; Moreland, D. W.; Czarnik, A. W. *Bioorg. Med. Chem. Lett.* **1995**, *5*, 2755–2760. See also: Mei, H.-Y.; Mack, D. P.; Galan, A. A.; Halim, N. S.; Heldsinger, A.; Loo, J. A.; Moreland, D. W.; Sannes-Lowery, K. A.; Sharmeen, L.; Truong, H. N.; Czarnik, A. W. *Bioorg. Med. Chem.* **1997**, *5*, 1173–1184.
33. Kirk, S. R.; Tor, Y. *Bioorg. Med. Chem.* **1999**, *7*, 1979–1991.
34. Michael, K.; Wang, H.; Tor, Y. *Bioorg. Med. Chem.* **1999**, *7*, 1361–1371.
35. Hermann, T.; Westhof, E. *J. Mol. Biol.* **1998**, *276*, 903–912.
36. Tor, Y. Hermann, T.; Westhof, E. *Chem. Biol.* **1998**, *5*, R277–R283.
37. Verhelst, S. H. L.; Michiels, P. J. A.; van der Marel, G. A.; van Boeckel, C. A. A.; van Boom, J. H. *ChemBioChem.* **2004**, *5*, 937–942.
38. Arts, E. J.; LeGrice, S. F. J., *Prog. Nucl. Acids. Res. Mol. Biol.* **1998**, *58*, 339–393.
39. Isel, C.; Ehresmann, C.; Keith, G.; Ehresmann, B.; Marquet, R., *J. Mol. Biol.* **1995**, *247*, 236–250.
40. Dvorin, J. D.; Malim, M. H. *Curr. Top. Microbiol. Immunol.* **2003**, *281*, 179–208.
41. Kim, S. H.; Sussman, J. L. *Nature* **1976**, *260*, 645–646.
42. Puglisi, E. V.; Puglisi, J. D. *Nature Struc. Biol.* **1998**, *5*, 1033–1036.
43. Robinson, H.; Wang, A. H.-J. *Nucleic Acids Res.* **1996**, *24*, 676–682.
44. Arya D. P.; Xue, L.; Willis, B. *J. Am. Chem. Soc.* **2003**, *125*, 10148–10149.
45. Campbell, E. M.; Hope, T. J. *Adv. Drug Deliv. Rev.* **2003**, *16*, 761–771.
46. Van Maele, B.; Debyser, Z. *AIDS Rev.* **2005**, *7*, 26–43.
47. Jones, K. A.; Peterlin, B. M. *Annu. Rev. Biochem. 1994*, *63*, 717–743.

48. Greene, W. C.; Peterlin, B. M. *Nat. Med.* **2002**, *8*, 673–680.

49. Strebel, K. *AIDS* ***2003***, *17*, 525–534.

50. Daelemans, D.; Afonina, E.; Nilsson, J.; Werner, G.; Kiems, J.; de Clercq, E.; Pavlakis, G. N.; VanDamme A. M. *Proc. Natl. Acad. Sci. USA* **2002**, *99*, 14440–14445.

51. Frankel, A. D.; Pabo, C. O. *Cell* **1988**, *55*, 1189–1193.

52. Green, M.; Loewenstein, P. M. *Cell* **1988**, *55*, 1179–1188.

53. Frankel, A. D. *Curr. Opin. Gen. Dev.* **1992**, *2*, 293–298.

54. Aboul-ela, F.; Karn, J.; Varani, G. *J. Mol. Biol.* **1995**, *253*, 313–332.

55. Hwang, S.; Tamilarasu, N.; Ryan, K.; Huq, I.; Richter, S.; Still, W. C.; Rana, T. M. *Proc. Natl. Acad. Sci. USA* **1999**, *96*, 12997–13002.

56. Mei, H.-Y.; Mack, D. P.; Galan A. A.; Halim, N. S.; Heldsinger, A.; Loo, J. A.; Moreland, D. W.; Sannes-Lowery, K. A.; Sharmeen, L.; Truong, H. N.; Czarnik, A. W. *Bioorg. Med. Chem.* **1997**, *5*, 1173–1184.

57. Froeyen, M.; Herdewijn, P. *Curr. Topi. Med. Chem.* **2002**, *2*, 1123–1145. Krebs, A,; Ludwig, V.; Boden, O.; Göbel, M. W. *ChemBioChem.* **2003**, *4*, 972–978. Baba, M. *Curr. Top. Med. Chem.* **2004**, *4*, 871–882.

58. See also: Mei, H.-Y.; Mack, D. P.; Galan, A. A.; Halim, N. S.; Heldsinger, A.; Loo, J. A.; Moreland, D. W.; Sannes-Lowery, K. A.; Sharmeen, L.; Truong, H. N.; Czarnik, A W. *Bioorg. Med. Chem.* **1997**, *5*, 1173–1184.

59. (a) Hamy, F.; Brondani, V.; Florsheimer, A.; Stark, W.; Blommers, M.J.; Klimkait, T. *Biochemistry* **1998**, *37*, 5086–5095. (b) Dassonneville, L.; Hamy, F.; Colson, P.; Houssier, C.; Bailly, C. *Nucleic Acids Res.* **1997**, *25*, 4487–4492. (c) Hamy, F.; Felder, E.R.; Heizmann, G.; Lazdins, J.; Aboul-ela, F.; Varani, G.; Karn, J.; Klimkait, T. *Proc. Natl. Acad. Sci. USA*, **1997**, *94*, 3548–3553. (d) Klimkait, T.; Felder E.R.; Albrecht, G.; Hamy F. *Biotechnol. Bioeng.* **1998–99**, *61*, 155–168. (e) Mei, H.-Y.; Cui, M.; Heldsinger, A.; Lemrow, S. M.; Loo, J. A.; Sannes-Lowery, K. A.; Sharmeen, L.; Czarnik, A.W. *Biochemistry* **1998**, *37*, 14204–14212. (f) Du, Z.; Lind, K. E.; James, T. L. *Chem. Biol.* **2002**, *9*,707–712.

60. Litovchick A.; Evdokimov, A. G.; Lapidot, A. *FEBS Lett.* **1999**, *445*, 73–79. Litovchick, A.; Evdokimov, A. G.; Lapidot A *Biochemistry* **2000**, *39*, 2838–2852. Litovchick, A,; Lapidot, A.; Eisenstein, M.; Kalinkovich, A.; Borkow, G. *Biochemistry* **2001**, *40*, 15612–15623.

61. Cabrera, C.; Gutierrez, A.; Barretina, J.; Blanco, J.; Litovchick, A.; Lapidot, A.; Clotet, B.; Este, J. A. *Antiviral Res.* **2002**, *53*, 1–8.

62. Riguet, E.; Tripathi, S.; Chaubey, B.; Désiré, J.; Pandey, V. N.; Décout, J.-L. *J. Med. Chem.* **2004**, *47*, 4806–4809.

63. Lee, J.; Kwon, M.; Lee, K. H.; Jeong, S.; Hyun, S.; Shin, K. J.; Yu, J. *J. Am. Chem. Soc.* **2004**, *126*, 1956–1957.

64. Gelus, N.; Bailly, C.; Hamy, F.; Klimkait, T.; Wilson, W. D.; Boykin, D. W. *Bioorg. Med. Chem.* **1999**, *7*, 1089–1096.

65. Gelus, N.; Hamy, F.; Bailly, C; *Bioorg. Med. Chem.* **1999**, *7*, 1075–1079.

66. For other low-molecular-weight TAR binders, see, for example: Yu X, Lin W, Li J, Yang M. *Bioorg. Med. Chem. Lett.* **2004**, *14*, 3127–3130. Renner, S.; Ludwig, V.; Boden, O.; Scheffer, U.; Göbel, M. W.; Schneider, G.; *Chembiochem.* **2005**, *6*, 1119–1125. Lee, C. W.; Cao, H.; Ichiyama, K.; Rana, T. M. *Bioorg. Med. Chem. Lett.* **2005**, *15*, 4243–4246.

67. Wang, X.; Huq, I.; Rana, T. M. *J. Am. Chem. Soc.* **1997**, *119*, 6444–6445. Tamilarasu, N.; Huq, I.; Rana, T. M. *J. Am. Chem. Soc.* **1999**, *121*, 1597–1598. Tamilarasu, N.; Huq, I.; Rana, T. M. *Bioorg. Med. Chem. Lett.* **2000**, *10*, 971–974. Tamilarasu, N.; Huq, I.; Rana, T. M. *Bioorg. Med. Chem. Lett.* **2001**, *11*, 505–507.
68. Blount, K. F.; Tor, Y. *Nucleic Acid Res.*, **2003**, *31*, 5490–5500.
69. Faber, C.; Sticht, H.; Schweimer, K.; Rosch, P. *J. Biol. Chem.* **2000**, *275*, 20660–20666.
70. Kierzek, R.; Li, Y.; Turner, D. H.; Bevilacqua, P. C. *J. Am. Chem. Soc.* **1993**, *115*, 4985–4992.
71. Fourmy, D., Recht, M., Blanchard, S., Dalquist, K., and Puglisi, J.D., *Science* **1996**, *274*, 1367–1371.
72. Vicens, Q.; Westhof, E. *Structure* **2001**, *9*, 647–658. See also: Vicens, Q.; Westhof, E. *Biopolymers* **2003**, *70*, 42–57.
73. Blount, K. F.; Zhao, F.; Hermann, T.; Tor, Y. *J. Am. Chem. Soc,* **2005**, *127*, 9818–9829.
74. Zhao, F.; Zhao, Q.; Blount, K. F.; Han, Q.; Tor, Y.; Hermann, T. *Angew. Chem.*, **2005**, *44*, 5329–5334.
75. Hammarskjöld, M.-L. *Cell Dev. Biol.* **1997**, *8*, 83–90.
76. Holland, S. M.; Chavez, M.; Gerstberger, S.; Venkatesan, S. *J. Virol.*, **1992**, *66*, 3699–3706. Tilley, L. S.; Malim, M. H.; Tewary, H. K.; Stockley, P. G.; Cullen, B. R. *Proc. Natl. Acad. Sci. USA* **1992**, *89*, 758–762.
77. Battiste, J. L.; Tan, R.; Frankel, A. D.; Williamson, J. R. *Biochemistry* **1994**, *33*, 2741–2747. Bartel, D. P.; Zapp, M. L.; Green, M. R.; Szostak, J. W. *Cell* **1991**, *67*, 529–536.
78. Schaal, H.; Klein, M.; Gehrmann, P.; Adams, O.; Scheid, A. *J. Virol.* **1995**, *69*, 3308–3314.
79. Phuphuakrat, A.; Auewarakul, P. *AIDS Res. Hum. Retrovir.* **2003**, *19*, 569–574.
80. Werstuck, G.; Zapp, M. L.; Green, M. R. *Chem. Biol.* **1996**, *3*, 129–137.
81. Lacourciere, K. A.; Stivers, J. T.; Marino, J. P. *Biochemistry* **2000**, *39*, 5630–5641.
82. Luedtke, N. W.; Tor, Y. *Angew. Chem. Int. Ed.* **2000**, *39*, 1788–1790.
83. Luedtke, N. W.; Tor, Y. *Biopolymers/Nucleic Acid Sci.* **2003**, 70, 103–119.
84. Luedtke, N. W.; Liu, Q.; Tor, Y. *Biochemistry* **2003**, *42*, 11391–11403.
85. Friedler, A.; Friedler, D.; Luedtke, N. W.; Tor, Y.; Loyter, A.; Gilon, C. *J. Biol. Chem.*, **2000**, *275*, 23783–23789.
86. Luedtke, N. W.; Hwang, J. S.; Glazer, E. C.; Gut, D.; Kol, M.; Tor, Y. *ChemBioChem.* **2002**, *3*, 766–771.
87. Park, W. K. C.; Auer, M.; Jaksche, H.; Wong, C.-H. *J. Am. Chem. Soc.* **1996**, *118*, 10150–10155.
88. Hamasaki, K.; Woo, M.-C.; Ueno, A. *Tetrahedron Lett.* **2000**, *41*, 8327–8332.
89. See also: Wang, Y.; Hamasaki, K.; Rando, R. R. *Biochemistry* **1997**, *36*, 768–779.
90. Ahn, D.-R.; Yu, J. *Bioorg. Med. Chem.* **2005**, *13*, 1177–1183.
91. Battiste, J. L.; Mao, H.; Rao, N. S.; Tan, R.; Muhandiram, D. R.; Kay, L. E.; Frankel, A. D.; Williamson, J. R. *Science*, **1996**, *273*, 1547–1551.
92. Wilson, W.B.; Ratmeyer, L.; Cegla, M.T.; Spychala, J.; Boykin, D.; Demeunynyck, M.; Lhomme, J.; Krishnan, G.; Kennedy, D.; Vinayak, R.; Zon, G. *New. J. Chem.* **1994**, *18*, 419–423.

93. Kirk, S. R.; Luedtke, N. W.; Tor, Y. *J. Am. Chem. Soc.* **2000**, *122*, 980–981.
94. Weiss, M. A.; Narayana, N. *Biopolymers/Nucleic Acid Sci.* **1999**, *48*, 167–180. See also: Bayer, T. S.; Booth, L. N.; Knudsen, S. M.; Ellington, A. D. *RNA* **2005**, *11*, 1848–1857.
95. Luedtke, N. W.; Baker, T. J.; Goodman, M.; Tor, Y. *J. Am. Chem. Soc.* **2000**, *122*, 12035–12036.
96. Baker, T. J.; Luedtke, N. W.; Tor, Y.; Goodman, M. *J. Org. Chem.* **2000**, 65 9054–9058.
97. Paillart, J.-C.; Shehu-Xhilaga, M.; Marquet, R.; Mak, J. *Nat. Rev. Microbiol.* **2004**, *2*, 461–472.
98. Berkhout, B.; van Wamel, J. L. *J. Virol.* **1996**, *70*, 6723–6732.
99. Huthoff, H.; Berkhout, B. *Biochemistry* **2002**, *41*, 10439–10445.
100. Muriaux, D.; De Rocquigny, H.; Roques, B.-P.; Paoletti, J. *J. Biol. Chem.* **1996**, *271*, 33686–33692.
101. Laughrea, M.; Jette, L.; Mak, J.; Kleiman, L.; Liang, C.; Wainberg, M. A. *J. Virol.* **1997**, *71*, 3397–3406. Shen, N.; Jette, L.; Liang, C.; Wainberg, M. A.; Laughrea, M. *J. Virol.* **2000**, *74*, 5729–5735.
102. Selected references: Mujeeb, A.; Clever, J. L.; Billeci, T. M.; James, T. L.; Parslow, T. G. *Nat. Struct. Biol.* **1998**, *5*, 432–436. Jossinet, F.; Paillart, J. C.; Westhof, E.; Hermann, T.; Skripkin, E.; Lodmell, J. S.; Ehresmann, C.; Ehresmann, B.; Marquet, R. *RNA* **1999**, *5*, 1222–1234. Lodmell, J. S.; Ehresmann, C.; Ehresmann, B. Marquet, R. *J. Mol. Biol.* **2001**, 311, 475–490.
103. McPike, M. P.; Goodisman, J.; Dabrowiak, J. C. *Bioorg. Med. Chem.* **2002**, *10*, 3663–3672. McPike, M. P.; Sullivan, J. M.; Goodisman, J.; Dabrowiak, J. C. *Nucleic Acids Res.* **2002**, *30*, 2825–2831. Sullivan, J. M.; Goodisman, J.; Dabrowiak, J. C. Bi*oorg. Med. Chem. Lett.* **2002**, *12*, 615–618.
104. Ennifar, E.; Walter, P.; Ehresmann, B.; Ehresmann, C.; Dumas, P. *Nat. Struct. Biol.* **2001**, *8*, 1064–1068.
105. Ennifar, E.; Paillart, J.-C.; Marquet, R.; Ehresmann, B.; Ehresmann, C.; Dumas, P.; Walter, P. *J. Biol. Chem.* **2003**, *278*, 2723–2730.
106. Hermann, T.; Tor, Y. *Exp. Opin. Ther. Pat.* **2005**, *15*, 49–62.
107. Tor, Y., *Biopolym./Nucleic Acid Sci.* **2003**, *70*, 1–3.
108. Hermann, T. *Biopolym./Nucleic Acid Sci.* **2003**, *70*, 4–18.
109. Induced fit is common in RNA–protein recognition, see: J. R. Williamson, *Nat. Struct. Biol.* **2000**, *7*, 834–837; N. Leulliot, G. Varani, *Biochemistry* **2001**, *40*, 7947–7956.
110. Nahvi, A.; Sudarsan, N.; Ebert, M. S.; Zou, X.; Brown, K. L.; Breaker, R. R. *Chem. Biol.* **2002**, *9*, 1043–1049.
111. See, for example: J. Boer, J.; Blount, K. F.; Luedtke, N. W.; Elson-Schwab, L.; Tor, Y. *Angew. Chem. Int. Ed.* **2005**, *44*, 927–932.
112. Ennifar, E.; Paillart, J. C.; Bodlenner, A.; Walter, P.; Weibel, J. M.; Aubertin, A. M.; Pale, P.; Dumas, P.; Marquet, R. *Nucleic Acids Res.* **2006**, *34*, 2328–2339.
113. Tam, V. K.; Kwong, D.; Tor, Y. *J. Am. Chem. Soc.* **2007**, *129*, in press; published online on Feb. 24, 2007 (DOI: 10.1021/ja0675797).

11

NOVEL TARGETS FOR AMINOGLYCOSIDES

DEV P. ARYA, NICHOLAS SHAW, AND HONGJUAN XI

Laboratory of Medicinal Chemistry, Department of Chemistry, Clemson University, Clemson, SC 29634

11.1. INTRODUCTION

Over the past decade, several nucleic acid structures other than the 16S rRNA A-site have been identified as aminoglycoside targets. Virtually all of these novel targets are RNA structures, and this affinity of aminoglycosides for various RNA structures has been the subject of numerous reviews.[1–3] Over the past few years, a deeper understanding of RNA recognition has been elucidated and the list of structures that bind aminoglycosides has been expanded to include DNA and even proteins. Several chapters within this book detail the recognition of the 16S rRNA A-site. This chapter presents targets apart from the 16S A-site, which have been discovered for their binding to aminoglycosides. Novel chemical approaches leading to selective recognition of such structures are also presented.

Since the discovery of aminoglycosides by Selman Waksman more than 50 years ago,[4] most attention has focused on their binding to rRNA[5] and, more recently, to other RNA structures.[6] The flexible and polycationic nature of the aminoglycoside antibiotics allows them to preferentially bind to prokaryotic ribosomal RNA but also allows binding to a variety of unrelated RNA and nucleic acids.[7] A summary of new nucleic acid and related structures that have been targeted using aminoglycosides will be presented here. These findings have suggested alternative pathways that can be used as novel targets for these drugs.[8–13] The literature of the past decade is replete with large numbers of different RNA structures that have been shown to bind to aminoglycosides. The reason for this RNA-centered development was understandable: Aminoglycosides exhibit their antibacterial action through rRNA binding and show high binding affinity (K_d in the nanomolar range) to such RNAs. Because RNA rapidly became a target for developing new drugs, exploring the potential of these functional RNAs was a logical extension of aminoglycoside based drug development research. What was remarkable, however, was the almost complete absence of reports on the non-RNA structures targeted by aminoglycosides

11.2. A-FORM NUCLEIC ACIDS: A BASIS FOR AMINOGLYCOSIDE SELECTIVITY TO NON-RNA TARGETS

The flexible and polycationic nature of the aminoglycoside antibiotics (Figure 11.1) allows them to preferentially bind to prokaryotic ribosomal RNA. Aminoglycosides are also able to bind to a variety of unrelated RNAs, group I introns,[9] a hammerhead ribozyme,[14] the RRE transcriptional activator region from HIV[15–17] (which contains the binding site for the Rev protein), the 5′-untranslated region of thymidylate synthase, targets for important enzymes such as ribonuclease H and reverse mRNA,[18] a variety of RNA aptamers from *in vitro* selection,[19,20] and human mRNAs.[21] These RNA structures are being investigated as novel targets for aminoglycoside-based therapeutic approaches. In 1996, Robinson and Wang reported an A-form DNA induction ability of neomycin {similar to spermine and $CO(NH_3)_6^{3+}$}.[22] We have recently shown that the A-form specificity of aminoglycosides explains their binding to various RNA forms as well as to other nucleic acid

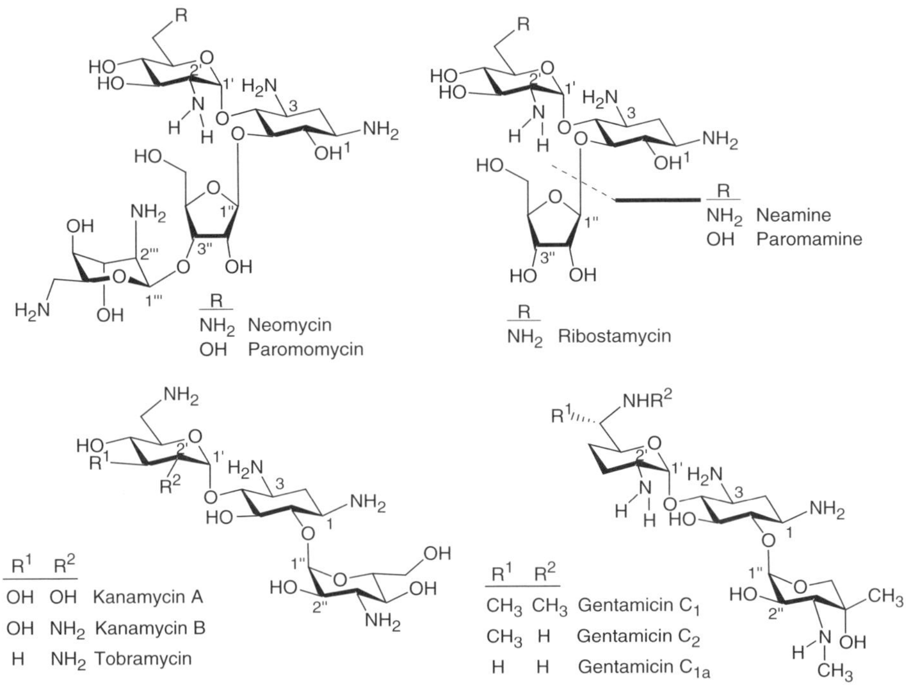

Figure 11.1. Representative chemical structures of aminoglycoside antibiotics.

targets previously reported by us.[7] Aminoglycoside specificity (neomycin, in high-nanomolar–low-micromolar-μM range) has been suggested to be for nucleic acid forms that show features characteristic of an A-type conformation (RNA triplex[23] DNA–RNA hybrid duplex,[24] RNA duplex,[25] DNA triplex,[23,26–28] A-form DNA duplex,[22] and DNA tetraplex),[29] rather than for just naturally occurring RNA. Groove recognition of these nucleic acids opens up possibilities of novel RNA and non-RNA therapeutic targets for aminoglycoside-based drug development in cancer, HIV, and antimicrobial therapy.

11.2.1. A-Form DNA: Structural Features and Biological Importance

The polymorphism of DNA was noticed early after the discovery of its ability to maintain a double-helical structure.[30] The conformations of DNA can be limited to two major distinctions, A-DNA and B-DNA, although other less-well-known structures like Z-DNA do exist.[31] Both structures consist of identical topology and hydrogen bonding patterns, but they differ largely in their overall shape (Figure 11.2). B-DNA, long believed to be the dominant biological conformation, implements water molecules and biological cations appropriately within its

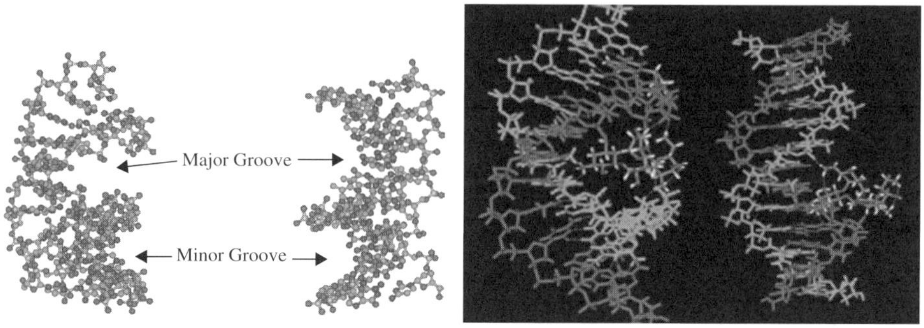

Figure 11.2. Globular conformations of an A-type duplex (left) and a B-type duplex (right), generally seen for RNA·RNA and DNA·DNA duplexes, respectively. Charge and shape complementarity of neomycin to the A-form major groove: Computer models of (left) neomycin docked in the major groove of A-form DNA, and (right) neomycin buried in the B-form major groove. See color plates.

structure. A-DNA, on the other hand, requires dehydrated conditions. The transition of B-A DNA is a reversible and cooperative process,[32] in which the A-form is considered the high-energy state. The underlying factors for this instability have been addressed, with little success.[33,34]

The pseudorotational circle[35] depicted in Figure 11.3 describes the possible nucleotide monomers. All distinct conformations (i.e., envelope, E, and twist, T, conformations) are separated by 18°, with a superscript designating an atom above the plane described by the remaining three or four atoms of the furanose ring and a subscript designating an atom below. The two dominating furanose conformations for DNA (B-form) and RNA (A-form) generally give rise to two different duplex forms, as depicted in Figure 11.1. Apart from having all furanose conformations in an S-type (*south*-type, C2-*endo*, ${}^{2}E$) conformation, the B-type helix, seen in solution for DNA, is characterized by (1) a 10-base-pair-per-helix turn almost perpendicular to the helix axis and (2) a distinct difference in the width of the major and the minor grooves.[31,36] In contrast, RNA adopts an A-type helix, in which the furanose conformations are all N-type (*north*-type, C3-*endo*, ${}^{3}E$), with (1) an 11-base-pair-per-helix turn tilted 20° with respect to the helix axis and (2) the minor and major grooves almost equally wide, but with different depths.[31,36]

The biological relevance of B–A transitions in DNA is of great importance.[37] Protein–DNA association results in an induction of partial deformation in the B-form helix to A-form helix. The cyclic AMP receptor protein is a specific example found to induce a B–A-like transition in its DNA target.[32] Perhaps the most attractive feature of A-DNA is its conformational similarity to duplex RNA. RNA–DNA hybrids, often formed during transcription, are also A-form.

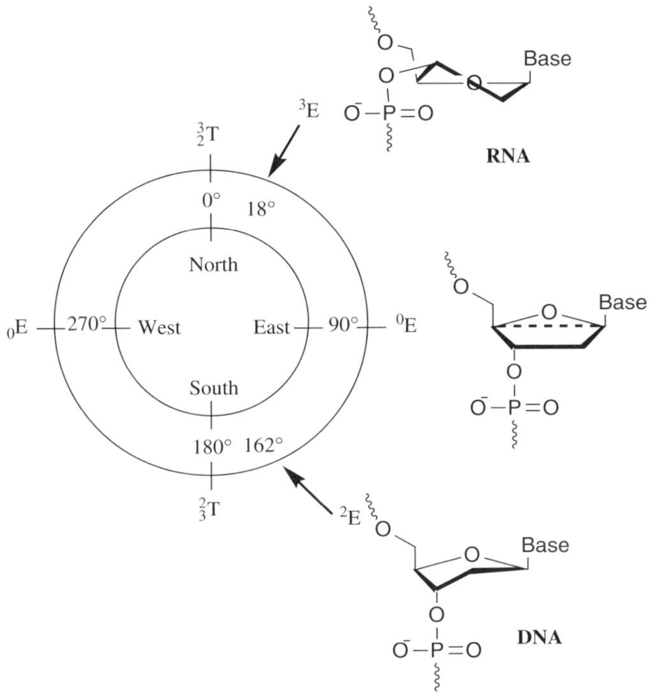

Figure 11.3. The pseudorotation cycle showing the relation between the pseudorotation angle *P* (0–360°); calculated from the five torsional angles of the furanose.

Therefore, the stabilization of these A-like conformations can be of critical importance for increased or decreased proliferation of DNA and RNA.

Native DNA, which comprises the genetic information of all known free organisms, adopts mostly B-form under physiological conditions because it is associated with high humidity in fibers or aqueous solutions. However, in living organisms it is important to switch B-DNA into A-form since constitutive conformation of double-stranded RNA is predominantly A-form. Because RNA probably preceded DNA in evolution,[38] the basic mechanisms of genetic information copying are likely to have evolved on A-form rather than B-form. In fact, the template DNA is induced by many polymerases into A-form at positions of genetic information copying in the microenvironment. Thus, DNA switching into A-form may influence replication and transcription of the existing genomes.

Nearly all structurally characterized DNA polymerases are found to induce B–A transition in DNA bases adjacent to nucleotide incorporation.[37] This induced A-conformation facilitates sufficient access for the protein to desirable contact points in the DNA minor groove. Negative base-pair displacement in the A-form helix results in a shallower minor groove, thus exposing contact atoms such as the O2 and N3 in pyrimidines and purines, respectively. With the helix now more compact, and somewhat stiffer than B-form, replication accuracy is ensured. The limited conformational freedom (higher energy) provides a natural check for base mispairing.[37] Proteins known to induce the A-form include HIV-1 reverse transcriptase,[39–41] Polymerase β,[42] Taq polymerase,[43] *Bacillus* polymerase I,[44] T7 polymerase,[45] DNase I,[46,47] I-PpoI homing endonuclease,[48] PvuII restriction endonuclease,[49–51] EcoRV endonuclease,[52–54] HhaI methyltransferase,[55] and the chromosomal protein Sac7d.[56]

Aminoglycoside A-form nucleic acid interactions have an underlying importance to both the area of drug development and the fundamental understanding of nucleic acid recognition. First, novel nucleic-acid therapeutic targets can be identified to better understand the thermodynamics of molecular recognition involved in aminoglycoside specificity. We have already initiated such a program for use in the development of sequence-specific antimicrobial agents targeting novel RNA and DNA sequences.[11–13] Second, as opposed to B-form DNA recognition, very few small molecules (multivalent cations)[32,37,63] are known to select for A-form structural features. Aminoglycosides present a novel scaffold for groove recognition of these A-form structures. A better understanding of aminoglycoside binding and selectivity can lead to the development of less toxic and more potent antibiotics.

As Table 11.1 shows, as canonical B-form DNA duplex poly(dA-dT)·poly(dA-dT) migrates to an A-form RNA duplex poly(rA)·poly(rU), there is an increase in K_a of approximately four orders of magnitude. The intermediate conformations, from GC-rich nucleic acids to DNA-RNA hybrids, show intermediate values of K_a. The higher observed affinities obtained for polynucleotides using calorimetric methods include the binding induced protonation in addition to the intrinsic enthalpies of drug binding. The nanomolar affinity of the ribosomal A-site was also obtained using similar calorimetric methods under similar salt and pH conditions. It is important to note that the variable low affinities of numerous other RNA structures (Table 11.1) that have been shown to bind aminoglycosides, since these values were obtained using different methods and under different salt conditions. Inclusion of $MgCl_2$ in most of these studies with RNA motifs is likely responsible for the lower affinities reported for different RNA structures and somewhat higher affinities with polynucleotides.

11.3. B-FORM DNA RECOGNITION BY NEOMYCIN CONJUGATES

To date, few carbohydrate interactions with DNA have involved interactions within the major groove. Among the classes of compounds known to exhibit DNA binding are enediyne antibiotics, anthracyclines, pluramycins, indolocarbazoles, and aureolic acids.[8] Of the limited number of carbohydrates known for DNA binding, only a select few have displayed major groove contacts. These

TABLE 11.1. Association Constants of Various Nucleic Acids with Neomycin. From top to bottom, various polynucleotides with their conformational preference (B- to A-form) are listed in 10 mM sodium cacodylate, 100 mM NaCl, 0.1 mM EDTA, pH 6.8. RNA targets that have previously been shown to bind neomycin are also listed. These targets are examples of RNA secondary structures that show high-affinity binding to aminoglycosides. Solution conditions for RNA targets vary as shown

	Nucleic Acid	$K_{a(20°C)}$ (M^{-1})
Polynucleotides (Top: B-form to bottom A-form)	1. poly(dA-dT)·poly(dA-dT) B-form DNA-DNA	$(3.9 \pm 0.4) \times 10^4$
	2. poly(dA)·poly(dT) DNA-DNA	$(9.3 \pm 0.7) \times 10^4$
	3. poly(dG-dC)·poly(dG-dC) DNA	$(1.0 \pm 0.5) \times 10^5$
	4. poly(dA)·2poly(dT) DNA-DNA-DNA	$(7.1 \pm 0.4) \times 10^6$
	5. A-form GC oligomer:5′-d($A_2G_{15}C_{15}T_2$)	$(1.2 \pm 0.2) \times 10^7$;
	6. poly(rA)·poly(dT) RNA-DNA	$(1.7 \pm 0.1) \times 10^6$
	7. poly(dA)·poly(rU) DNA-RNA	$(3.1 \pm 0.2) \times 10^7$
	8. poly(rA)·poly(rU) A-form RNA-RNA	$(2.9 \pm 0.2) \times 10^8$
RNA secondary structures with pockets that favor aminoglycoside binding.	9. TAR (HIV)[57] 10 mM Tris-HC1, pH 7.5, 50 mM NaC1, 5 mM MgC12,	1.09×10^6
	10. RRE (HIV)[58] 140 mM NaCl, 5 mM KCl, 1 mM MgC12, 1 mM CaC12, and 20 mM HEPES (pH 7.4)	4.0×10^6
	11. Hammerhead ribozyme (HH16)[14] 50 mM Tris-HCl buffer, pH 7.3, 10 mM MgC12	7.41×10^4
	12. $tRNA^{Phe}$[59] 50 mM Tris-HCl buffer (pH 7.0) 100 mM NaCl	$IC_{50} = 100$ μM
	13. RNase P RNA[60] 50 mM Tris-HCl, pH 7.2, 10 mM spermidine,10 mM MgC12	$K_i = 35$ μM
	14. Hepatitis delta virus ribozyme[61] 40 mM Tris-HCl pH 7.2 and 0.4 mM spermidine	$K_i = 28$ μM
	15. Group I intron[62]	$K_i = 0.5$ μM
	16. 16S A-site RNA highest affinity site	$(9.1 \pm 1.44) \times 10^8$

interactions are often assisted by an intercalative structure, as displayed in DNA cleavage agents (neocarzinostatin), alkylating agents (altromycin B), or tandem intercalative-groove binding ligands (nogalamycin, respinomycin, NB-506). We have shown that neomycin conjugated to minor groove binders can be used to target B-DNA. In order to investigate the effect of neomycin binding to B-DNA,

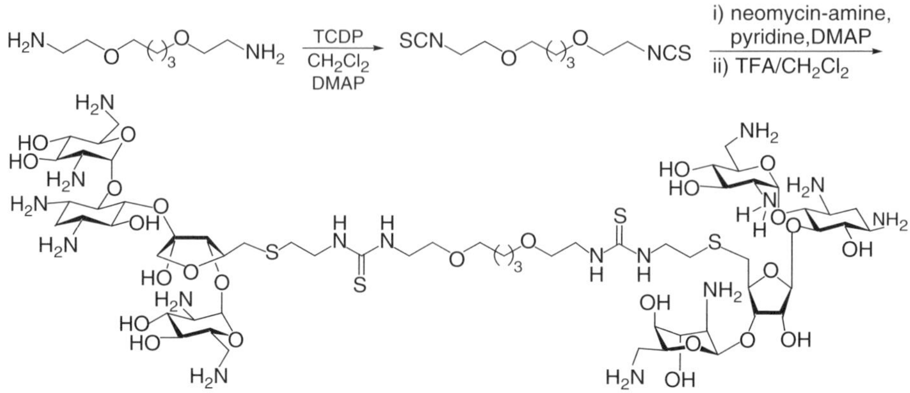

Figure 11.4. A novel neomycin–neomycin dimer, a potent DNA duplex groove binder.

we have covalently linked neomycin with Hoechst 33258, a well-known minor groove binding ligand that has particular affinity for A-T base pairs known to adopt B-form structure. The design, synthesis, and spectroscopic studies of novel neomycin–Hoechst 33258 and neomycin–neomycin conjugates for recognition of B-DNA has been reported.[10] A neomycin dimer (Figure 11.4) has previously been suggested to bind in the DNA major groove with high affinities[8,64]. These results illustrated the preliminary successes in extending aminoglycoside recognition to include the major groove of B-DNA.

NH1 and NH2 (Figure 11.5) were shown to significantly enhance the thermal stability of B-DNA. NH1 was found to exhibit a more significant enhancement than both NH2 and Hoechst 33258 alone. UV melting experiments with both polymeric and oligomeric DNA indicated significant shifts in T_m when compared with samples in the absence of ligand. Triplex DNA inhibition by both NH1 and NH2 suggests dual groove binding. UV melting experiments of NH1 and NH2 under triplex-forming conditions indicated an inhibition of TFO binding, likely due to neomycin binding in the major groove. Furthermore, viscosity experiments indicated neomycin (in NH1 and NH2) binding to B-DNA involves a groove binding mode. The linker between Hoechst and neomycin plays an important role in delivering both binding moieties to the grooves. Both UV melting and fluorescence data indicate significant enhancement of DNA binding for NH1 over NH2. Conjugates of different minor groove binders and linker sizes can then perhaps be designed to target a structure of preference and should aid in the development of even more selective and potent conjugates. Since neomycin affinity is much higher for A-form than B-form DNA, this work is expected to lead to the design of dual-groove binding A-form DNA specific molecules.

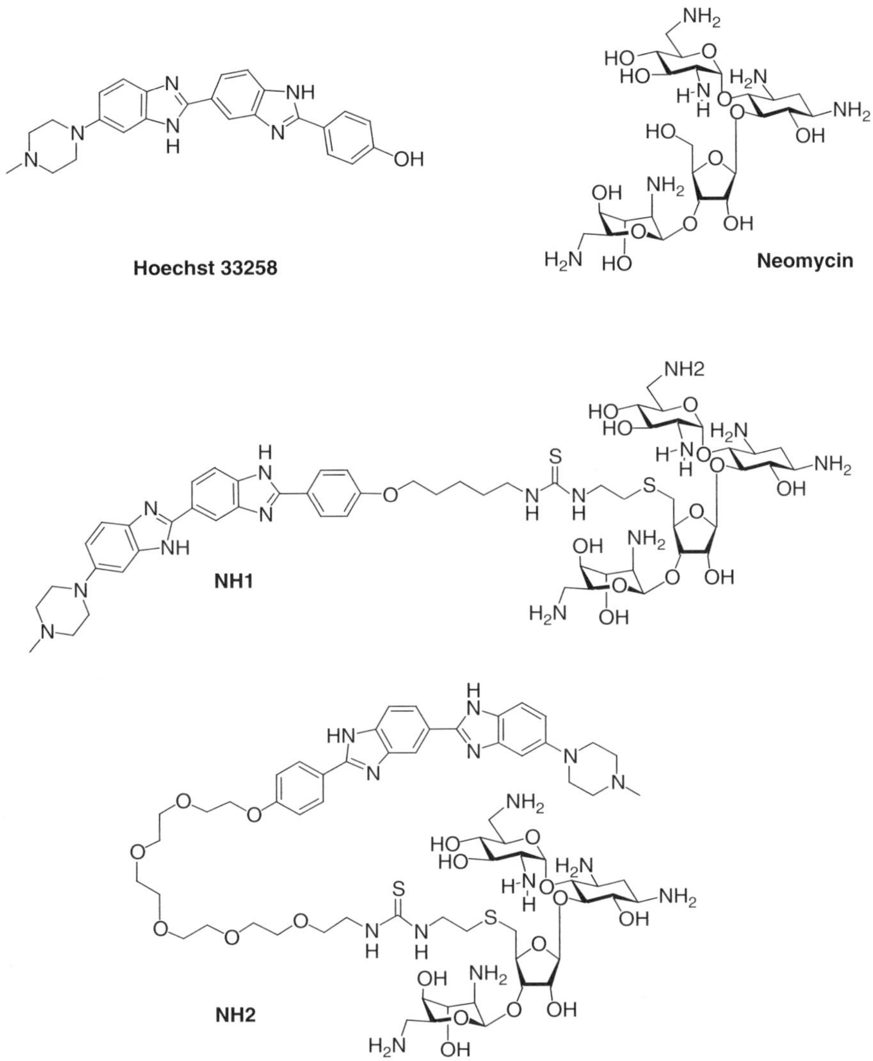

Figure 11.5. Structures of Hoechst 33258, neomycin, and conjugates NH1 and NH2.

11.4. DNA TRIPLEX

The association of homopyrimidine·homopurine stretches of duplex DNA are known as receptors for triplex formation by major groove association of a TFO.[65] TFO recognition of duplex DNA can be exploited by inducing transcription inhibition, site-directed mutagenesis, or recombination. Another attractive feature of triplex DNA is the feature of H–DNA, an intramolecular-forming triplex, found in biological systems. H–DNA formation is found within mirror repeats of homopyrimidine·homopurine stretches in plasmid DNA, in which triplex formation requires a negative supercoiling (dissociation of symmetrical duplex stretch with folding back of a single strand to form triplex).[66] Because the constrained, bent DNA conformation that occurs upon H–DNA formation is often observed in observation with regulatory proteins, the formation of such structures may represent a form of molecular switch in controlling gene expression. The targeting of triplex DNA is thus of obvious interest. However, triplex formation is thermodynamically and kinetically less favorable than duplex–TFO dissociation. Therefore, the driving force for utilization of TFO-based recognition for therapeutic purposes is the development or discovery of ligands that stabilize and kinetically favor the formation of triplex structures in a specific fashion.

Neomycin, among a series of aminoglycoside antibiotics studied, has been shown to significantly stabilize DNA triplexes.[23,24,26,27,67] Neomycin was also shown to enhance the rate of TFO–duplex association.[23] The binding and stabilization of DNA triplexes by neomycin is unique among other triplex-stabilizing ligands in that very weak DNA duplex binding occurs, leading to 100–1000 fold differences in affinities, under identical conditions. Molecular modeling has suggested neomycin binding within the Watson–Hoogsteen groove and that it is neomycin's charge and shape complementarity that drives triplex recognition over duplex (Figure 11.6).[26] All previously discovered triplex-stabilizing ligands have also displayed a large degree of duplex stabilization. Moreover, neomycin is the

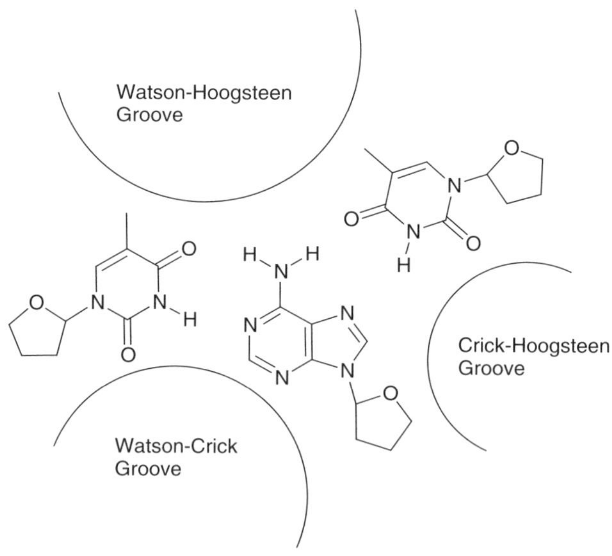

Figure 11.6. A: TAT triplet showing the positioning of the three grooves and the neomycin binds to the larger triplex Watson–Hoogsteen (W–H) groove.

first groove-binding ligand to exhibit DNA triplex stabilization (the absence of a fused, planar ring system eliminates the structural possibilities for intercalation).

11.4.1. Dual Recognition of DNA Triple Helix by BQQ-Neomycin

Both intercalators and groove binders can stabilize the DNA, RNA, or hybrid triplex to a different degree. The most potent triplex specific intercalators and groove binders have been combined to obtain more selective and potent triplex-binding ligands. Pentacyclic BQQ-based structures, designed by the late Claude Hélène, have been shown to be one of the most potent and specific intercalators over the past few decades.[68] BQQ, designed and synthesized in 1998 for use in developing a more potent triplex intercalator, was shown to be the best triplex-stabilizing ligand in existence. BQQ stabilized a 14 T/C mix-based triple helix with a 51°C increase in melting temperature. It was used to inhibit restriction enzyme cleavage by inducing and stabilizing a short (9-mer) triplex.[68] Hélène also used BQQ to achieve triple helix-specific DNA cleavage by covalently linking it to a ethylenediaminotetraacetic acid (EDTA) which was positioned by BQQ in the Watson–Crick minor groove.[69] By combining BQQ with neomycin, a more stable and specific triplex stabilization effect has been achieved, with BQQ intercalating into the triplex and neomycin binding to the Watson-Hoogsteen groove. The 5′-OH on ring III was selected to provide the linkage to the intercalating unit. The intercalator (Scheme 11.1, BQQ-amine[69]) was linked to neomycin isothiocyanate **3** (prepared in four steps from neomycin)[13] in the presence of 4-(dimethylamino) pyridine (DMAP) as a catalyst. The synthesis rested on the selective conversion of ring III 5′-OH of neomycin into a good leaving group TPS as shown in Schemes 11.1 and 11.2.[70] The displacement of TPS by aminoethanethiol, the conversion into isothiocyanate, followed by coupling with the primary amine in BQQ,[69] with subsequent deprotection with HCl yielded conjugate **4** (Scheme 11.1).

Using individual solutions containing 4 μM of neomycin, 4 μM of BQQ, and 4 μM of BQQ-neomycin, it was found that BQQ-neomycin induced a triplex with a melting temperature as high as 82°C. CD melts further confirmed these results, which showed a negative transition at 49°C (2 → 1 melting) and a positive transition at 82°C (3 → 1 melting). Similar results were obtained with the mixed base 22-mer triplex. Molecular modeling studies of the conjugate binding to DNA triplex further support the dual recognition of DNA triple helix by the conjugate BQQ-neomycin. As BQQ "stacks" between the triplex bases, neomycin is buried in the larger triplex W-H groove. This design of such dual recognition ligands opens a new paradigm for recognition of triplex nucleic acids which is expected to aid in developing even more selective and potent conjugates.

Previous work[7] has suggested that aminoglycoside specificity may occur in nucleic acid forms that display features characteristic of an A-type conformation (RNA triplex,[23] DNA–RNA hybrid duplex,[24] RNA duplex,[25] DNA triplex,[26] A-form DNA duplex,[22] and DNA tetraplex[29]), rather than in naturally occurring RNA. However, conflicting results have been reported regarding the conformation of the triplex and of the Watson-Crick duplex within these triplexes. Both

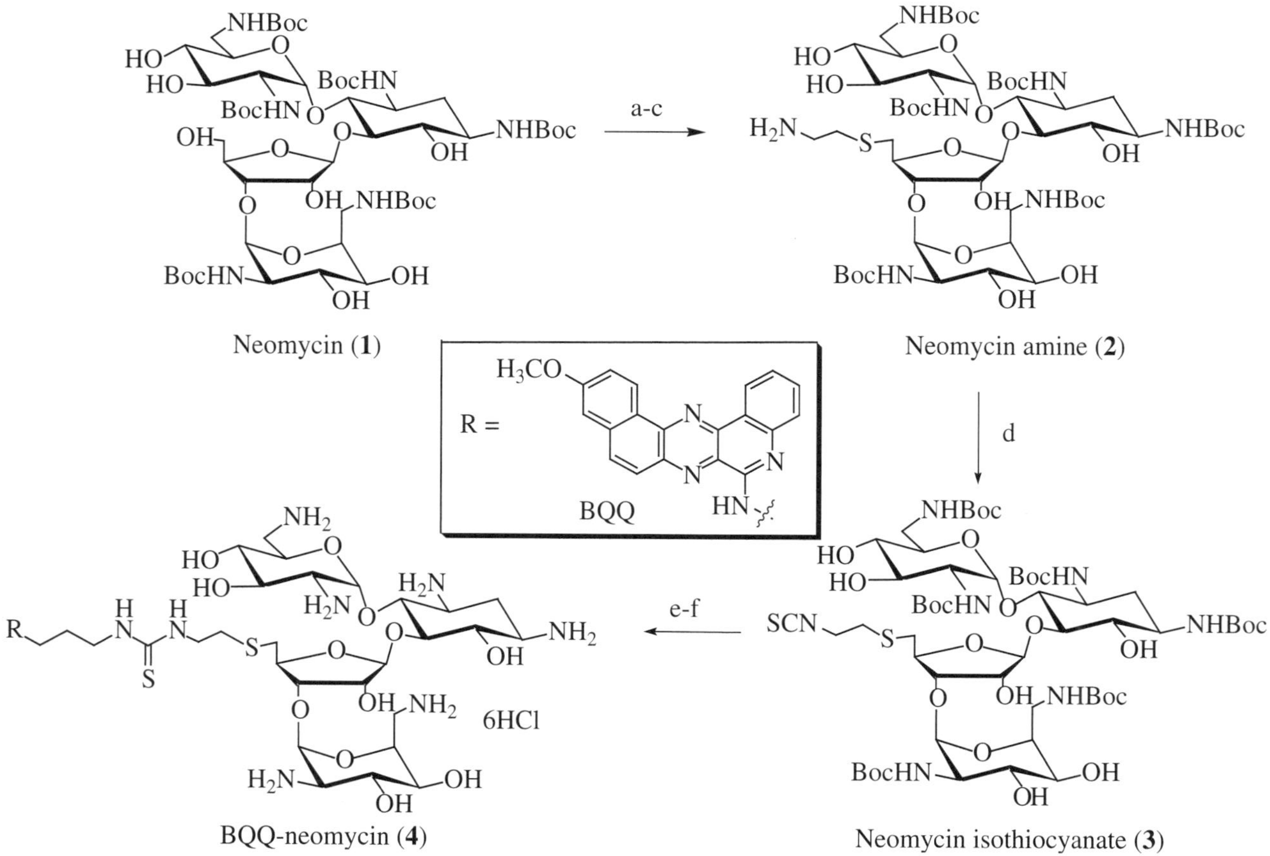

Scheme 11.1. Synthesis of BQQ-neomycin conjugate. Reagents and conditions: (a) $(Boc)_2O$, DMF, H_2O, Et_3N, 60° C, 5 h, 60%; (b) 2,4,6-triisopropylbenzenesulfonyl chloride, pyridine, room temperature, 40 h, 50%; (c) $HSCH_2CH_2$ $NH_2 \cdot HCl$, NaOEt/EtOH, room temperature, 18 h, 50%; (d) 1,1′-thiocarbonyldi-2(1*H*)-pyridone, CH_2Cl_2, room temperature, overnight, 86%; (e) BQQ, DMF, DMAP, room temperature, 7 h, 83%; (f) 4 M HCl/dioxane, $HSCH_2CH_2SH$, room temperature, 5 min, 58%.

A-like[71–74] and B-like[75–77] geometry have been described in the literature. Arnott and Selsing first reported A-like geometry and 3′-endo (N-type) sugar pucker in poly(dT)-poly(dA)*poly(dT) by fiber diffraction studies.[71] Later, studies on FT-IR,[78,79] fiber diffraction, and solution NMR[75,77,80–82] with different DNA triplex (inter-, intra-) suggested that the structure of DNA triple helices have more characteristics of B-form than of A-form. In contrast to the results reported by Arnott et al., B-like geometry and 2′-endo (S type) sugar pucker were observed. Thus far, most studied pyrimidine-purine-pyrimidine triplexes have an axial rise similar to B-DNA, an X-displacement intermediate between A- and B-DNA, a low helical twist similar to that in A-DNA, and a predominance of S-type sugar puckers which are found in B-DNA. Since DNA triplexes have both A-like and B-like characteristics, others have used terms such as Ψ-DNA and P-DNA to describe DNA triplex conformation.[83,84] Given the available data, it can be conclusively inferred that DNA triplexes have some A-form characteristics. The lower association constant observed in 2-deoxystreptamine containing aminoglycoside-triplex interactions (when compared to RNA triplex, RNA duplex and RNA·DNA hybrids)[7] can be explained by the fact that the DNA triplex lacks all of the A-form characteristics present in RNA duplexes and triplexes. Based on this hypothesis, as the number of A-form features in a set of nucleic acid structures increases, the association constant of aminoglycoside–nucleic acid also increases. The available set of aminoglycoside–polynucleotide acid interactions seems to support this trend.[7]

11.5. DNA–RNA HYBRIDS

11.5.1. Hybrid Duplex

DNA–RNA hybrid duplexes are biologically relevant due to their recognition by such enzymes as RNase H and reverse transcriptase.[85,86] Small-molecule recognition of such structures therefore has potential in antiviral applications. The importance of RNase H in such applications is well recognized. Earlier mutational studies indicate that genetic deactivation of the RNase H activity of HIV-1 reverse transcriptase (RT) results in noninfectious virus particles.[87] Targeting crucial RNase H-based interactions is a pathway for developing anti-HIV agents, and is even more attractive when considering that RNA–DNA hybrids formed during the reverse transcription process are not associated with a high mutation frequency, as are RT[88] and protease inhibitors[89–92] (both AIDS therapeutic agents).

Of those small molecules best able to bind DNA–RNA hybrids, very few have been found. At last count, the number was less than ten. For comparison, the number of molecules known to target duplex DNA is perhaps in the thousands. Few reported attempts have been made to target the unique structure of the DNA–RNA hybrids, and consequently only a handful of molecules have been identified.[24,93,94] A few years ago, we identified aminoglycosides as the antibiotics most likely to stimulate groove-based recognition of A-form nucleic acid structures. These major groove binders were then later shown to bind DNA–RNA hybrids in the 100 nM range and to also inhibit the RNA processing enzymes, RNase H and RNase A, by binding to RNA–DNA hybrids.[95] Recent reports indicate that ethidium bromide is the clearly preferred catalyst to elucidate this DNA–RNA hybrid recognition.[94] The binding was shown to be in the low-micromolar range.

With the conjugation of neomycin using a specific DNA–RNA binding intercalator such as methidium, new classes of molecules with high DNA–RNA hybrid affinity can be identified. A neomycin–methidium conjugate **6** has been synthesized by forming an amide linkage between neomycin and methdium carboxylic acid (methidium) (Scheme 11.2). Conjugate **6** is more potent in stabilizing either DNA–RNA hybrids than neomycin, ethidium, or a combination of both. In fact, conjugate affinity exceeds the highest affinity ever identified for aminoglcyoside–nucleic acid recognition. The intercalator 6-(4-carboxyphenyl)-3,8-diamino-5-methylphen-anthridinium chloride **5** (Scheme 11.1, methidium carboxylic acid[69] was linked to neomycin amine **2** (prepared in three steps from neomycin)[13] using DCC-mediated activation of the acid. TFA deprotection yielded the conjugate methidium–neomycin **6** in good yields.

CD and UV melting profiles in the presence of the ligands were carried out to confirm the presence of a DNA–RNA hybrid duplex in the presence of the ligands. The addition of 2.2 μM of neomycin and ethidium bromide results in a T_m increase of 12°C, whereas the addition of **6** results in a T_m increase of 20°C. Increased concentration of **6** caused a gradual increase in the hybrid melting temperature. As the concentration of NM was increased from 0 μM to 3 μM, the melting temperature of the hybrid increases from 48.9°C to 70.1°C, a 21.2°C increase overall.

If a ligand molecule A and ligand molecule B bind to a receptor, ligands A–B are expected to bind so that binding constant of A–B equals the product of two individual binding constants. It is expected that K_d appears in the subnanomolar range (Table 11.1) when observing the product of the K_d values for methidium and neomycin. There are few methods that can accurately determine such large binding constants. Fortunately, a DSC- and UV-based method has been previously used to determine K_{app} in this range,[96] enabling us to calculate the neomycin-binding constants for methidium and neomycin–methidium **6**. The data of these calculations, shown in Table 11.2, illustrate the power of conjugation of two ligands that bind to a receptor at different sites. Neomycin binds to the hybrid duplex with 100 nM affinity and the conjugate shows a 1000-fold improvement, with an affinity of 0.1 nM. The conjugate reflects the first example of an aminoglycoside ligand that binds to a nucleic acid target with affinity higher than the nanomolar affinities shown for the eubacterial A-site. The work should facilitate aminoglycoside-based recognition of novel nucleic acid structures with therapeutic applications.

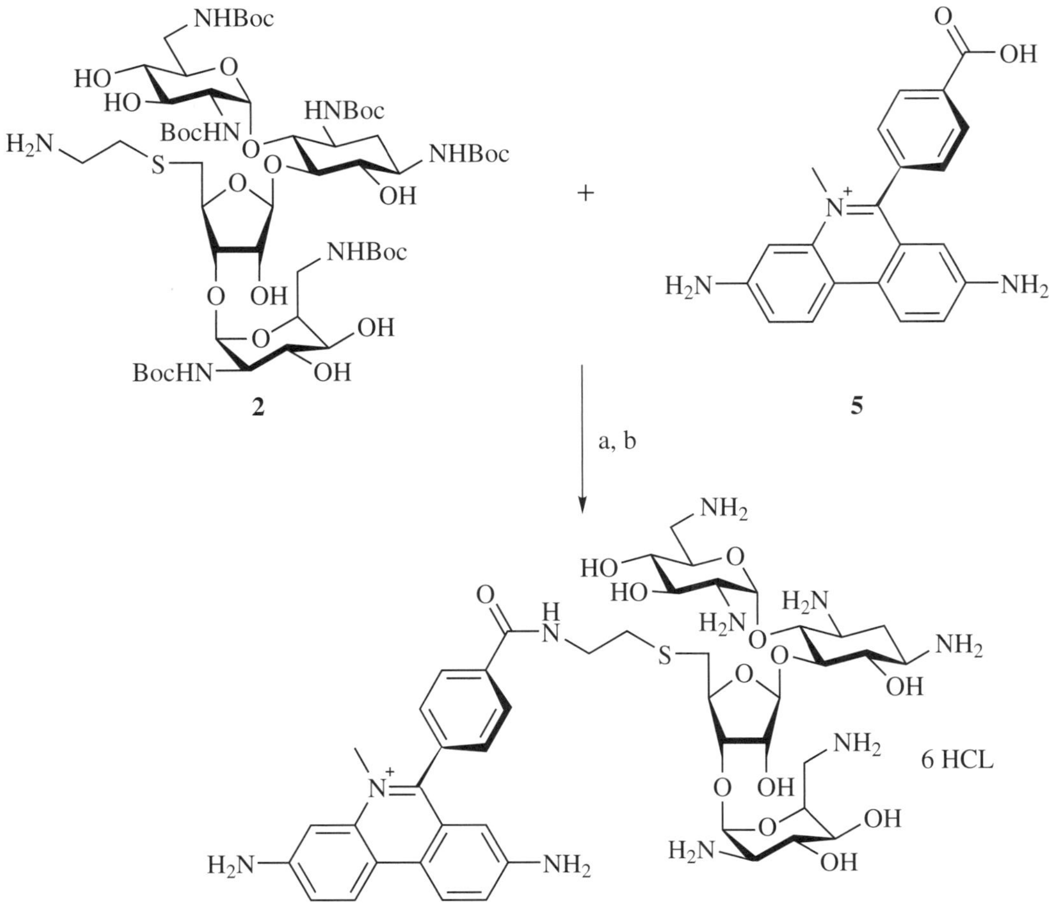

2

5

6

Scheme 11.2. Synthesis of a neomycin–methidium conjugate. Reagents and conditions: (a) DCC, DMF, rt, 7 h, 83%; (b) 4 M HCl/dioxane, $HSCH_2CH_2SH$, rt, 5 min, 88%.

TABLE 11.2. Thermodynamic Profile of poly(dA)·poly(rU) Interaction with Neomycin–methidium 6, Neomycin and Ethidium Bromide in 10 mM Sodium Cacodylate, 0.5 mM EDTA, 100 mM NaCl, and pH 6.8

Parameter	NM (3)	Neomycin	Ethidium Bromide
ΔH_{wc} (kcal/mol)^a	5.27	5.27	5.27
T_{m0} (°C)^b	48.9	48.9	48.9
T_m (°C)^c	70.9	64.2	50.3
N^c	9.3	6.0	4.4
$\Delta H_{(10°C)}$ (kcal/mol)	-11.2 ± 0.1	-2.01 ± 0.1	-7.34 ± 0.1
$\Delta H_{(20°C)}$ (kcal/mol)	-16.2 ± 0.1	-6.2 ± 0.1	-7.7 ± 0.1
$\Delta H_{d(10°C)}$ (kcal/mol)	-1.2 ± 0.1	-0.9 ± 0.1	-1.3 ± 0.1
$\Delta H_{d(20°C)}$ (kcal/mol)	-0.7 ± 0.1	-0.7 ± 0.1	-1.1 ± 0.1
ΔC_p (cal/mol · K)	-501 ± 22	-427 ± 14	-35 ± 13
$K_{T(20°C)}$ (M^{-1})	$(3.2 \pm 0.5) \times 10^{10}$	$(3.1 \pm 0.21) \times 10^7$	$(5.6 \pm 0.2) \times 10^4$

[a]Data obtained from DSC melting profiles.
[b]Data obtained from CD melting profiles.
[c]Data obtained from CD titrations.
[d]Data obtained from ITC excess titration. ΔH is binding heat. ΔH_d is drug dilution heat.
[e]The ΔC_p calculated from equation $\Delta C_p = [\Delta H(10°C) - \Delta H(20°C)]/(10°C - 20°C)$.

These findings clearly show that the conjugation of the methidium moiety results in increased affinity for the DNA–RNA hybrid. Because methdium and neomycin are able to respectively bind with micromolar and 100 nM affinities, linker modification is expected to result in ligands with much higher specificities and affinities, approaching or even surpassing the picomolar range. These affinities were derived using calorimetric methods, and as discussed in Table 11.1, binding induced drug protonation contributes to the observed enthalpies. At low pH (5.5), where all aminoglycoside amines are protonated, an even higher affinity is observed for binding of **6** to DNA–RNA hybrid duplex.

11.5.2. Hybrid Triplex

DNA–RNA hybrid duplexes can be targeted by TFOs to form hybrid triplexes structure. The hybrid triplex TFO consists of a DNA or RNA strand complementary to either strand of the duplex[97] [consider the existing examples poly(rA) · 2poly(dT) and 2poly(rA) · poly(dT)].[98] As with small ligands that bind hybrid duplex, TFOs may produce similar results concerning the prevention of key biological events involving hybrid structures. In fact, stable hybrid triplex formation has been shown to inhibit RNA polymerase,[97] RNase,[99] and DNase I.[99] However, the formation of such triplex structures requires molar salt concentrations. Recent studies have circumvented this requirement by introducing neomycin. Neomycin was shown to induce the hybrid triplex structures poly(rA) · 2poly(dT) and 2poly(rA) · poly(dT) using a series of spectroscopic techniques.[24] The induction and binding of this groove-binding ligand occurred at low micromolar neomycin concentrations and low millimolar sodium concentrations.

11.5.3. RNA Triplex

Though reported as the first observed triplex nucleic acid structure,[100] RNA triplex has received little attention when compared with other RNA structures or DNA triplexes, as discussed below. With the renewed interest in RNA duplex binding proteins, the introduction of a third strand to the duplex, to form a triplex, has important implications for inhibiting protein function at their recognition sites. Single-stranded RNA can also be targeted by circular or foldback triplex forming oligonucleotides (TFOs), which intramolecularly form duplex structures.[101–104] Though triplex formation is limited to homopyrimidine or homopurine stretches, which limit its therapeutic applicability, it still has great potential because of new knowledge about the RNA primary sequence. One example of an important RNA sequence for TFO targeting has been the 5′-noncoding region of hepatitis C viral RNA, which has been shown to form a triplex structure in the presence of Mg^{2+} and the polyamine spermidine.[105]

More recently, aminoglycosides were shown to significantly stabilize RNA triplex structures. Among these aminoglycoside studies, and many other RNA-binding studies, neomycin was found to be the most significant RNA triplex stabilizing aminoglycoside.[23]

11.6. TARGETING RNA USING DNA–RNA HYBRIDS

11.6.1. Carbohydrate (Neomycin)-Mediated Oligonucleotide Delivery

Covalent attachment of a ligand to an oligonucleotide increases the complex stabilization inferred by DNA binding agents such as spermine,[106,107] acridine,[108] Hoechst 33258,[109] and psoralen.[110] Conjugation of an aminoglycoside to an oligonucleotide increases the stabilization incurred by these nucleic acid stabilizing agents while also assisting in the delivery of the aminoglycoside to specific DNA–RNA targets. Various methods have been attempted to use carbohydrate–protein interactions to enhance the cellular uptake and targeting of drug delivery vehicles by coating them with carbohydrate ligands.[111] Examples of carbohydrate ligands used to increase antisense oligonucleotides uptake[112] include mannosylated polylysine[113] and galactosylated polylysine[114] in receptor-mediated endocytosis of antisense oligonucleotides. We have recently shown that neomycin assists in the lipid-mediated delivery of plasmid DNA and oligonucleotides.[115] Delivery of oligonucleotides has been a major obstacle in the development of nucleic acid-based drugs. Surprisingly, neomycin, when combined with a cationic lipid preparation such as DOTAP, enhances transfection efficiency of both reporter plasmids and oligonucleotides, resulting in a significant increase in transgene expression.[115]

11.6.2. RNA Sequence-Specific Aminoglycoside-ODN Conjugates

To advantageously use these unique properties of aminoglycoside–nucleic acid interaction, aminoglycoside–nucleic acid conjugates have been developed.

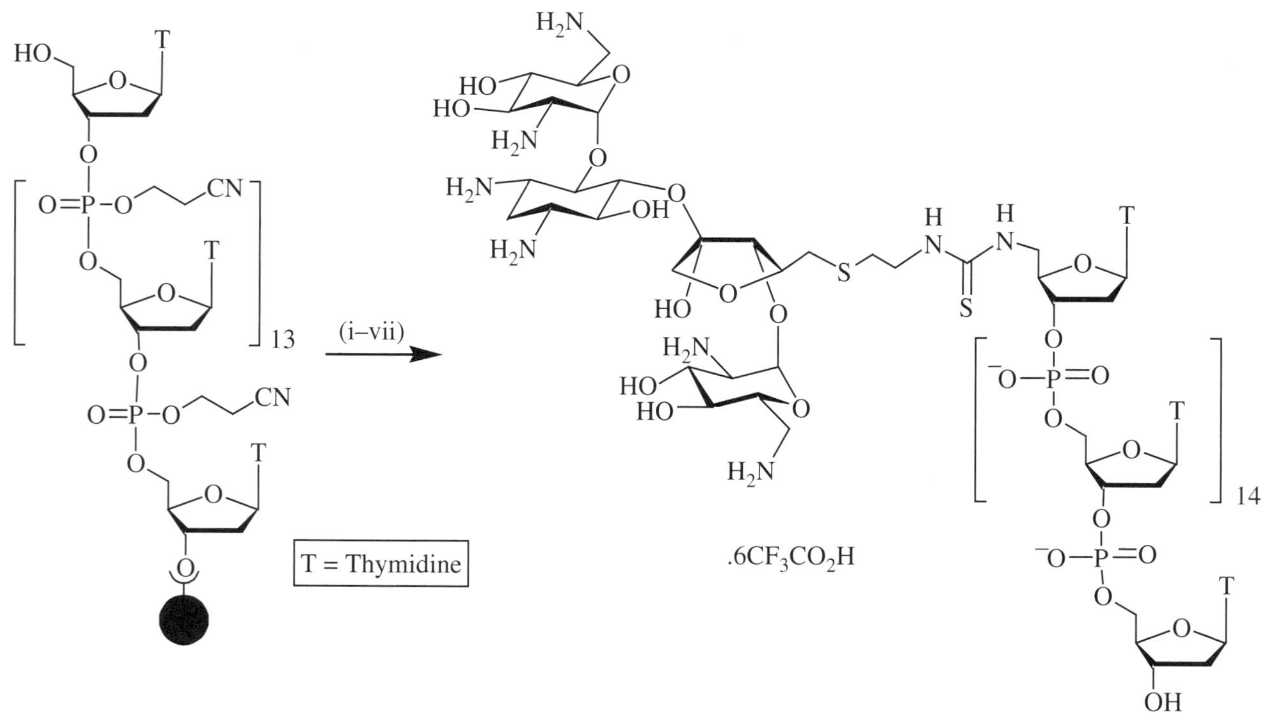

Scheme 11.3. Preparation of neomycin–DNA conjugates.[11] *Reaction conditions:* (i) 1*H*-tetrazole and CH_3CN; (ii) capping of unreacted 5′-hydroxyl group; (iii) oxidation of P(III) to P(V) with I_2, H_2O/pyridine/THF; (iv) deprotection of p-methoxyphenyldiphenylmethyl group from the 5′-amino group with 4% CCl_3CO_2H in CH_2Cl_2; (v) neomycin isothiocyanate (1) 3, 4-*N*,*N*-dimethylaminopyridine and pyridine; (vi) β-elimination followed by deprotection from the solid support using conc. NH_4OH; (vii) 1,4-dioxane solution containing 5% CF_3CO_2H and 1% *m*-cresol (v/v/v %).

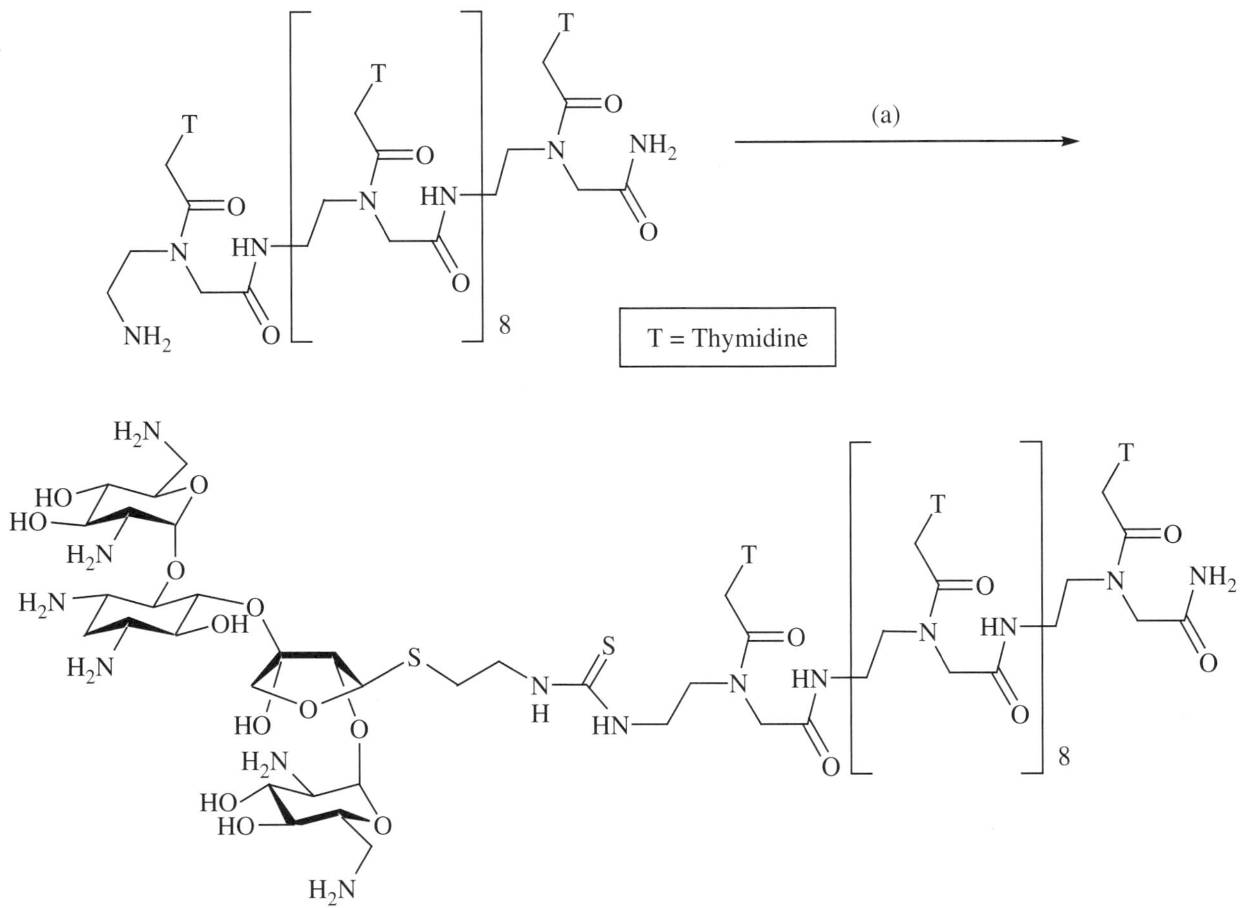

Scheme 11.4. Preparation of PNA–neomycin conjugates.[11] *Reaction conditions*: (a) (i) neomycin isothiocyanate, pyridine, cat. DMAP; (ii) 1M HCl/dioxane, 1,2-ethanedithiol.

Aminoglycosides (neomycin in particular) can induce DNA–RNA hybrid duplex/triplex formation,[24] suggesting that neomycin–DNA conjugates[13] can be effective models for targeting RNA in a sequence-specific manner. Biologically important hybrid duplexes can be constructed from single-stranded RNA and single-stranded DNA.[116] Ribonucleoproteins can act as repressors, with mRNA providing the sequence specificity for duplex strand formation.[117] Recently, the synthesis of neomycin conjugates was reported in which neomycin was conjugated at the 5′-end of the nucleic acid analogs (DNA/PNA) via thiourea linkage (Scheme 11.3, 11.4).[11,13] The application of such conjugates, however, is limited as the drug can be added only at the 5′-end of the oligomer. The applicability of conjugating a drug molecule to any given site within an oligonucleotide can be accomplished if the drug is attached to the base or sugar using an appropriate linker, leaving 5′- and 3′-ends free for suitable oligomer elongation. With this methodology, the addition of drug-phosphoramidite conjugates to the growing strand during oligonucleotide synthesis can be monitored using a trityl monitor. Recently, an efficient phosphoramidite-based method for the synthesis of neomycin attached to the C5-position of 2′-deoxyuridine through a hexynyl linker, its extension to oligonucleotide-neomycin synthesis (Scheme 11.5), and its hybridization with a complementary RNA structure has been performed.[12] These conjugates were referred to as nucleo-aminoglycoside conjugates (N-Ag-C), of which the first example, nucleo-neomycin conjugate (N-Neo-C), was reported.

The synthesis permits the incorporation of an aminoglycoside (neomycin) at any given site in an oligonucleotide (ODN) where a thymidine (or uridine) is present. Incorporating this modified base into an oligonucleotide, which was complementary to a 7-base-long α-sarcin loop RNA sequence, results in enhanced duplex hybridization. The increase in T_m for this duplex ($\Delta T_m = 6^\circ$C) is similar to the increase observed when adding neomycin to the unmodified duplex in a 1:1 ratio. CD spectroscopy indicates an A-type confirmation adopted by the modified duplex. ITC measurements indicate the additive effects of ODN and neomycin binding to the RNA target ($K_a = 4.5 \times 10^7$ M^{-1}). The enhanced stability of the hybrid duplex from this neomycin–ODN conjugate originates primarily from the enthalpic contribution of neomycin $\{\Delta\Delta H_{obs} = -7.21\ \text{kcal/mol}(\Delta H_{\text{neomycin conjugated}} - \Delta H_{\text{nonconjugated}})\}$ binding to the hybrid duplex. The short linker length allows for selective stabilization of the hybrid duplex over the hybrid triplex. The results described here open up new avenues in the design and synthesis of nucleo-aminoglycoside-conjugates (N-Ag-C), where the inclusion of any number of aminoglycoside (neomycin) molecules per oligonucleotide can be accomplished.

This report was the first of its type to show the ability of such a large covalently linked carbohydrate (neomycin) to aid in RNA–DNA hybrid complexation. N-Neo-conjugate and a 7-base-long RNA sequence present in α-sarcin loop can form a stable hybrid duplex. The stability of a hybrid duplex with addition of one neomycin equivalent is comparable to the stability of a hybrid duplex where the DNA has a covalent bond to neomycin. The presence of a single mismatch decreases the melting temperature of both neomycin-conjugated

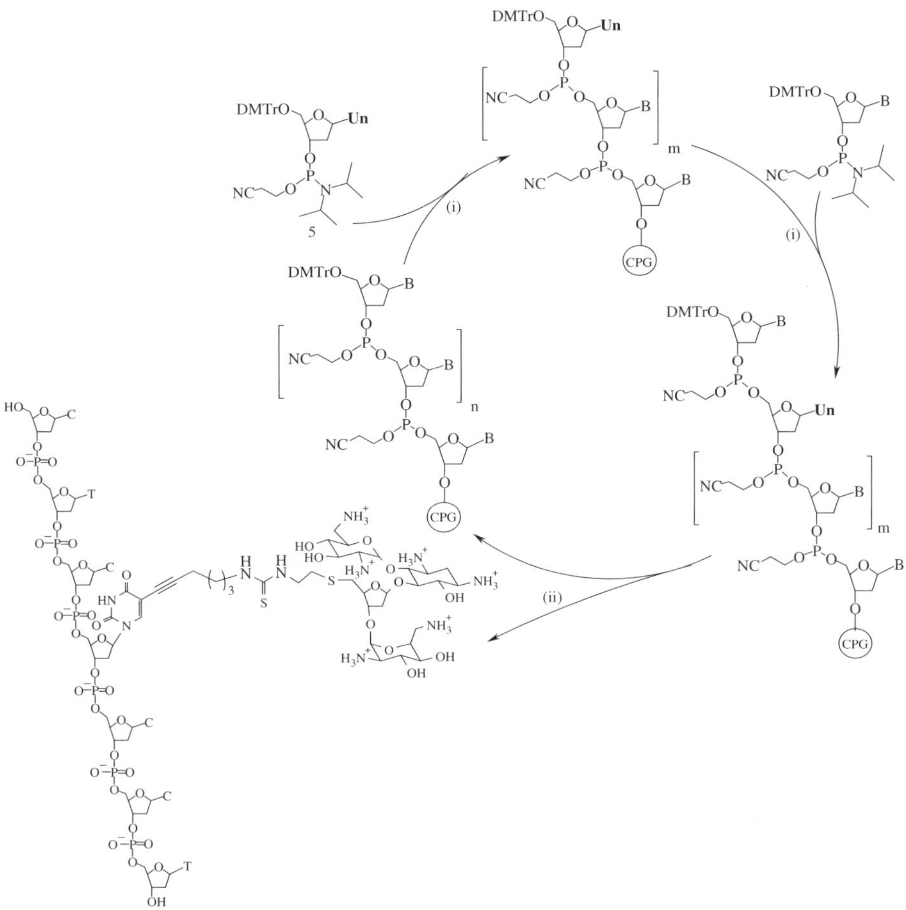

Scheme 11.5. Synthesis of 7-mer-neo-conjugate N-Neo-C. (i) Deprotection with 4% trichloroacetic acid in CH_2Cl_2, and coupling with **3** in the presence of 1*H*-tetrazole followed by capping with acetic anhydride in pyridine/THF solution, and oxidation with I_2 in THF/H_2O/pyridine solution; (ii) β-elimination followed by deprotection from the solid support using conc. NH_4OH; (c) 1,4-dioxane solution containing 3% CF_3CO_2H and 1% m-cresol (v/v/v %). Reprinted with permission from *Bioconj. Chem.* **2007**, *18*, 160–169.

and -nonconjugated hybrid duplexes. This important finding supports the finding that conjugate association is driven by DNA–RNA base-pair association rather than the non-sequence-specific electrostatic interactions between neomycin and the phosphodiester linkages. The linker length, which does not induce short triplex formation, is thus specific for RNA targeting as opposed to duplex DNA targeting. Potential applications of these conjugates include drug and oligonucleotide delivery to site-specific RNA, since the formation of such hybrid duplexes can control gene expression by inhibiting translation steps, slicing, or activation of RNase H leading to mRNA degradation.

These nucleoneomycin conjugates (N-Neo-C) are efficiently synthesized and chemically stable, and they behave similar to an unmodified DNA strand in solution. The synthetic methodology allows for inclusion of any number of aminoglycoside (neomycin) molecules per oligonucleotide. These conjugates can provide a platform for the development of a new form of hybrid biomaterials. They combine the unique specificity of nucleic acids and the recognition/binding ability of the aminoglycosides. N-Neo-Cs could have enormous potential in oligonucleotide therapeutic strategies (duplex as well as triplex approaches) and in the engineering of novel DNA-based materials.

11.7. OTHER AMINOGLYCOSIDE TARGETS: THE ANTHRAX LETHAL FACTOR

A recent discovery in the Wong laboratory involves the inhibition of the Anthrax Lethal Factor protein, one of three plasmid-encoded proteins responsible for anthrax disease development.[118] Among a library of 3000 studied compounds, neomycin was found to be the most potent in inhibitory activity ($K_i = 7$ nM). Interestingly, a further comparison of dimeric neomycin derivatives indicated that neomycin was still the most potent, although dimeric aminoglycosides bind the 16S A-site more strongly. These interesting findings further illustrate the potential for utilizing aminoglycoside-based structures to target not just RNA, but DNA and proteins as well. As with the rest of the structural targets for neomycin, both charge and shape complementarity to the protease active site contribute to its binding.

The development of novel assays that can be used to screen for aminoglycoside binding to various proteins and nucleic acids will be extremely important in deciphering drug specificity because it relates to biological targets as well as targets that lead to nonspecific effects.[7] Though few such assays currently exist, progress has been reported.[7,119] Aminoglycosides, as positively charged nucleic acid binders, offer a unique template for molecular recognition of A- and B-form nucleic acid as well as protein targets.[119] It is now becoming clear that aminoglycosides binding to numerous nucleic acid and protein targets offer physiologically relevant basis for further study of the basis for their interactions and the development of novel and more effective drugs. Developments in aminoglycoside chemistry, biochemistry, and structure biology will lead to the development of novel sugars that have immense benefits in therapy and rational design of ligands for macromolecular recognition.

ACKNOWLEDGMENTS

Financial support for this work was provided by NSF-CAREER (CHE/MCB-0134972).

REFERENCES

1. Arya, D. P. In *Topics in Current Chemistry: DNA Binders;* Chaires, J. B.; Waring, M. J., Eds.; Heidelburg: Springer Verlag: 2005; pp. 149–178.
2. Magnet, S.; Blanchard, J. S. *Chem. Rev.* **2005,** *105*, 477–497.
3. Walter, F.; Vicens, Q.; Westhof, E. *Curr. Opin. Chem. Biol.* **1999,** *3*, 694–704.
4. Schatz, A.; Bugie, E.; Waksman, S. A. *Proc. Soc. Exp. Biol. Med.* **1944,** *55*.
5. Fourmy, D.; Recht, M. I.; Blanchard, S. C.; Puglisi, J. D. *Science* **1996,** *274*, 1367–1371.
6. Walter, F.; Vicens, Q.; Westhof, E. *Current Opinion in Chemical Biology* **1999,** *3*, 694–704.
7. Arya, D. P.; Xue, L.; Willis, B. *J. Am. Chem. Soc.* **2003,** *125*, 10148–10149.
8. Willis, A., III; Arya, D. P. *Curr. Org. Chem.* **2006,** *10*, 663–673.
9. Willis, A., III; Arya, D. P. In *Adv. Carb. Chem. Biochem.;* Horton, D., Ed.; New York: Elsevier, 2006; pp. 251–302.
10. Willis, B.; Arya, D. P. *Biochemistry* **2006,** *45*, 10217–10232.
11. Charles, I.; Arya, D. P. *J. Carb. Chem.* **2005,** *24*, 145–160.
12. Charles, I.; Xi, H.; Arya, D. P. *Bioconj. Chem.* **2007,** *18*, 160–169.
13. Charles, I.; Xue, L.; Arya, D. P. *Bioorg. Med. Chem. Lett.* **2002,** *12*, 1259–1262.
14. Stage, T. K.; Hertel, K. J.; Uhlenbeck, O. C. *RNA* **1995,** *1*, 95–101.
15. Wang, Y.; Hamasaki, K.; Rando, R. R. *Biochemistry* **1997,** *36*, 768–779.
16. Cho, J.; Rando, R. R. *Biochemistry* **1999,** *38*, 8548–8554.
17. Park, W. K. C.; Auer, M.; Jaksche, H.; Wong, C.-H. *J. Am. Chem. Soc.* **1996,** *118*, 10150–10155.
18. Tok, J. B. H.; Cho, J.; Rando, R. R. *Biochemistry* **1999,** *38*, 199–206.
19. Hamasaki, K.; Killian, J.; Cho, J.; Rando, R. R. *Biochemistry* **1998,** *37*, 656–663.
20. Hamasaki, K.; Rando, R. R. *Biochemistry* **1997,** *36*, 12323–12328.
21. Hermann, T. *Angew. Chem. Int. Ed. Engl.* **2000,** *39*, 1890–1905.
22. Robinson, H.; Wang, A. H. J. *Nucleic Acids Res.* **1996,** *24*, 676–82.
23. Arya, D. P.; Coffee, R. L., Jr.; Willis, B.; Abramovitch, A. I. *J. Am. Chem. Soc.* **2001,** *123*, 5385–5395.
24. Arya, D. P.; Coffee, R. L., Jr.; Charles, I. *J. Am. Chem. Soc.* **2001,** *123*, 11093–11094.
25. Chen, Q.; Shafer, R. H.; Kuntz, I. D. *Biochemistry* **1997,** *36*, 11402–11407.
26. Arya, D. P.; Micovic, L.; Charles, I.; Coffee, R. L., Jr.; Willis, B.; Xue, L. *J. Am. Chem. Soc.* **2003,** *125*, 3733–3744.
27. Arya, D. P.; Xue, L.; Tennant, P. *J. Am. Chem. Soc.* **2003,** *125*, 8070–8071.
28. Arnott, S.; Bond, P. J.; Selsing, E.; Smith, P. J. *Nucleic Acids Res.* **1976,** *3*, 2459–2470.
29. Kypr, J.; Fialova, M.; Chladkova, J.; Tumova, M.; Vorlickova, M. *Eur. Biophys. J.* **2001,** *30*, 555–558.
30. Franklin, R. E.; Goslin, R. G. *Acta Crystallogr.* **1953,** *6*, 673–677.
31. Saenger, W. *Principles of Nucleic Acid Structure*; New York: Springer-Verlag, 1984.
32. Ivanov, V. I.; Minchenkova, L. E. *Molekulyarnaya Biologiya (Moscow)* **1995,** *28*, 780–788.

33. Calladine, C. R.; Drew, H. R. *J. Mol. Biol.* **1984,** *178*, 773–782.
34. Marky, N. L.; Olson, W. K. *Biopolymers* **1994,** *34*, 121–142.
35. Altona, C.; Sundaralingam, M. *J. Am. Chem. Soc.* **1972,** *94*, 8205–8212.
36. Egli, M. *Antisense Nucleic Acid Drug Dev.* **1998,** *8*, 123–128.
37. Lu, X.-J.; Shakked, Z.; Olson, W. K. *J. Mol. Biol.* **2000,** *300*, 819–840.
38. Jeffares, D. C.; Poole, A. M.; Penny, D. *J. Mol. Evol.* **1998,** *46*, 18–36.
39. Ding, J.; Hughes, S. H.; Arnold, E. *Biopolymers* **1997,** *44*, 125–138.
40. Ding, J.; Das, K.; Hsiou, Y.; Sarafianos, S. G.; Clark, A. D., Jr.; Jacobo-Molina, A.; Tantillo, C.; Hughes, S. H.; Arnold, E. *J. Mol. Biol.* **1998,** *284*, 1095–1111.
41. Huang, H.; Chopra, R.; Verdine, G. L.; Harrison, S. C. *Science* **1998,** *282*, 1669–1675.
42. Pelletier, H.; Sawaya, M. R.; Kumar, A.; Wilson, S. H.; Kraut, J. *Science* **1994,** *264*, 1891–1903.
43. Eom, S. H.; Wang, J.; Steitz, T. A. *Nature* **1996,** *382*, 278–281.
44. Kiefer, J. R.; Mao, C.; Braman, J. C.; Beese, L. S. *Nature* **1998,** *391*, 304–307.
45. Doublie, S.; Tabor, S.; Long, A. M.; Richardson, C. C.; Ellenberger, T. *Nature* **1998,** *391*, 251–258.
46. Carver, T. E., Jr.; Millar, D. P. *Biochemistry* **1998,** *37*, 1898–1904.
47. Li, Y.; Korolev, S.; Waksman, G. *EMBO Journal* **1998,** *17*, 7514–7525.
48. Galburt, E. A.; Chevalier, B.; Tang, W.; Jurica, M. S.; Flick, K. E.; Monnat, R. J., Jr.; Stoddard, B. L. *Nat. Struct. Biol.* **1999,** *6*, 1096–1099.
49. Horton, J. R.; Bonventre, J.; Cheng, X. *Biol. Chem.* **1998,** *379*, 451–458.
50. Horton, J. R.; Nastri, H. G.; Riggs, P. D.; Cheng, X. *J. Mol. Biol.* **1998,** *284*, 1491–1504.
51. Cheng, X.; Balendiran, K.; Schildkraut, I.; Anderson, J. E. *EMBO J.* **1994,** *13*, 3927–3935.
52. Thomas, M. P.; Brady, R. L.; Halford, S. E.; Sessions, R. B.; Baldwin, G. S. *Nucleic Acids Res.* **1999,** *27*, 3438–3445.
53. Horton, N. C.; Perona, J. J. *J. Biol. Chem.* **1998,** *273*, 21721–21729.
54. Horton, N. C.; Perona, J. J. *J. Mol. Biol.* **1998,** *277*, 779–787.
55. O'Gara, M.; Horton, J. R.; Roberts, R. J.; Cheng, X. *Nat. Struct. Biol.* **1998,** *5*, 872–877.
56. Robinson, H.; Gao, Y.-G.; McCrary, B. S.; Edmondson, S. P.; Shriver, J. W.; Wang, A. H. J. *Nature* **1998,** *392*, 202–205.
57. Mei, H. Y.; Mack, D. P.; Galan, A. A.; Halim, N. S.; Heldsinger, A.; Loo, J. A.; Moreland, D. W.; Sannes-Lowery, K. A.; Sharmeen, L.; Truong, H. N.; Czarnik, A. W. *Bioorg Med. Chem.* **1997,** *5*, 1173–1184.
58. Lacourciere, K. A.; Stivers, J. T.; Marino, J. P. *Biochemistry* **2000,** *39*, 5630–5641.
59. Kirk, S. R.; Tor, Y. *Bioorg Med Chem* **1999,** *7*, 1979–1191.
60. Mikkelsen, N. E.; Brannvall, M.; Virtanen, A.; Kirsebom, L. A. *Proc. Natl. Acad. Sci U S A* **1999,** *96*, 6155–6160.
61. Rogers, J.; Chang, A. H.; von Ahsen, U.; Schroeder, R.; Davies, J. *J. Mol. Biol.* **1996,** *259*, 916–925.
62. von Ahsen, U.; Davies, J.; Schroeder, R. *Nature* **1991,** *353*, 368.

63. Robinson, H.; Gao, Y.; Sanishvili, R.; Joachimiak, A.; Wang, A. H. J. *Nucleic Acids Res.* **2000,** *28*, 1760–1766.

64. Arya, D. P.; Coffee, R. L.; Xue, L. *Bioorg. Med. Chem. Letts.* **2004,** *14*, 4643–4646.

65. Frank-Kamenetskii, M. D.; Mirkin, S. M. *Annu. Rev. Biochem* **1995,** *64*, 65–95.

66. Htun, H.; Dahlberg, J. E. *Science* **1989,** *243*, 1571–1576.

67. Arya, D. P.; Coffee, R. L., Jr. *Bioorg. Med. Chem. Letts.* **2000,** *10*, 1897–1899.

68. Escude, C.; Nguyen, C. H.; Kukreti, S.; Janin, Y.; Sun, J.-S.; Bisagni, E.; Garestier, T.; Helene, C. *Proc. Natl. Acad. Sci. USA* **1998,** *95*, 3591–3596.

69. Zain, R.; Marchand, C.; Sun, J.; Nguyen, C. H.; Bisagni, E.; Garestier, T.; Helene, C. *Chem. Biol.* **1999,** *6*, 771–777.

70. Michael, K.; Tor, Y. *Chem. Eur. J.* **1998,** *4*, 2091–2098.

71. Arnott, S.; Selsing, E. *J. Mol. Biol.* **1974,** *88*, 509–521.

72. Betts, L.; Josey, J. A.; Veal, J. M.; Jordan, S. R. *Science* **1995,** *270*, 1838–1841.

73. Umemoto, K.; Sarma, M. H.; Gupta, G.; Luo, J.; Sarma, R. H. *J. Am. Chem. Soc.* **1990,** *112*, 4539–4545.

74. Weerasinghe, S.; Smith, P. E.; Mohan, V.; Cheng, Y.-K.; Pettitt, B. M. *J. Am. Chem. Soc.* **1995,** *117*, 2147–2158.

75. Macaya, R. F.; Schultze, P.; Feigon, J. *J. Am. Chem. Soc.* **1992,** *114*, 781–783.

76. Radhakrishnan, I.; Gao, X.; Santos, C. D. L.; Live, D.; Patel, D. J. *Biochemistry* **1991,** *30*, 9022–9030.

77. Koshlap, K. M.; Schultze, P.; Brunar, H.; Dervan, P. B.; Feigon, J. *Biochemistry* **1997,** *36*, 2659–2668.

78. Dagneaux, C.; Liquier, J.; Taillandier, E. *Biochemistry* **1995,** *34*, 16618–16623.

79. Howard, F. B.; Miles, H. T.; Liu, K.; Frazier, J.; Raghunathan, G.; Sasisekharan, V. *Biochemistry* **1992,** *31*, 10671–10677.

80. Bartley, J. P.; Brown, T.; Lane, A. N. *Biochemistry* **1997,** *36*, 14502–14511.

81. Radhakrishnan, I.; Patel, D. J.; Veal, J. M.; Gao, X. *J. Am. Chem. Soc.* **1992,** *114*, 6913–6915.

82. Wang, G.; Seidman, M. M.; Glazer, P. M. *Science* **1996,** *271*, 802–805.

83. Saminathan, M.; Antony, T.; Shirahata, A.; Sigal, L. H.; Thomas, T.; Thomas, T. J. *Biochemistry* **1999,** *38*, 3821–3830.

84. Radhakrishnan, I.; Patel, D. J. *Biochemistry* **1994,** *33*, 11405–11416.

85. Kohlstaedt, L. A.; Wang, J.; Friedman, J. M.; Rice, P. A.; Steitz, T. A. *Science* **1992,** *256*, 1783–1790.

86. Stein, C. A.; Cohen, J. S. *Cancer. Res.* **1988,** *48*, 2659–2668.

87. Tisdale, M.; Schulze, T.; Larder, B. A.; Moelling, K. *J. Gen. Virol.* **1991,** *72*, 59–66.

88. Larder, B. A. *Cold Spring Harbor Monograph Ser.* **1993,** *23*, 205–222.

89. Richman, D. D. *Nature* **2001,** *410*, 995–1001.

90. Nabel, G. J. *Nature* **2001,** *410*, 1002–1007.

91. Root, M. J.; Kay, M. S.; Kim, P. S. *Science* **2001,** *291*, 884–888.

92. Cervia, J. S.; Smith, M. A. *Clin. Infect. Dis.* **2003,** *37*, 1102–1106.

93. Glukhov, A. I.; Zimnik, O. V.; Gordeev, S. A.; Severin, S. E. *Biochem. Biophys. Res. Commun.* **1998,** *248*, 368–371.

94. Ren, J.; Chaires, J. B. In *Methods in Enzymology*; New York: Academic Press, 2001; pp. 99–108.
95. Barbieri, C. M.; Li, T. K.; Guo, S.; Wang, G.; Shallop, A. J.; Pan, W.; Yang, G.; Gaffney, B. L.; Jones, R. A.; Pilch, D. S. *J. Am. Chem. Soc.* **2003,** *125*, 6469–6477.
96. Leng, F.; Priebe, W.; Chaires, J. B. *Biochemistry* **1998,** *37*, 1743–1753.
97. Morgan, A. R.; Wells, R. D. *J. Mol. Biol.* **1968,** *37*, 63–80.
98. Steely, H. T., Jr.; Gray, D. M.; Ratliff, R. L. *Nucleic Acids Res.* **1986,** *14*, 10071–10090.
99. Murray, N. L.; Morgan, A. R. *Can. J. Biochem.* **1973,** *51*, 436–449.
100. Felsenfeld, G.; Davies, D. R.; Rich, A. *J. Am. Chem. Soc.* **1957,** *79*, 2023–2024.
101. Wang, S.; Kool, E. T. *J. Am. Chem. Soc.* **1994,** *116*, 8857–8858.
102. Kool, E. T. *New J. Chem.* **1997,** *21*, 33–45.
103. Wang, S.; Kool, E. T. *Nucleic Acids Res.* **1994,** *22*, 2326–2333.
104. Prakash, G.; Kool, E. T. *Chem. Commun.* **1991**, 1161–1163.
105. Carmona, P.; Molina, M. *Nucleic Acids Res.* **2002,** *30*, 1333–1337.
106. Rajeev, K. G.; Jadhav, V. R.; Ganesh, K. N. *Nucleic Acids Res.* **1997,** *25*, 4187–4193.
107. Tung, C. H.; Breslauer, K. J.; Stein, S. *Nucleic Acids Res.* **1993,** *21*, 5489–5494.
108. Stonehouse, T. J.; Fox, K. R. *Biochim. Biophys. Acta* **1994,** *1218*, 322–330.
109. Robles, J.; McLaughlin, L. W. *J. Am. Chem. Soc.* **1997,** *119*, 6014–6021.
110. Perkins, B. D.; Wensel, T. G.; Vasquez, K. M.; Wilson, J. H. *Biochemistry* **1999,** *38*, 12850–12859.
111. Sihorkar, V.; Vyas, S. P. *J. Pharm. Pharmaceut. Sci.* **2001,** *4*, 138–158.
112. Akhtar, S.; Hughes, M. D.; Khan, A.; Bibby, M.; Hussain, M.; Nawaz, Q.; Double, J.; Sayyed, P. *Adv. Drug Delivery Rev.* **2000,** *44*, 3–21.
113. Liang, W.; Shi, X.; Deshpande, D.; Malanga, C. J.; Rojanasakul, Y. *Biochim. Biophys. Acta* **1996,** *1279*, 227–234.
114. Biessen, E. A. L.; Vietsch, H.; Rump, E. T.; Fluiter, K.; Kuiper, J.; Bijsterbosch, M. K.; Berkel, T. J. C. V. *Biochem. J.* **1999,** *340*, 783–792.
115. Napoli, S.; Carbone, G. M.; Catapano, C.; Shaw, N.; Arya, D. P. *Bioorg. Med. Chem. Lett.* **2005,** *15*, 3467–3469.
116. Felsenfeld, G.; Rich, A. *Biochim. Biophys. Acta* **1957,** *26*, 457–468.
117. Miller, J. H.; Sobell, H. M. *Proc. Natl. Acad. Sci. USA* **1966,** *55*, 1201–1205.
118. Lee, L. V.; Bower, K. E.; Liang, F.-S.; Shi, J.; Wu, D.; Sucheck, S. J.; Vogt, P. K.; Wong, C.-H. *J. Am. Chem. Soc.* **2004,** *126*, 4774–4775.
119. Liang, F.-S.; Greenberg, W.; Hammond, J. A.; Hoffmann, J.; Head, S. R.; Wong, C.-H. *Proc. Natl. Acad. Sci. USA* **206,** *103*, 12311–12316.

INDEX